Thomas Grombein

Gravity forward modeling with a tesseroid-based Rock-Water-Ice approach

Theory and applications in the context of the GOCE mission and height system unification

Karlsruher Institut für Technologie
Schriftenreihe des Studiengangs Geodäsie und Geoinformatik
2017, 1

Eine Übersicht aller bisher in dieser Schriftenreihe erschienenen Bände finden Sie am Ende des Buches.

Gravity forward modeling with a tesseroid-based Rock-Water-Ice approach

Theory and applications in the context of the GOCE mission and height system unification

by
Thomas Grombein

KIT Scientific Publishing

Karlsruher Institut für Technologie
Geodäsie und Geoinformatik

Gravity forward modeling with a tesseroid-based Rock-Water-Ice approach – Theory and applications in the context of the GOCE mission and height system unification

Zur Erlangung des akademischen Grades eines Doktor-Ingenieurs
von der KIT-Fakultät für Bauingenieur-, Geo- und Umweltwissenschaften
des Karlsruher Instituts für Technologie (KIT) genehmigte Dissertation

von Dipl.-Ing. Thomas Grombein aus Mönchengladbach

Tag der mündlichen Prüfung: 20. Oktober 2016
Referent: Prof. Dr.-Ing. Dr. h.c. Bernhard Heck
Korreferent: Univ.-Prof. Dr.-Ing. Dr. h.c. mult. Reiner Rummel

Diese Arbeit ist gleichzeitig veröffentlicht in:
Veröffentlichungen der DGK, Reihe C, Band 798
Ausschuss Geodäsie der Bayerischen Akademie der Wissenschaften
Verlag der Bayerischen Akademie der Wissenschaften, München, 2017
ISBN 978-3-7696-5210-9

Impressum

KIT Scientific Publishing

Karlsruher Institut für Technologie (KIT)
KIT Scientific Publishing
Straße am Forum 2
D-76131 Karlsruhe

KIT Scientific Publishing is a registered trademark
of Karlsruhe Institute of Technology.
Reprint using the book cover is not allowed.

www.ksp.kit.edu

Print on Demand 2017 – Gedruckt auf FSC-zertifiziertem Papier

ISSN 1612-9733
ISBN 978-3-7315-0655-3
DOI 10.5445/KSP/1000068500

"I do not know what I may appear to the world, but to myself I seem to have been only like a boy playing on the sea-shore, and diverting myself in now and then finding a smoother pebble or a prettier shell than ordinary, whilst the great ocean of truth lay all undiscovered before me."

(Sir Isaac Newton, 1642 – 1727)

Brewster, D. (1855): Memoirs of the life, writings, and discoveries of Sir Isaac Newton, vol. 2, p. 407.

Abstract

Due to the increasing availability of global high-resolution digital terrain models (DTMs), it has nowadays become possible to obtain a detailed image of the Earth's topography. This enables to precisely determine the gravitational effect of the topographic masses on the Earth's gravity field. The central technique for this aim is *gravity forward modeling* (GFM), which is based on Newton's law of universal gravitation, and allows to convert topographic heights along with suitable density assumptions into corresponding values of the gravitational potential and its derivatives. This topographic gravity forward modeling attracts a growing interest in various areas of geodetic gravity field determination and geophysical studies of the Earth's composition and structure (e.g., solid-earth sciences).

However, previous GFM methods have proven unsuitable for the increasing accuracy requirements stemming from an improved precision of geodetic measurements. This is due to commonly used simplifications and approximations, such as (i) the use of condensed heights for water and ice masses (rock-equivalent heights), (ii) mass discretizations or arrangements based on planar and spherical approximations, and (iii) assumptions regarding the spectral consistency between band-limited topographic heights and induced gravity, as in residual terrain modeling (RTM) techniques.

This thesis contributes to state-of-the-art GFM in the space domain by providing effective techniques and refinements that overcome these limitations. More concretely, the theory of the *Rock-Water-Ice (RWI) approach* is developed that encompasses a more realistic modeling of the Earth's topographic and isostatic masses, i.e., the masses of the continents, oceans, lakes, ice sheets and shelves, as well as their deeper lying (isostatic) compensation masses in the Earth's interior. The RWI method is characterized by a three-layer decomposition of the Earth's topography that accounts for a rigorous separate modeling of the rock, water, and ice masses with variable density values. Furthermore, a modified Airy-Heiskanen isostatic concept is applied that is enhanced by additional geophysical information in terms of a seismologically derived depth model of the Mohorovičić discontinuity, i.e., the boundary surface between the Earth's crust and mantle.

To counteract the increased computational demand of the more complex modeling, an efficient numerical algorithm is needed for the forward modeling. For space domain GFM, it has become more and more customary to use a mass discretization based on *tesseroids*, which are mass bodies bounded by geocentric spherical coordinate lines, and hence are directly linked to the curvature of the Earth. Several studies have demonstrated their superiority over classical prism methods with respect to precision and computation time. However, for global applications based on high-resolution DTMs, any computational speed-up with respect to a single mass body leads to a massive improvement in the overall computation time. This thesis presents a considerable optimization of previously used tesseroid formulas, where the gravitational field of a tesseroid and its derivatives up to second-order are represented in a compact and computationally attractive form. This allows an efficient numerical evaluation that reduces the overall runtime by about 20 to 55 %, depending on the evaluated gravity field functional. Additionally, to correctly locate topographic masses in space, tesseroids are arranged on an ellipsoidal reference surface.

Within this thesis, the novel tesseroid-based RWI approach is applied to different topographic input data and is used for various gravity field functionals in two main applications. Both are connected to ESA's *satellite mission GOCE* (Gravity field and steady-state Ocean Circulation Explorer) that measured the second-order derivatives of the gravitational potential, commonly known as gravity gradients.

In the first application, RWI-based topographic-isostatic effects are calculated along the orbit of the GOCE satellite and are subtracted from the gravity gradient observations. In this way, the measurement signal is smoothed so that interpolation and prediction tasks, such as harmonic downward continuation of the gradients from satellite altitude to the Earth's surface, can be executed with an improved numerical stability. While in previous studies such a concept was applied to simulated gravity gradients, this thesis presents the application to real GOCE measurements. As the smoothing effect strongly depends on the variability of the topography crossed by the satellite, this procedure is particularly suitable for regional applications. For a time series when the satellite passed the Himalayan region, a comparison of the observed gradients to the reduced ones reveals significant smoothing effects that are quantified by analyses in the space and frequency domain.

The second application contributes to the task of *height system unification*, which aims to connect the different locally defined reference levels, conventionally used for national height systems. This is achieved by a satellite-based method which employs global geopotential models derived from data of the GOCE mission, whose limited spectral resolution is extended by high-frequency topographic effects of the RWI approach. To extract these high-frequency signals, a novel (residual) gravity forward modeling method is proposed that allows to perform the required high pass filtering directly in the gravity domain, thus, avoiding the above-mentioned assumption (iii) of the RTM method. By using three representative study areas in Germany, Austria, and Brazil, the benefit and importance of high-frequency topography-implied gravity signals for an accurate estimation of height datum offsets is demonstrated.

As a highlight of this thesis, the RWI approach is utilized to generate a series of topographic-isostatic gravity field models. These RWI models provide a high-resolution representation of the Earth's topographic-isostatic gravitational potential in terms of spherical harmonics expanded up to degree and order 1800 (Release 2012), and 2190 (Release 2015). The spherical harmonic coefficients of these models are obtained from a spherical harmonic analysis of global gridded potential values, which have been calculated by massive parallel computing on high-performance computer systems. By using spherical harmonic synthesis, the RWI model can be used to efficiently calculate various functionals of the topographic-isostatic potential in different heights. For this purpose, the RWI models are publicly available via the database of the International Centre for Global Earth Models (ICGEM) and have already been used in a wide range of studies by other research groups.

This article-based (cumulative) thesis consists of a detailed introductory chapter and five chapters associated with the following peer-reviewed publications (four original articles in international journals and one proceedings contribution to an IAG Scientific Assembly):

Grombein, T., Seitz, K., and Heck, B. (2013): Optimized formulas for the gravitational field of a tesseroid. *Journal of Geodesy* 87(7):645–660. DOI: 10.1007/s00190-013-0636-1.

Grombein, T., Luo, X., Seitz, K., and Heck, B. (2014): A wavelet-based assessment of topographic-isostatic reductions for GOCE gravity gradients. *Surveys in Geophysics* 35(4):959–982. DOI: 10.1007/s10712-014-9283-1.

Grombein, T., Seitz, K., and Heck, B. (2016a): The Rock-Water-Ice topographic gravity field model RWI_TOPO_2015 and its comparison to a conventional rock-equivalent version. *Surveys in Geophysics* 37(5):937–976. DOI: 10.1007/s10712-016-9376-0.

Grombein, T., Seitz, K., and Heck, B. (2017): On high-frequency topography-implied gravity signals for height system unification using GOCE-based global geopotential models. *Surveys in Geophysics* 38(2):443–477. DOI: 10.1007/s10712-016-9400-4.

Grombein, T., Seitz, K., and Heck, B. (2016b): Height system unification based on the fixed GBVP approach. In: Rizos, C., and Willis, P. (eds.) IAG 150 years. Proceedings of the IAG Scientific Assembly, Postdam, Germany, Sept. 1–6, 2013. *International Association of Geodesy Symposia*, vol. 143. Springer Berlin Heidelberg, pp. 305–311. DOI: 10.1007/1345_2015_104.

Zusammenfassung

Mit globalen, hochauflösenden digitalen Geländemodellen (DGMs) stehen heutzutage detaillierte Informationen über die Topographie der Erde zur Verfügung. Diese erlauben es den gravitativen Einfluss der Topographie auf das Erdschwerefeld hochgenau zu bestimmen. Die zentrale Methode hierfür ist die *Vorwärtsmodellierung* (gravity forward modeling), die auf dem Newtonschen Gravitationsgesetz beruht und es ermöglicht, topographische Höhen, zusammen mit geeigneten Dichteannahmen, in Schwerepotentialwerte umzurechnen. Diese topographische Vorwärtsmodellierung spielt eine zunehmend größere Rolle in verschiedenen Anwendungen der geodätischen Schwerefeldmodellierung sowie in geophysikalischen Studien zum Aufbau und zur Struktur der Erde.

Durch immer genauere und sensitivere Beobachtungsverfahren sind auch die Genauigkeitsanforderungen an die Vorwärtsmodellierung gestiegen. Bisher verwendete Vereinfachungen und Approximationen müssen daher kritisch hinterfragt werden. Dies betrifft unter anderem (i) die Verwendung von kondensierten Höhen für Wasser- und Eismassen (sogenannte rock-equivalent heights), (ii) die planare bzw. sphärische Diskretisierung und Anordnung topographischer Massen, und (iii) Annahmen bezüglich der spektralen Übereinstimmung von Höhen und Schwerewerten, wie im Fall der weit verbreiteten RTM-Methode (residual terrain modeling).

Die vorliegende Arbeit trägt maßgeblich zur aktuellen Weiterentwicklung der Vorwärtsmodellierung im Ortsbereich bei, indem neue Methoden und Verfeinerungen bereitgestellt werden, mit denen es gelingt, die oben genannten Einschränkungen zu überwinden. Hierzu wird die Theorie des *Rock-Water-Ice (RWI)-Ansatzes* entwickelt, der eine realistischere Modellierung der topgraphischen und isostatischen Erdmassen erlaubt. Diese setzen sich aus den Massen der Kontinente, Ozeane, Seen, Eisflächen und dem Schelfeis zusammen, sowie deren tiefer liegenden (isostatischen) Kompensationsmassen. Die RWI-Methode ist durch eine Zerlegung der Erdtopographie in drei Schichten gekennzeichnet, die es erlaubt, eine strenge, separate Modellierung der Gesteins-, Wasser-, und Eismassen mit variablen Dichtewerten durchzuführen. Des Weiteren wird ein modifiziertes isostatisches Airy-Heiskanen-Konzept angewandt, welches durch die Verwendung von zusätzlichen geophysikalischen Informationen verbessert wurde. In diesem Zusammenhang werden seismisch bestimmte Tiefen der Mohorovičić-Diskontinuität, d. h. der Grenzfläche zwischen Erdkruste und Erdmantel, in das isostatische Konzept eingeführt.

Um dem gesteigerten Rechenaufwand durch die komplexere Modellierung entgegenzuwirken, sind effiziente Berechnungsmethoden für die Vorwärtsmodellierung notwendig. Für die Modellierung im Ortsbereich setzt sich hierbei immer mehr die Verwendung von *Tesseroiden* durch. Diese Massenkörper sind durch geozentrisch-sphärische Koordinatenlinien begrenzt und berücksichtigen daher direkt die Krümmung der Erde. In verschiedenen Studien konnte bereits der Vorteil von Tesseroiden gegenüber der herkömmlich verwendeten Quaderformel gezeigt werden, sowohl hinsichtlich einer gesteigerten Genauigkeit als auch einer verringerten Rechenzeit. In Hinblick auf globale Anwendungen und hochauflösende DGMs ist es dennoch notwendig die Rechenzeit weiter zu reduzieren. In diesem Zusammenhang werden in der vorliegenden Arbeit optimierte Berechnungsformeln präsentiert, wodurch das Schwerepotential eines Tesseroids inklusive seiner ersten und zweiten Ableitungen in einer kompakten und einfach auszuwertenden Form dargestellt werden kann. Dies ermöglicht eine numerisch effiziente Vorwärtsmodellierung, wobei der Gesamtrechenaufwand gegenüber bisher verwendeten Tesseroidformeln je nach Schwerefunktional um 20 bis 55 % sinkt. Im Sinne einer verbesserten räumlichen Anordnung der topographischen Massen werden Tesseroide zudem auf eine ellipsoidische Referenzfläche aufgesetzt.

In Rahmen dieser Arbeit wird der neu entwickelte RWI-Ansatz unter Verwendung verschiedener topographischer Eingangsdaten auf unterschiedliche Funktionale des Schwerefeldes in zwei Anwendungen eingesetzt. Beide Anwendungen sind mit der *ESA-Satellitenmission GOCE* (Gravity field and steady-state Ocean Circulation Explorer) verbunden, welche die zweiten Ableitungen des Gravitationspotentials gemessen hat, die auch als Gravitationsgradienten bezeichnet werden.

In der ersten Anwendung werden topographisch-isostatische Effekte mittels des RWI-Ansatzes entlang des GOCE-Satellitenorbits berechnet und als Reduktionen an die gemessenen Gradienten angebracht. Hierdurch kann eine Glättung des Messsignals bewirkt werden, wodurch sich die numerische Stabilität bei Interpolationsaufgaben und Feldtransformationen verbessert, wie z. B. bei einer harmonischen Fortsetzung der Gradienten nach unten, vom Satellitenorbit zur Erdoberfläche. Während diese Vorgehensweise in früheren Untersuchungen auf simulierte Gradienten angewendet wurde, werden in der vorliegenden Arbeit gemessene GOCE-Gradienten topographisch-isostatisch reduziert. Da der erreichte Glättungseffekt stark von der Variabilität der Topographie abhängt, die der Satellit überfliegt, ist diese Vorgehensweise besonders für regionale Anwendungen geeignet. Durch den Vergleich von gemessenen und reduzierten Gradienten für eine Zeitreihe, in der der Satellit die Himalaya-Region überquert hat, konnten durch eine Analyse im Orts- und Frequenzraum signifikante Glättungseffekte beobachtet und quantifiziert werden.

Die zweite Anwendung trägt zur aktuellen Fragestellung der *Vereinheitlichung von Höhensystemen* bei. Hierbei sollen die meist lokal definierten Bezugsflächen unterschiedlicher nationaler Höhensysteme miteinander verbunden werden. In diesem Zusammenhang wird ein satellitenbasiertes Verfahren vorgeschlagen, welches auf der Verwendung von globalen Schwerefeldmodellen der GOCE-Mission beruht. Um die spektral begrenzte Auflösung dieser Modelle zu erweitern, werden hochfrequente topographische Effekte aus der RWI-Methode eingesetzt. Hierzu wird ein neuer Ansatz zur residualen Vorwärtsmodellierung entwickelt und getestet. Dieser erlaubt es die notwendige Hochpassfilterung direkt auf Schwerebasis durchzuführen und ist daher nicht auf die oben erwähnte Annahme (iii) der RTM-Methode angewiesen. Am Beispiel von drei repräsentativen Untersuchungsgebieten in Deutschland, Österreich und Brasilien konnte die Bedeutung von hochfrequenten topographischen Schwerefeldinformationen für die genaue Berechnung von Höhenoffsets eindrucksvoll gezeigt werden.

Den Höhepunkt der vorliegenden Arbeit bildet eine Reihe topographisch-isostatischer Schwerefeldmodelle, die unter Verwendung des RWI-Ansatzes generiert wurden. Diese RWI-Modelle liefern eine hochauflösende Darstellung des topographisch-isostatischen Gravitationspotentials in Form von Kugelfunktionsentwicklungen bis Grad und Ordnung 1800 (Version 2012), bzw. 2190 (Version 2015). Die Kugelfunktionskoeffizienten dieser Modelle werden über eine sphärisch-harmonische Analyse aus globalen Gittern von Potentialwerten abgeleitet. Für die Generierung der Gitterwerte wird eine stark parallelisierte Berechnung auf Hochleistungsrechnern eingesetzt. Mittels sphärisch-harmonischer Synthese können die RWI-Modelle zur effizienten Berechnung verschiedener Funktionale des topographisch-isostatischen Potentials in unterschiedlichen Auswertehöhen verwendet werden. Zu diesem Zweck wurden die Koeffizienten der RWI-Modelle über die Datenbank des International Centre for Global Earth Models (ICGEM) frei verfügbar gemacht und sind bereits vielfältig in Untersuchungen internationaler Forschungsgruppen zum Einsatz gekommen.

Diese kumulative Dissertation besteht aus einem ausführlichen Einleitungskapitel und fünf Kapiteln, die die folgenden, begutachteten Publikationen beinhalten (vier Artikel in internationalen Zeitschriften und ein Tagungsbeitrag zu einer Konferenz der International Association of Geodesy):

Grombein, T., Seitz, K. und Heck, B. (2013): Optimized formulas for the gravitational field of a tesseroid. *Journal of Geodesy* 87(7):645–660. DOI: 10.1007/s00190-013-0636-1.

Grombein, T., Luo, X., Seitz, K. und Heck, B. (2014): A wavelet-based assessment of topographic-isostatic reductions for GOCE gravity gradients. *Surveys in Geophysics* 35(4):959–982. DOI: 10.1007/s10712-014-9283-1.

Grombein, T., Seitz, K. und Heck, B. (2016a): The Rock-Water-Ice topographic gravity field model RWI_TOPO_2015 and its comparison to a conventional rock-equivalent version. *Surveys in Geophysics* 37(5):937–976. DOI: 10.1007/s10712-016-9376-0.

Grombein, T., Seitz, K. und Heck, B. (2017): On high-frequency topography-implied gravity signals for height system unification using GOCE-based global geopotential models. *Surveys in Geophysics* 38(2):443–477. DOI: 10.1007/s10712-016-9400-4.

Grombein, T., Seitz, K. und Heck, B. (2016b): Height system unification based on the fixed GBVP approach. In: Rizos, C., und Willis, P. (Hrsg.) IAG 150 years. Proceedings of the IAG Scientific Assembly, Postdam, Germany, 1.–6. Sep. 2013. *International Association of Geodesy Symposia*, vol. 143. Springer Berlin Heidelberg, S. 305–311. DOI: 10.1007/1345_2015_104.

Contents

„Daß ich erkenne, was die Welt
im Innersten zusammenhält“

“So that I may perceive whatever holds
the world together in its inmost folds”

(Johann Wolfgang von Goethe, Faust I,
V. 382f., transl. by George M. Priest)

I. Introductory chapter

1. Introduction

1.1. Motivation and background

Ever since, humanity desires to know “*whatever holds the world together in its inmost folds*”. Although the answer to this more general and philosophical question is sought in various disciplines, from the perspective of geodesy, the decisive role is certainly played by gravity – and, more concretely, the phenomenological and quantitative effect of gravitation on our planet Earth.

In this context, the primary task of geodesy is to determine the figure of the Earth and its time-variable external gravity field, where both aspects are closely related to each other (Torge and Müller, 2012, p. 2). The Earth's gravity field is composed by a gravitational and a centrifugal component, which result from the attraction of the Earth's masses and the Earth's rotation, respectively. Mathematically, the gravity field is a conservative vector field that is usually represented by a scalar potential, namely the gravity potential W. The geometry of the gravity field can be illustrated by equipotential surfaces, i.e., surfaces of constant gravity potential. A distinguished equipotential surface with a fundamental meaning is the geoid, which in first approximation can be associated with the mean sea level (MSL) that is conceptionally extended below the continents. Conventionally, the geoid is defined by the potential value $W = W_0$.

The gravity potential W results from the sum of the gravitational potential V and the centrifugal potential Z, i.e., $W = V + Z$. The centrifugal potential Z at an Earth's surface point P can be calculated analytically as a function of the orthogonal distance of the point to the Earth's rotational axis, and the Earth's angular velocity ω, which can both be measured very precisely. Hence, the main challenge of gravity field modeling is the accurate determination of the gravitational potential V that depends on the mass density distribution in the Earth's interior. As this information is not globally known with sufficient accuracy and spatial resolution, gravity field determination is usually related to the solution of the Geodetic Boundary Value Problem (GBVP, Heiskanen and Moritz, 1967, p. 36f.).

Depending on whether the theory of Stokes or Molodensky is used, the geoid or the Earth's surface is employed as the boundary surface, respectively. For the determination of the required boundary values, various measurement approaches can be applied, which are based on the observation of different gravity field functionals on or close to the Earth's surface. These include:

Terrestrial gravity measurements. Based on the principle of gravimetry, gravity values g can be measured at the Earth's surface, where $g(P) = |\operatorname{grad} W(P)|$ denotes the modulus of the gradient of the gravity potential at a point P. These terrestrial gravity measurements provide highly precise information and are the basic observables within the solution of the GBVP. However, they can only be determined pointwise and require time-consuming and expensive field work. For this reason, gravity data is generally inhomogeneously distributed and of varying quality. Moreover, for larger parts of the Earth, the access is restricted or no data is available. Due to these gaps in the global gravity coverage, a gravity field determination solely based on terrestrial data might only be possible on a regional scale.

Satellite altimetry. For the coverage of the oceans, information derived from satellite altimetry can be utilized. In this case, the height of the sea surface above a reference ellipsoid is measured by means of radar techniques. Differences between the observed sea surface and the geoid can be attributed to the sea surface topography. In order to derive precise gravity field data, i.e., geoidal heights, from satellite altimetry, a suitable separation between the geoid information and the sea surface topography needs to be performed.

Satellite-based gravity field missions. In contrast to terrestrial gravity data, the main benefit of satellite-based gravity field observations is that they provide independent and homogeneously distributed information with a consistent accuracy and a nearly global coverage. However, due to the decay of gravity with distance, the measurement sensitivity in satellite altitude decreases with an increasing distance from the Earth's center of mass. Therefore, gravity field information recovered from satellite-based missions is generally restricted in terms of its spatial and spectral resolution. The most prominent gravity field missions are CHAMP (CHAllenging Minisatellite Payload), GRACE (Gravity Recovery And Climate Experiment), and GOCE (Gravity field and steady-state Ocean Circulation Explorer), see Rummel et al. (2002) for an overview about their respective measurement concepts and objectives.

In addition to these measurements, the demand for a precise knowledge of the gravitational effect of the Earth's topography gained an increased interest within current gravity field modeling and has become more and more important in recent years. Analyzing the spectral composition of the gravity field, the high-frequency components can mainly be attributed to the gravitational attraction of the Earth's topographic and isostatic masses. These encompass the rock, water, and ice masses of the continents, oceans, lakes, and ice sheets, as well as their deeper lying (isostatic) compensation masses in the Earth's interior.

By using the information of global high-resolution digital terrain models (DTMs), the topographic-isostatic potential and related functionals can be determined by gravity forward modeling (GFM), see Sect. 2.1, and utilized for various applications in geodesy and geophysics. While Table 1 provides a general overview of the main applications of topographic-isostatic GFM and their purposes, they are summarized in the following:

Table 1. Main applications of topographic-isostatic gravity forward modeling.

Application	Purpose
Mass reductions	Solution of the GBVP according to the theory of Stokes
	Interpretation of observations in geophysical studies
Smoothing	Numerical stabilization of interpolation tasks
	Numerical stabilization of field transformations
Spectral extension	Augmentation of band-limited measurements
	Reduction of the omission error of GGMs
Gravity prediction	Improvement of the spatial resolution of terrestrial data sets
	Fill-in information for the construction of combined GGMs
Validation	Independent evaluation of satellite-based GGMs
	Verification of new methods, theories, and algorithms

1) *Mass reduction.* In physical geodesy, topographic-isostatic mass reductions are utilized to obtain mass-free boundary values at geoid level as required for the solution of the GBVP according to the theory of Stokes (Heiskanen and Moritz, 1967, Chapter 3). In geophysical studies and solid-earth sciences, topographic mass reductions of gravity field observations are needed to interpret crustal structures (Ebbing et al., 2001; Tenzer et al., 2012), to analyze mass anomalies in the Earth's interior (Wieczorek, 2007), or to perform a general lithospheric modeling (Bouman et al., 2015).

2) *Smoothing.* Many calculation methods in gravity field modeling are embedded in a remove-compute-restore framework (Forsberg and Tscherning, 1997), where topographic-isostatic information is used to smooth gravity field observations. In this context, the main purpose is to achieve a numerical stabilization of interpolation tasks or field transformations (Forsberg, 1984, Sect. 4). Such a procedure is frequently used for harmonic downward continuation of airborne and satellite-based measurements (Novák et al., 2003; Makhloof and Ilk, 2008; Wild and Heck, 2008; Janák et al., 2014), but is also applied for a gravimetric geoid determination (Omang and Forsberg, 2000).

3) *Spectral extension.* In recent years, one of the most important applications of topographic GFM is the spectral extension of band-limited gravity field observations or global geopotential models (GGMs). This is particularly important in the case of satellite-derived GGMs, which are generally restricted to long and medium wavelengths according to the maximum degree of their spherical harmonic representation. To determine the remaining signal above this maximum degree, known as the omission

error of the GGM, topographic GFM is usually applied in terms of residual terrain modeling (Forsberg and Tscherning, 1997).

4) *Gravity prediction.* Topographic-isostatic gravity information can be utilized to refine the spatial resolution of terrestrial data sets. This is of special interest for regions where no or only limited gravity data is available, like in Africa (Abd-Elmotaal et al., 2016). In this context, Hirt et al. (2016) recently presented a combination of satellite-based gravity data with topographic information in order to improve the spatial resolution of the gravity field information for Antarctica. Moreover, topographic-isostatic gravity data is needed as fill-in information for the compilation of combined global high-resolution gravity field models, like EGM2008 (Pavlis et al., 2012); see also Hirt et al. (2013) and Fecher et al. (2015, 2016) for other examples.

5) *Validation.* As an independent source of gravity field information, topographic-isostatic gravity data can be used to assess the performance of GGMs (Tsoulis and Patlakis, 2013). Recently, this has particularly been done to evaluate the improvements gained from different releases of GOCE satellite-based GGMs (Hirt et al., 2012, 2015). Moreover, topographic-isostatic information is applied to construct a synthetic Earth gravity model (Kuhn and Featherstone, 2005), which can be used for testing and verification of new methods, theories, and algorithms (Baran et al., 2006).

1.2. Outline

The rest of this introductory chapter is organized as follows: in Sect. 2, the general principle of GFM is introduced considering space and frequency domain techniques. Additionally, background information is provided for the two central applications related to the GOCE gravity field mission and height system unification. In Sect. 3, the limitations of current GFM approaches are outlined, and, based on this consideration, the research objectives of this thesis are defined. While Sect. 4 summarizes the main contributions of this thesis, Sect. 5 provides an overview of the related publications as presented and reprinted in Chapters II – VI. An overall conclusion of the thesis including an outlook to future research can be found in Sect. 6.

2. Foundations

2.1. Gravity forward modeling

Gravity forward modeling (GFM) is generally based on the evaluation of Newton's integral for the gravitational potential of a solid body $\Omega \subset \mathbb{R}^3$ (Heiskanen and Moritz, 1967, p. 3):

$$V(P) = G \iiint\limits_{\Omega} \frac{\rho}{\ell} \, \mathrm{d}\Omega, \tag{1}$$

where G denotes Newton's gravitational constant, $\rho = \rho(Q)$ is the location-dependent density function, and $\ell = \ell(P, Q)$ is the Euclidean distance between the attracted computation point $P \in \mathbb{R}^3$ and the running integration mass point $Q \in \Omega$. Depending on the specific application, the integration domain Ω can extend over quite different mass distributions, with the primary example being the Earth's topographic-isostatic masses. However, in each case, information on the geometry and density of the mass distribution is required.

In the frequency domain, Newton's integral in Eq. (1) can alternatively be represented by a series of solid spherical harmonics. In terms of geocentric spherical coordinates (r, φ, λ), this spherical harmonic expansion is expressed by (Heiskanen and Moritz, 1967, p. 59):

$$V(P) = \frac{GM}{r} \sum_{n=0}^{\infty} \left(\frac{R}{r}\right)^n \sum_{m=0}^{n} \left(\overline{C}_{nm} \cos m\lambda + \overline{S}_{nm} \sin m\lambda\right) \overline{P}_{nm}(\sin\varphi), \tag{2}$$

where GM is the geocentric gravitational constant, R is the radius of a reference sphere, $\overline{P}_{nm}$ denotes the fully normalized associated Legendre functions (ALFs) of degree n and order m, and $\overline{C}_{nm}$ and $\overline{S}_{nm}$ are the fully normalized spherical harmonic (SH) coefficients.

According to the representations in Eqs. (1) and (2), different GFM methods have been proposed that can be classified as space and frequency domain techniques (e.g., Kuhn and Seitz, 2005). In Fig. 1, a general processing scheme for GFM in both domains is presented and is described in the following two paragraphs.

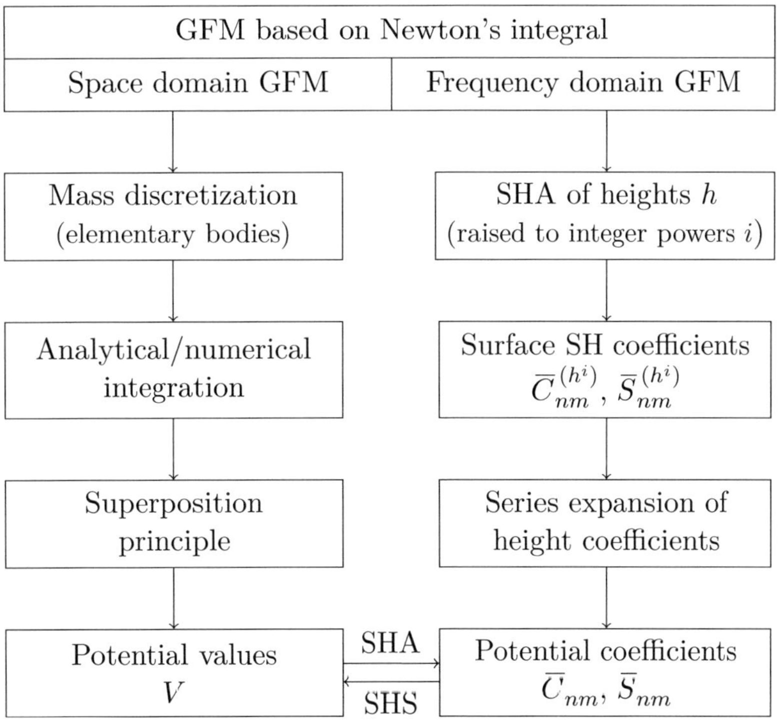

Fig. 1. General processing scheme for gravity forward modeling (GFM) in the space and frequency domain (Grombein et al., 2016a). The results of both GFM methods, i.e., potential values or coefficients, can be converted to each other by means of spherical harmonic analysis (SHA) and synthesis (SHS).

In the case of *space domain GFM*, Newton's integral in Eq. (1) is evaluated by a mass discretization. In the first step, the integration domain Ω is decomposed into elementary mass bodies $\Omega_i^* \subset \Omega$ with assumed constant density values ρ_i, where $\Omega = \bigcup \Omega_i^*$. Depending on the type of application (local, regional, global) and the accuracy requirements, various mass bodies or suitable approximations are frequently used, see Kuhn (2000, Chapter 6) and Wild-Pfeiffer (2008) for a detailed overview.

In a second step, the gravitational potential V_i^* of each individual mass body is calculated by analytical or numerical integration. In the third step, the potential V of the whole mass distribution is then approximated by the sum of the potential values V_i^* according to the superposition principle:

$$V \approx \sum_i V_i^* = \sum_i G\rho_i \iiint\limits_{\Omega_i^*} \frac{1}{\ell}\, \mathrm{d}\Omega. \tag{3}$$

Analogously, this procedure can be applied to the derivatives of V.

For *frequency domain GFM*, Newton's integral in Eq. (1) is evaluated through a transformation into the frequency domain (Rummel et al., 1988; Wieczorek, 2007; Hirt and Kuhn, 2012). In the first step, global gridded DTM heights h are raised to integer powers, i.e., h^i with $i \in \mathbb{N}$, and the corresponding surface SH coefficients $(\overline{C}_{nm}^{(h^i)}, \overline{S}_{nm}^{(h^i)})$ are calculated via spherical harmonic analysis (SHA). In a second step, the solid SH coefficients $(\overline{C}_{nm}, \overline{S}_{nm})$ of the potential are calculated based on a series expansion of the inverse distance, involving the surface SH coefficients $(\overline{C}_{nm}^{(h^i)}, \overline{S}_{nm}^{(h^i)})$ obtained from the DTM heights. In the third step, the derived set of SH coefficients $(\overline{C}_{nm}, \overline{S}_{nm})$ can be used to calculate potential values and functionals via spherical harmonic synthesis (SHS) according to Eq. (2).

2.2. GOCE gravity gradiometry

As the first core mission of the Living Planet Programme, initiated by the European Space Agency (ESA), the satellite mission GOCE observed the Earth's static gravity field between November 2009 and October 2013 (ESA, 1999). Based on the concept of satellite gravity gradiometry (Rummel and Colombo, 1985), GOCE measured in-situ the second-order derivatives of the Earth's gravitational potential at an altitude of about 250 km. These derivatives are specified in a local Cartesian coordinate frame (x_1, x_2, x_3) and can be arranged in the symmetric Marussi tensor $\mathbf{M}$, where

$$M_{ij} = \frac{\partial^2 V(x_1, x_2, x_3)}{\partial x_i\, \partial x_j} = M_{ji}, \quad i, j \in \{1, 2, 3\} \tag{4}$$

are its physical components, known as gravity gradients. In the context of GOCE gradiometry, gravity gradients are commonly represented in two different coordinate frames: (i) the instrument-fixed Gradiometer Reference Frame (GRF), where the x_1-axis points in the flight direction, the x_2-axis is orthogonal to the instantaneous orbit plane, and the x_3-axis points upwards in the geocentric radial direction, and (ii) the local north-oriented frame (LNOF), where the x_1-axis points north, the x_2-axis west, and the x_3-axis upwards in the geocentric radial direction. Within the measurement bandwidth of the GOCE gradiometer

(5 to 100 mHz), the main diagonal tensor components M_{11}, M_{22}, M_{33} and the off-diagonal component M_{13} were observed with a precision of about 1 to $2\,\mathrm{mE}/\sqrt{\mathrm{Hz}}$ ($1\,\mathrm{E} = 10^{-9}\,\mathrm{s}^{-2}$) in the GRF, while the other off-diagonal components M_{12} and M_{23} are less accurately measured with a precision of about $1\,\mathrm{E}/\sqrt{\mathrm{Hz}}$ (cf. Rummel et al., 2011).

A primary geodetic mission objective of GOCE is to contribute to global height system unification by providing a precise global geoid with an accuracy of 1–2 cm (Rummel, 2002). Due to the measurement at satellite altitude and the limited bandwidth of the GOCE gradiometer, the derived geoid is restricted to a spatial resolution of about 100 km. This corresponds to a SH expansion approximately up to degree and order (d/o) 200 in terms of a derived GGM.

2.3. Height system unification

By using tide gauge observations, national height reference systems have conventionally been linked to the local MSL. Due to variations in the sea surface topography, different tide gauges do not realize the same height reference level in terms of a global equipotential surface. Therefore, hundreds of different national height systems exist worldwide that are realized by their own local vertical datum. The resulting discrepancies cause height datum offsets of about ±1–2 m at a global scale (Heck, 1990; Gerlach and Rummel, 2013; Sánchez, 2015), making it impossible to directly compare the physical heights of different countries.

This can cause practical problems for regional applications, when height information from neighboring countries needs to be combined, e.g., in the case of international engineering projects or flooding control. Additionally, for monitoring global geodynamic and climatological processes, such as sea level rise, a consistent physical height reference level is required. Furthermore, height system unification is also relevant for the realization of the Global Geodetic Observing System (Ihde and Sánchez, 2005).

3. Problem statements and research objectives

Due to increasing accuracy requirements for different applications and an improved sensitivity and precision of terrestrial and space geodetic measurements, it is required to scrutinize current approximations and simplifications in gravity forward modeling. If necessary, appropriate refinements of the theory and modifications of commonly used methods have to be taken into consideration.

3.1. State-of-the-art of gravity forward modeling

In geodesy, rectangular prisms (Mader, 1951; Nagy et al., 2000) have traditionally been utilized for space domain GFM, and are still being applied in the case of the widely-used *GRAVSOFT* program routine *TC* (Forsberg and Tscherning, 2008). While rectangular prisms are suitable for local applications, they cannot generally be recommended for applications at a regional or global scale due to their underlying planar approximation. To improve the rough approximation, prisms with curved surfaces, e.g., topped by a bilinear surface (Smith et al., 2001; Tsoulis et al., 2003), as well as general polyhedral bodies (Tsoulis, 2012; D'Urso, 2013) have been proposed.

With an increasing number of global applications, tesseroid mass bodies, as introduced by Anderson (1976, p. 48ff.), have become more and more popular in recent years, as they directly take the curvature of the Earth into account. As shown in Fig. 2, tesseroids are bounded by a pair of concentric spheres (r_1 = const., r_2 = const.), a pair of meridional planes (λ_1 = const., λ_2 = const.), and a pair of coaxial circular cones, defined by the parallels φ_1 = const., φ_2 = const. In terms of precision and computation time, various studies have demonstrated the high numerical efficiency when utilizing tesseroids instead of conventional rectangular prisms, see, e.g., Heck and Seitz (2007), Wild-Pfeiffer (2008), and Grombein et al. (2010a, Chapter 7). However, even in the case of tesseroids, one of the main issues of space domain GFM is the still enormous computational demand, particularly when using information of global high-resolution DTMs.

Although this property also holds true for frequency domain GFM, a major advantage is that the gravitational potential is directly represented in terms of SH coefficients (e.g., Rummel et al., 1988; Balmino et al., 2012; Hirt and Kuhn, 2012), allowing to efficiently calculate various gravity field functionals by SHS. However, frequency domain methods suffer from several drawbacks, which are pointed out in the following. Due to the nature of spherical harmonics, frequency domain GFM is generally limited to a spherical approximation, i.e., topographic masses are arranged on a spherical reference surface. As this is insufficient for current high-resolution applications, Claessens and Hirt (2013) proposed a new spectral approach taking into account an ellipsoidal approximation; see also Novák and Grafarend (2005) for the use of an expansion in ellipsoidal harmonics. Another issue is that almost all GFM approaches in the frequency domain are currently restricted to the use of one mass layer of constant density.

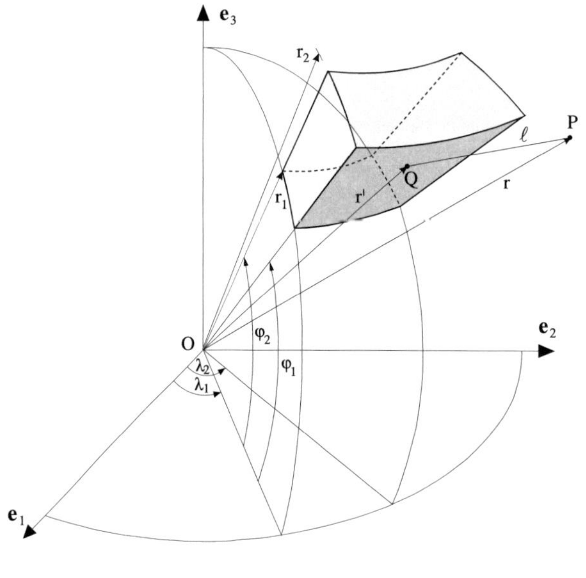

Fig. 2. Geometry of a tesseroid used for the space domain GFM (Heck and Seitz, 2007).

A common way to account for density variations of different terrain types is the concept of rock-equivalent heights (REQ), also called rock-equivalent topography (e.g., Rummel et al., 1988; Kuhn and Seitz, 2005; Hirt et al., 2015). By applying a mass conservation condition, the DTM columns of water and ice masses are condensed such that their associated mass elements correspond to an equivalent mass of rock density. Hence, topographic masses can be modeled with a constant density value, which considerably simplifies the GFM calculation. However, the shortcoming of this simplification is a changed geometry with considerable mass displacements, as noted by Tsoulis and Kuhn (2007). Due to the distance dependency, this has also an impact on any derived gravity field functional.

The load of the topographic masses is generally compensated by deeper lying mass distributions in the Earth's interior. In order to consider these isostatic compensation masses, two concepts are commonly employed in geodesy: the isostatic concept of Airy-Heiskanen and the one of Pratt-Hayford (Göttl and Rummel, 2009). Both models rely on the simplified hypothesis of a local and column-based mass compensation with respect to a particular normal compensation depth D. In the Airy-Heiskanen isostatic concept (Heiskanen and Moritz, 1967, p. 135ff.), the thickness d' of an isostatic mass column varies, while its density value $\Delta\rho$ is kept constant. Following this idea, the depth $D + d'$ is then assumed to correspond to the boundary surface between the Earth's crust and mantle, known as Mohorovičić discontinuity (abbreviated as Moho), and $\Delta\rho$ represents the crust-mantle density contrast. On the contrary, the Pratt-Hayford isostatic concept (Heiskanen and Moritz, 1967, p. 134ff.) assumes a constant thickness of an isostatic mass column along with laterally varying density values, depending on the topographic height.

However, both classically applied concepts have their disadvantages and do not provide reasonable results when applied globally. For example, the Airy-Heiskanen isostatic concept fails over deep ocean trenches, where d' may rise above the ocean bottom (Wild and Heck, 2005). Therefore, several studies suggested to apply a combination of different isostatic concepts, e.g., the use of the Airy-Heiskanen concept over land areas and the Pratt-Hayford model over the oceans (Wild and Heck, 2005; Göttl and Rummel, 2009; Hirt et al., 2012).

Based on a refinement of Airy's theory, the Vening-Meinesz model (Moritz, 1990, p. 222ff.; Abd-Elmotaal, 1995) accounts for a regional, instead of local isostatic compensation. Although such a compensation scheme is more realistic, from a mathematical point of view, the Vening-Meinesz model is much more complicated, making it hardly applicable on a global scale. Moreover, as pointed out by Kaban et al. (2004), additional information about the Earth's interior and its crustal structure needs to be integrated, in order to achieve a more geophysically meaningful isostatic compensation.

For the spectral extension of band-limited measurements or GGMs, the results of gravity forward modeling cannot be used directly, as the topography generally contributes to all spectral scales of the gravity field. In addition, some kind of high-pass filtering has to be performed. For this purpose, residual terrain modeling (RTM, Forsberg and Tscherning, 1997) is a widely used tool that has been utilized in various studies and for different kinds of applications (e.g., Hirt et al., 2010; Hirt, 2013; Šprlák et al., 2015). In this approach, gravity forward modeling is applied to a residual topography between a high-resolution DTM and a smoothed reference topography, whose spectral information is assumed to be already contained in the measurements or GGM that should be augmented. In this way, the RTM method presupposes a spectral consistency of topographic heights and implied gravity. However, as pointed out by Hirt and Kuhn (2014), a band-limited topography can also generate a full spectrum gravity field. Therefore, the basic assumption of the RTM approach cannot be considered valid in general. However, this issue has not received due attention so far.

3.2. Research objectives

The general research objective of this thesis is to develop a new, more realistic GFM approach that allows to efficiently calculate topographic-isostatic effects for different gravity field functionals and overcomes the current limitations of other methods as specified above. Besides this methodological aspect, it is intended to conduct different kind of studies in order to evaluate the performance and see the benefit of the newly developed method. In detail, the following questions and aspects have been addressed within the thesis:

- Is it possible to optimize the currently used tesseroid formulas in order to achieve a significant reduction of the computation time for space domain GFM?
- How does the approximation error of the commonly used rock-equivalent modeling affect different gravity field functionals and which magnitudes can be expected for terrestrial, airborne and satellite-based applications?
- Is it possible to enhance the topographic-isostatic modeling by a rigorous separate modeling of different terrain types with variable density values?
- Can space domain GFM approaches overcome current limitations of global GFM in the frequency domain?
- Is it possible to find an alternative approach to residual terrain modeling that allows to perform the required high-pass filtering directly in the gravity domain?
- Is it feasible to effectively smooth measured GOCE gravity gradients by means of topographic-isostatic reductions?
- Can topography-implied gravity signals profitably contribute to the problem of global height system unification?

4. Contributions

In this section, the main contributions of the thesis are summarized and references are provided with respect to the attached publications. Succinctly stated, these are:

- Elaboration of optimized tesseroid formulas that reduce the computational demand by about 20 to 55 %, depending on the respective gravity field functional of interest, see Grombein et al. (2013) in Chapter II.

- Development of the advanced Rock-Water-Ice (RWI) approach that is based on a three-layer decomposition of the Earth's topography with respect to its rock, water, and ice mass constituents, and a modified Airy-Heiskanen concept that incorporates Moho depths, see Grombein et al. (2014a, 2016a) in Chapters III and IV.
- Release of several Rock-Water-Ice gravity field models (RWI models) that provide a high-resolution representation of the Earth's topographic-isostatic potential in terms of SH coefficients, see Grombein et al. (2014a, 2016a) in Chapters III and IV.
- Analyses of the benefit of the RWI approach in comparison to the conventional rock-equivalent technique, see Grombein et al. (2016a) in Chapter IV.
- Application of the RWI approach for the topographic-isostatic reduction of measured GOCE gravity gradients, see Grombein et al. (2014a) in Chapter III.
- Development of an alternative to residual terrain modeling based on GFM using the RWI approach and the corresponding RWI models, see Grombein et al. (2017) in Chapter V.
- Application of this strategy for the spectral extension of satellite-based GOCE gravity field models in the context of global height system unification, see Grombein et al. (2017) in Chapter V.
- Analyses of a newly developed fixed GBVP approach for height system unification, see Grombein et al. (2016b) in Chapter VI.

4.1. Optimized tesseroid formulas

Gravity forward modeling is generally a very time consuming task, where the runtime increases quadratically with the spatial resolution of the used topography model. To handle the enormous computational demand in the case of global high-resolution input data, it is indispensable to use techniques of parallel computation. However, as a first step, it is reasonable to evaluate and optimize existing formulas.

Investigations in this context have concentrated on formulas for the gravitational potential V^* of a homogeneous spherical tesseroid, as well as its first- and second-order derivatives that are defined with respect to a topocentric Cartesian coordinate frame (x_1, x_2, x_3), i.e., $\partial V^*/\partial x_i$ and $\partial^2 V^*/(\partial x_i\,\partial x_j)$ with $i, j \in \{1, 2, 3\}$, respectively. Instead of the local north-oriented frame as defined above and used for the GOCE gradiometry, a conventional left-handed frame is applied, where the x_2-axis points east instead of west.

Previously published tesseroid formulas are based on integral kernels with respect to geocentric spherical coordinates (e.g., Heck and Seitz, 2007; Wild-Pfeiffer, 2007, 2008). Thus, in order to calculate the first- and second-order derivatives in the moving Cartesian frame, additional transformations have to be applied that show polar singularities (Tscherning, 1976). In contrast to these approaches, first ideas of optimized tesseroid formulas based on Cartesian integral kernels have been formulated in Grombein et al. (2010a), and further elaborated and generalized within this thesis (cf. Grombein et al., 2013). These formulas avoid the explicit transformation and thus allow to evaluate the first- and second-order derivatives directly in the local Cartesian frame for any position on the globe. Moreover, this leads to a compact representation of the gravitational field functionals of a tesseroid that can be described consistently by the general formula

$$\left\{\begin{array}{c} V^*(P) \\ \frac{\partial}{\partial x_i} V^*(P) \\ \frac{\partial^2}{\partial x_i \, \partial x_j} V^*(P) \end{array}\right\} = G\rho \int_{\lambda_1}^{\lambda_2} \int_{\varphi_1}^{\varphi_2} \int_{r_1}^{r_2} \frac{1}{\ell^3} \left\{\begin{array}{c} \Delta x_i \Delta x_i \\ \Delta x_i \\ \frac{3\Delta x_i \Delta x_j}{\ell^2} - \delta_{ij} \end{array}\right\} r'^2 \cos\varphi' \, \mathrm{d}r' \, \mathrm{d}\varphi' \, \mathrm{d}\lambda', \tag{5}$$

where

$$\begin{aligned} \Delta x_1 &= r'\left[\cos\varphi \sin\varphi' - \sin\varphi \cos\varphi' \cos(\lambda' - \lambda)\right], \\ \Delta x_2 &= r' \cos\varphi' \sin(\lambda' - \lambda), \\ \Delta x_3 &= r'\left[\sin\varphi \sin\varphi' + \cos\varphi \cos\varphi' \cos(\lambda' - \lambda)\right] - r, \end{aligned} \tag{6}$$

denote the coordinate differences between the computation point $P(r, \varphi, \lambda)$ and the running integration point $Q(r', \varphi', \lambda')$,

$$\ell = \sqrt{\Delta x_1{}^2 + \Delta x_2{}^2 + \Delta x_3{}^2} = \sqrt{\Delta x_i \Delta x_i} \tag{7}$$

is the corresponding Euclidean distance, and δ_{ij} is the Kronecker delta, i.e., $\delta_{ij} = 1$ if $i = j$, and $\delta_{ij} = 0$ otherwise.

Since volume integrals linked to tesseroids cannot be solved analytically, a numerical evaluation of Eq. (5) is achieved by a Taylor series approach with a fourth-order error that has been adapted from Heck and Seitz (2007), see also Deng et al. (2016). As the structure of the Cartesian integral kernel is substantially simplified, Taylor coefficients can be represented in a compact and computationally attractive form that allows an efficient numerical evaluation. Thus, the main benefit of using the optimized tesseroid formulas is a significant speed-up of the calculation process, which has been verified by conducting a realistic numerical experiment (cf. Grombein et al., 2013). In comparison to previously published tesseroid implementations, the computation time is decreased by about 20 % for the potential, 28 % for the first-order derivatives, and 56 % for the second-order derivatives.

Particularly, due to studies related to GOCE gravity gradients, the optimized tesseroid formulas for the second-order derivatives have been rapidly spread and used for various applications in the field of forward and inverse modeling (e.g., Álvarez et al., 2014; Guerri et al., 2015; Bouman et al., 2016; Fuchs et al., 2016). Moreover, the elaborated formulas provide the basis for the software package *Tesseroids* (Uieda et al., 2011), which is used mainly in the geophysical community. Recently, the optimized tesseroid formulas have even been adapted to provide efficient evaluation rules for the magnetic potential and its first- and second-order derivatives (cf. Du et al., 2015).

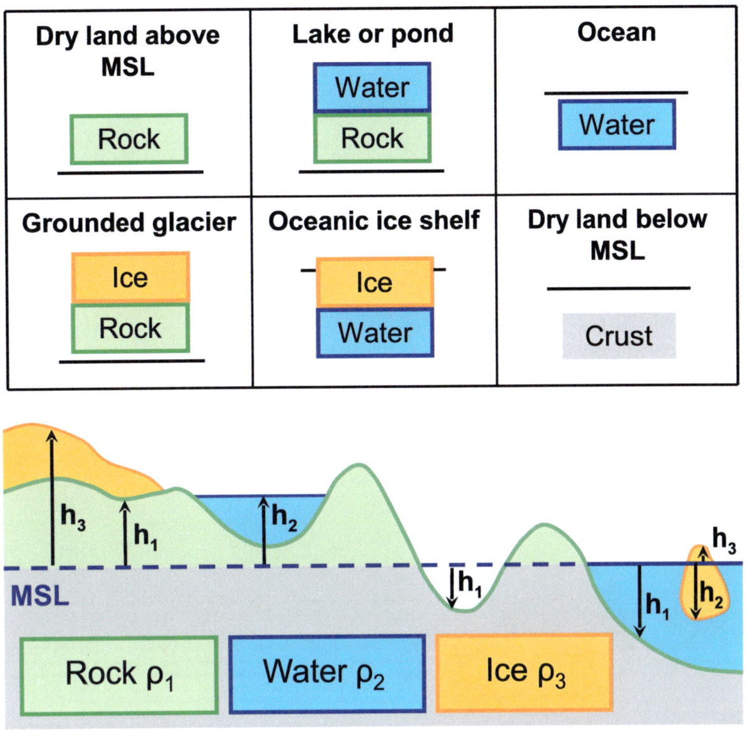

Fig. 3. Schematic representation of the RWI-based topographic model and the composition of different terrain types by rock, water, and ice masses (Grombein et al., 2016a). Note that a planar illustration is only used for the purpose of simplification, whereas all formulas of the RWI approach correspond to a spherical or ellipsoidal approximation.

4.2. The Rock-Water-Ice (RWI) approach

To address the above specified limitations of current GFM and to establish a more realistic topographic-isostatic modeling, the Rock-Water-Ice (RWI) approach has been developed within this thesis. This method relies on a three-layer decomposition of the Earth's topography that enables a rigorous separate GFM of the rock, water, and ice masses with variable density values. This prevents geometry changes and mass displacements, occurring in conventional rock-equivalent methods. Furthermore, a modified Airy-Heiskanen isostatic concept is applied that incorporates a depth model of the Mohorovičić discontinuity (Moho), i.e., the boundary surface between the Earth's crust and mantle.

While a detailed description and further refinements of the RWI approach are provided in the two attached publications Grombein et al. (2014a, 2016a), additional information is presented in Grombein et al. (2010b, 2011, 2014b,c). In the following, the topographic and isostatic models associated with the RWI approach are summarized.

The basic idea of the RWI approach is that the topographic masses of each terrain type, e.g., bedrock, oceans, lakes, ice sheets and ice shelves, can be composed by rock, water, and ice proportions. As schematically shown in Fig. 3, the sequence of these three masses is the same for each terrain type, however, some mass types may be lacking. For this

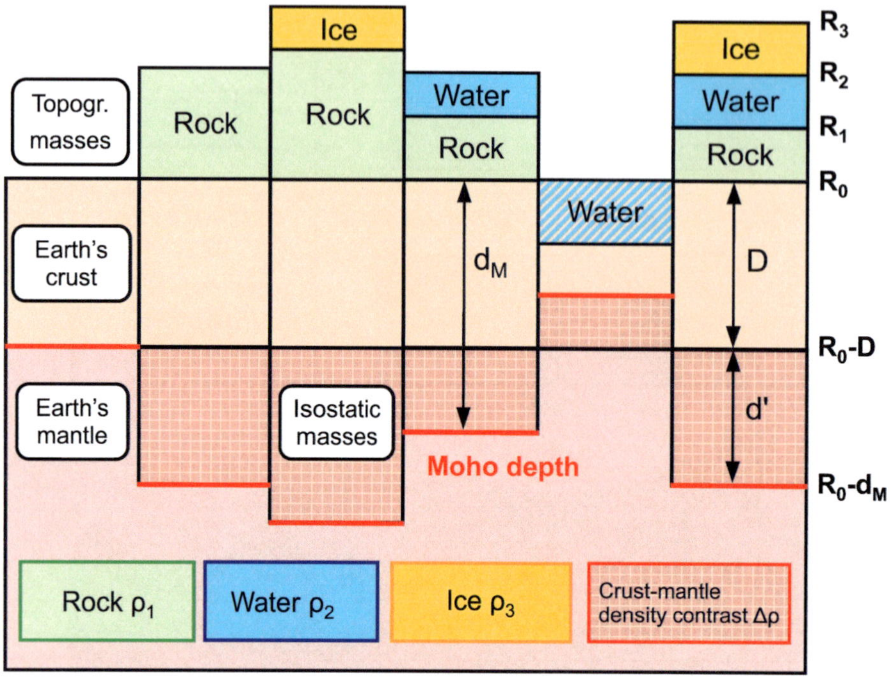

Fig. 4. Schematic representation of the modified Airy-Heiskanen isostatic concept (Grombein et al., 2014a). Note that a planar illustration is only used for the purpose of simplification, whereas all formulas correspond to a spherical or ellipsoidal approximation.

reason, the Earth's topographic masses can consistently be represented by a vertically arranged (1) rock, (2) water, and (3) ice layer. For the geometrical description of these layers, a global DTM is needed that provides the MSL heights (h_1, h_2, h_3), corresponding to the upper boundary surfaces of the rock, water, and ice masses, see Fig. 3. To correctly locate the topographic masses in space, an ellipsoidal reference surface is used, which is parameterized by a position-dependent geocentric radius $R_0(\varphi, \lambda)$. The geocentric radii of the rock, water, and ice masses are then approximated by $R_s = R_0 + h_s$, for $s \in \{1, 2, 3\}$. By using layer-specific density values (ρ_1, ρ_2, ρ_3), the gravitational effect of each mass layer is calculated separately, applying space domain GFM based on tesseroid mass bodies. For an efficient numerical calculation, the elaborated optimized tesseroid formulas are utilized.

To account for deeper lying compensation masses, the classical Airy-Heiskanen isostatic concept has been adapted to the RWI approach, see Fig. 4. This means that the isostatic masses are quantified from the topographic load, i.e., the sum of the rock, water, and ice masses, by applying a column-based mass equality condition with respect to a fixed normal compensation depth D. To enhance this simplified concept, additional geophysical information is introduced in terms of seismic Moho depths d_M, which are used to define the thickness d' of an isostatic mass column (cf. Fig. 4). Thus, in contrast to the original idea of Airy-Heiskanen, the geometry of the isostatic masses is fixed, allowing to account for variable density values. To this end, a location-dependent crust-mantle density contrast $\Delta\rho$ is estimated by the above specified mass equality condition, see Grombein et al. (2014a)

Table 2. Overview of the generated gravity field models of Release 2012 and 2015.

Release	Model name	Gravitational potential
Release 2012 (Grombein et al., 2014a)	RWI_TOPO_2012	Topographic
	RWI_ISOS_2012	Isostatic
	RWI_TOIS_2012	Topographic-isostatic
Release 2015 (Grombein et al., 2016a)	RWI_TOPO_2015	Topographic
	REQ_TOPO_2015	Topographic (rock-equivalent)

for the applied formulas in spherical approximation. While the gravitational effect of the isostatic masses is analogously calculated by tesseroid-based GFM, the combined topographic-isostatic effect results from the addition of both constituents.

Within the studies performed in this thesis, the RWI approach has been applied to different input data and was used for various gravity field functionals, such as gravity gradients (Grombein et al., 2014a) and height anomalies (Grombein et al., 2017). Moreover, a series of topographic-isostatic gravity field models has been generated based on the RWI approach, as will be outlined in the next subsection.

4.3. Topographic-isostatic gravity field models

In the framework of this thesis, several topographic-isostatic gravity field models have been generated and made publicly available in two releases, 2012 and 2015, associated with the publications Grombein et al. (2014a) and Grombein et al. (2016a), respectively. These RWI models provide a high-resolution representation of the Earth's topographic, isostatic, and topographic-isostatic gravitational potential in terms of SH coefficients. By using SHS, the models' coefficients can be used to efficiently calculate various functionals of the potential, and evaluate them in different heights or on different surfaces (e.g., Barthelmes, 2013).

The SH coefficients of the RWI models are obtained in a two-step sequence. In the first step, GFM with the tesseroid-based RWI approach is performed in the space domain, resulting in a high-resolution global grid of potential values. In a second step, these global gridded values are then transformed to the frequency domain by using SHA. While Table 2 provides a general overview of the derived gravity field models, their differences with respect to Release 2012 and 2015 are specified in Table 3 and are outlined in the following.

Release 2012. The RWI models of Release 2012 are based on the topographic information of the $5' \times 5'$ global DTM2006.0 model (Pavlis et al., 2007). Furthermore, standard density values, $\rho_1 = 2670\,\mathrm{kg\,m^{-3}}$, $\rho_2 = 1000\,\mathrm{kg\,m^{-3}}$, $\rho_3 = 920\,\mathrm{kg\,m^{-3}}$, and the reference surface of the GRS80 ellipsoid (Moritz, 1980) have been used to define and locate the rock, water, and ice masses. For the determination of the isostatic masses, Moho depths were obtained

Table 3. Specifications and differences between the models of Release 2012 and 2015.

		Release 2012	Release 2015
Input	Topography model	DTM2006.0 ($5' \times 5'$)	Earth2014 ($1' \times 1'$)
	Moho model	CRUST 2.0 ($2° \times 2°$)	–
Density values [$\text{kg}\,\text{m}^{-3}$]	Rock	2670	2670
	Water (Land/Ocean)	1000	1000 / 1030
	Ice	920	917
Space domain GFM	Mass discretization	Tesseroids	Tesseroids
	Reference surface	Ellipsoid	Ellipsoid + Geoid
	Calculation grid	Spherical grid ($5' \times 5'$)	Ellipsoidal grid ($4' \times 4'$)
SHA	Method	Least-squares technique	Least-squares technique
	Maximum d/o	1800	2190
	ALF algorithm	Thong (1989)	Holmes and Featherstone (2002)

from the $2° \times 2°$ global CRUST 2.0 model (Bassin et al., 2000) and accommodated to the resolution of the topography by means of SHA and SHS.

By applying the RWI approach to these input data, topographic and isostatic potential values were calculated on a $5' \times 5'$ global spherical grid with radius $R = 6\,398\,137\,\text{m}$. As the computations for each grid point are independent of each other, techniques of parallel computing can be used, in order to handle the enormous computational demand. To this end, a GFM software optimized for massive parallel computing on high-performance computer systems has been developed, using the message passing interface (MPI). For the transformation of the potential values to the frequency domain, SHA based on a least-squares technique for global gridded data has been applied (cf. Abd-Elmotaal et al., 2014). The used software implementation for this SHA is originally based on Seitz and Heck (1991) and has been further developed by K. Seitz. For the required calculation of the ALFs, an algorithm according to Thong (1989) has been utilized.

As a result of the computations, Release 2012 consists of three sets of SH coefficients up to degree and order (d/o) 1800:

- RWI_TOPO_2012 (topographic potential)
- RWI_ISOS_2012 (isostatic potential)
- RWI_TOIS_2012 (combined topographic-isostatic potential).

In Fig. 5a, the dimensionless signal degree variances of these models are shown.

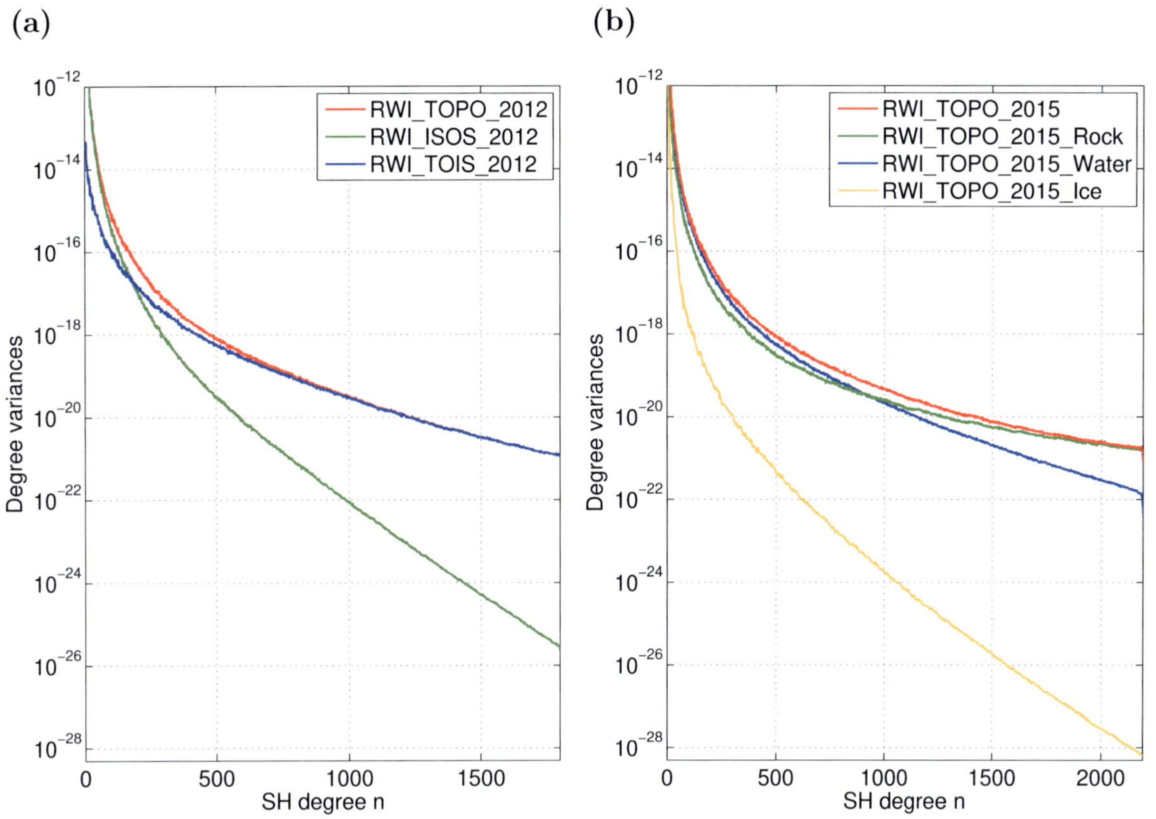

Fig. 5. Dimensionless signal degree variances of the SH coefficients from **(a)** Release 2012 models (Grombein et al., 2014a) – RWI_TOPO_2012 (*red curve*), RWI_ISOS_2015 (*green curve*), RWI_TOIS_2012 (*blue curve*) – and **(b)** Release 2015 models (Grombein et al., 2016a) – RWI_TOPO_2015 (*red curve*), RWI_TOPO_2015_Rock (*green curve*), RWI_TOPO_2015_Water (*blue curve*), RWI_TOPO_2015_Ice (*yellow curve*).

Release 2015. In the case of Release 2015, updated topographic information of the $1' \times 1'$ Earth2014 model (Hirt and Rexer, 2015) have been used to characterize the rock, water, and ice masses. Moreover, the applied layer-specific density values (ρ_1, ρ_2, ρ_3) were slightly modified, accounting for the difference in salt- and fresh-water density (see Table 3), and the used ellipsoidal reference surface has been refined by an additional geoid model.

Generally, the same software implementations have also been used for the processing of Release 2015. However, two main changes made it possible to extend the representation in spherical harmonics up to d/o 2190. Firstly, the space domain tesseroid-based GFM has been performed on an ellipsoidal grid with an increased resolution of $4' \times 4'$ that is approximately 20 km above the GRS80 surface; secondly, to improve the numerical stability in calculating the ALFs of higher degrees, a new algorithm based on Holmes and Featherstone (2002) has been implemented.

Release 2015 additionally features separate SH coefficients for the rock, water, and ice proportions. Furthermore, to see the additional benefit of the RWI approach, consistent rock-equivalent (REQ) versions of the RWI models have been generated, in which the heights of water and ice masses are condensed to the constant rock density. This allows a detailed comparison of both approaches, which has been conducted in the publication Grombein et al. (2016a). When evaluated on the surface of the GRS80 ellipsoid, the differences of the RWI- and REQ-based models reach maximum amplitudes of about 1 m, 50 mGal, and 20 mE in terms of height anomaly, gravity disturbance, and the radial-radial gravity gradient, respectively, see Fig. 6 for global variations. Although these differences are damped with increasing ellipsoidal height, significant magnitudes can even be detected in satellite altitudes of current gravity field missions (cf. Grombein et al., 2016a). Summarized, Release 2015 consist of the following models, each expanded up to d/o 2190,

- RWI_TOPO_2015 (topographic potential)
- RWI_TOPO_2015_Rock (rock proportion)
- RWI_TOPO_2015_Water (water proportion)
- RWI_TOPO_2015_Ice (ice proportion)

and their associated rock-equivalent versions

- REQ_TOPO_2015 (rock-equivalent topographic potential)
- REQ_TOPO_2015_Water (condensed water proportion)
- REQ_TOPO_2015_Ice (condensed ice proportion).

A separate REQ model for the rock proportion has not been generated, as it is identical to RWI_TOPO_2015_Rock. In Fig. 5b, the signal degree variances of the models are shown.

Availability and application of the models. The SH coefficients of the RWI models are freely available from the website https://www.gik.kit.edu/rwi_model.php and via the database of the International Centre for Global Earth Models (ICGEM, http://icgem.gfz-potsdam.de/ICGEM). Besides the conducted application in this thesis, the RWI models have already been used in a wide range of studies by different international research groups.

For generating the combined gravity field model GOCO05c, Fecher et al. (2015, 2016) use topographic-isostatic gravity anomalies of the RWI model to fill gaps in the global gravity coverage. To analyze the structure and state of stress of the Chilean subduction zone, Gutknecht et al. (2014) utilize the RWI models to derive topographically reduced gravity gradients of the satellite-based GOCO03s model (Mayer-Gürr et al., 2012). Bouman et al. (2016) make use of the RWI models to derive topographic mass reductions for GOCE gravity gradient grids, as used in geophysical applications. Furthermore, in some studies the RWI models have been applied as a reference for an independent validation (e.g., Novák and Tenzer, 2013; Pitonák et al., 2016).

(a)

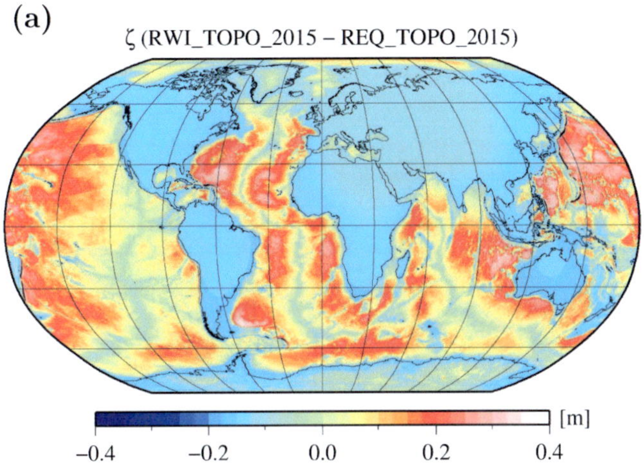

Statistics in [m]	
Min	−0.195
Max	0.987
Mean	0.000
STD	0.131

(b)

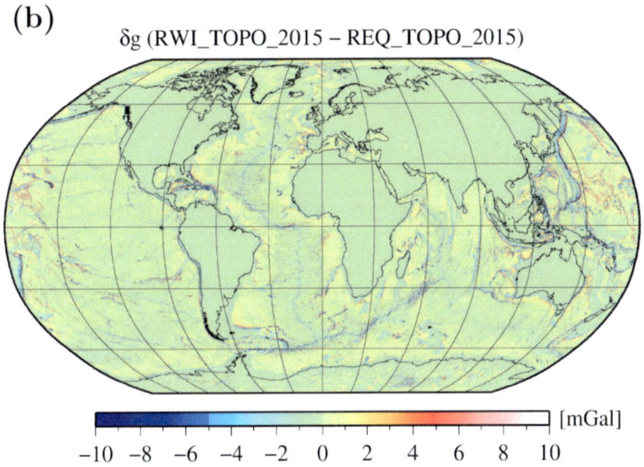

Statistics in [mGal]	
Min	−46.723
Max	36.410
Mean	0.000
STD	1.916

(c)

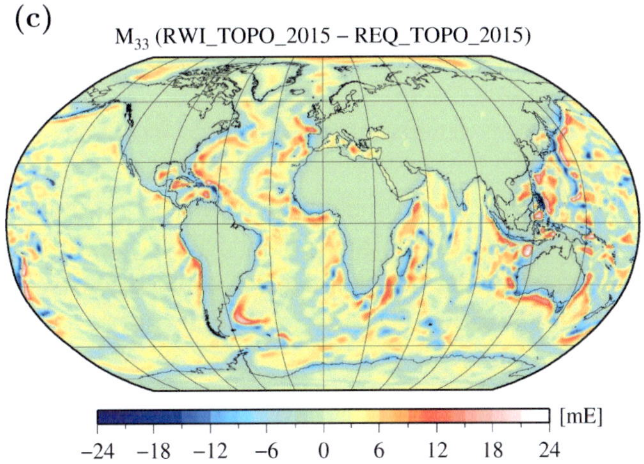

Statistics in [mE]	
Min	−19.336
Max	22.589
Mean	0.000
STD	3.833

Fig. 6. Difference of RWI_TOPO_2015 and REQ_TOPO_2015 in terms of **(a)** height anomalies ζ and **(b)** gravity disturbances δg evaluated on the surface of the GRS80 ellipsoid, as well as **(c)** the radial-radial gravity gradient M_{33} evaluated at a satellite altitude of about 250 km (Grombein et al., 2016a). Robinson projection centered at 0° longitude.

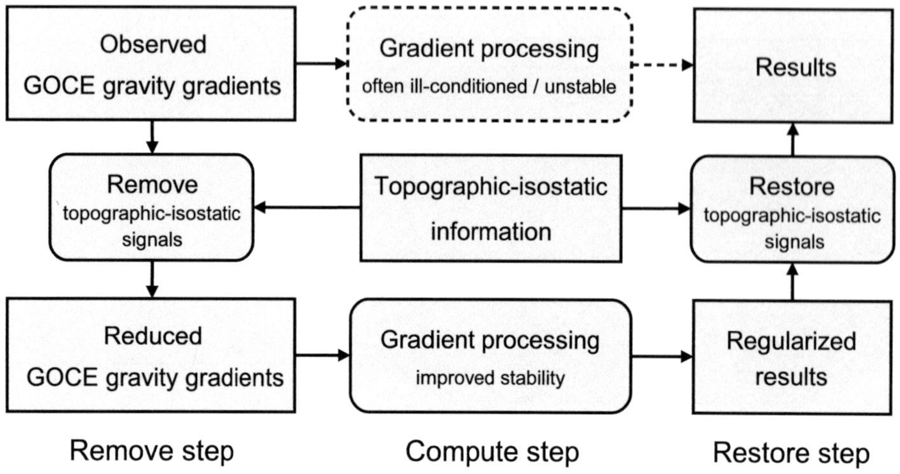

Fig. 7. Processing of GOCE gravity gradients by using topographic-isostatic information in a remove-compute-restore concept (Grombein et al., 2014c).

4.4. Smoothing of GOCE gravity gradients

By analyzing observed GOCE gravity gradients, high- and mid-frequency signals can be detected that are primarily caused by the attraction of the Earth's topographic and isostatic masses. Due to the induced rough measurement signal, interpolation tasks or field transformations, such as harmonic downward continuation of the gradients, can generally be considered an ill-conditioned processes. One possibility to improve the numerical stability is to smooth the observed gradients with respect to topographic-isostatic effects using a remove-compute-restore technique (Forsberg and Tscherning, 1997), as schematically shown in Fig. 7. In the remove step, topographic-isostatic signals are reduced from the observed GOCE gradients. Thereby, the high- and mid-frequency components of the gradients are mitigated, and mainly low-frequency constituents remain in the residual signal. In the compute step, the smoothed observation signals can be processed with an improved stability to gain regularized results. In the restore step, consistent topographic-isostatic signals are reconsidered to obtain the final results.

While previous studies in this context were based on simulated gravity gradients (Wild-Pfeiffer, 2007, 2008; Janák and Wild-Pfeiffer, 2010; Janák et al., 2012), the publication Grombein et al. (2014a) of this thesis provides the first application of topographic-isostatic reductions to real GOCE gravity gradient measurements. For this purpose, the RWI approach has been used to calculate topographic-isostatic signals in terms of gravity gradients along the orbit of the GOCE satellite, using the measurement positions as computation points. Note that in this study, the input data of Release 2012 was used, as specified in Table 3.

The RWI-based topographic and isostatic signals calculated along the orbit of the GOCE satellite, both reach extreme values of about $\pm 8\,\mathrm{E}$ and largely cancel each other out. Hence, the combined topographic-isostatic signal is nearly one order of magnitude smaller and ranges between $\pm 1\,\mathrm{E}$. To get an impression on global variations, Fig. 8 visualizes the

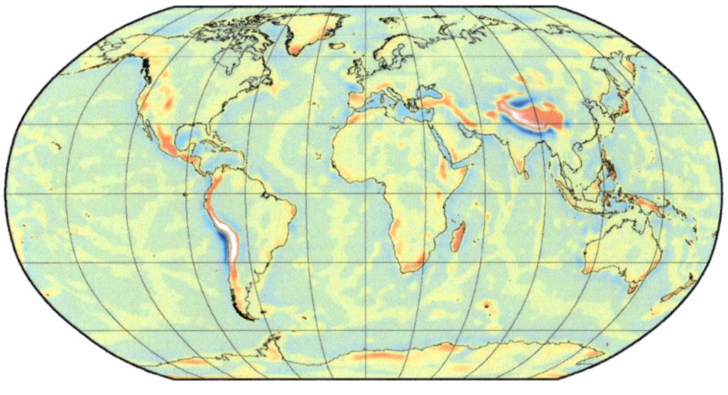

Fig. 8. RWI-based topographic-isostatic signal in terms of the radial-radial gravity gradient M_{33}, evaluated on a spherical grid at a mean GOCE satellite altitude (254.9 km) by using RWI_TOIS_2012 (Grombein et al., 2014a). Robinson projection centered at 0° longitude.

RWI-based topographic-isostatic effects for the radial-radial gravity gradient component M_{33}, evaluated on a spherical grid at a mean GOCE satellite altitude of 254.9 km.

Using a data set in the Himalayan region, the smoothing impact of topographic-isostatic signals on GOCE measurements has been analyzed. For this purpose, observed gravity gradients have been compared to the corresponding reduced ones within the measurement bandwidth of the gradiometer. This comparison has been performed in both, the space and the frequency domain. In the former, the degree of smoothing has been quantified by percentage changes in the standard deviation and range of the gradients before and after reduction. For spectral analysis, a continuous wavelet transform has been applied to the measured and reduced GOCE gravity gradients, allowing to simultaneously examine the signal structure in both time and frequency domains. By using percentage changes of the amplitude range, the degree of smoothing can then be estimated separately for different frequency values. In Fig. 9, the obtained results are exemplarily shown for the M_{33} gravity gradient signal, which can be reduced by about 50 %.

Generally, the analyses in both domains demonstrate that the RWI-based topographic-isostatic reductions lead to significant smoothing effects on the measurements for the M_{11}, M_{22}, M_{33}, and M_{13} components. For the less accurately observed gradients M_{12} and M_{23}, smoothing effects can only be detected in the spectral analysis and were invisible in the space domain, as in this case all frequencies are superposed (cf. Grombein et al., 2014a).

4.5. Spectral extension in the framework of height system unification

For the purpose of height system unification, a satellite-based method is proposed in the publication Grombein et al. (2017), using GGMs derived from the GOCE mission. These GOCE GGMs provide global independent and homogeneous gravity field information that is not affected by any height datum offset and, hence, can be used for a consistent connection of different height systems (Rummel, 2002; Gatti et al., 2013). In this context, height

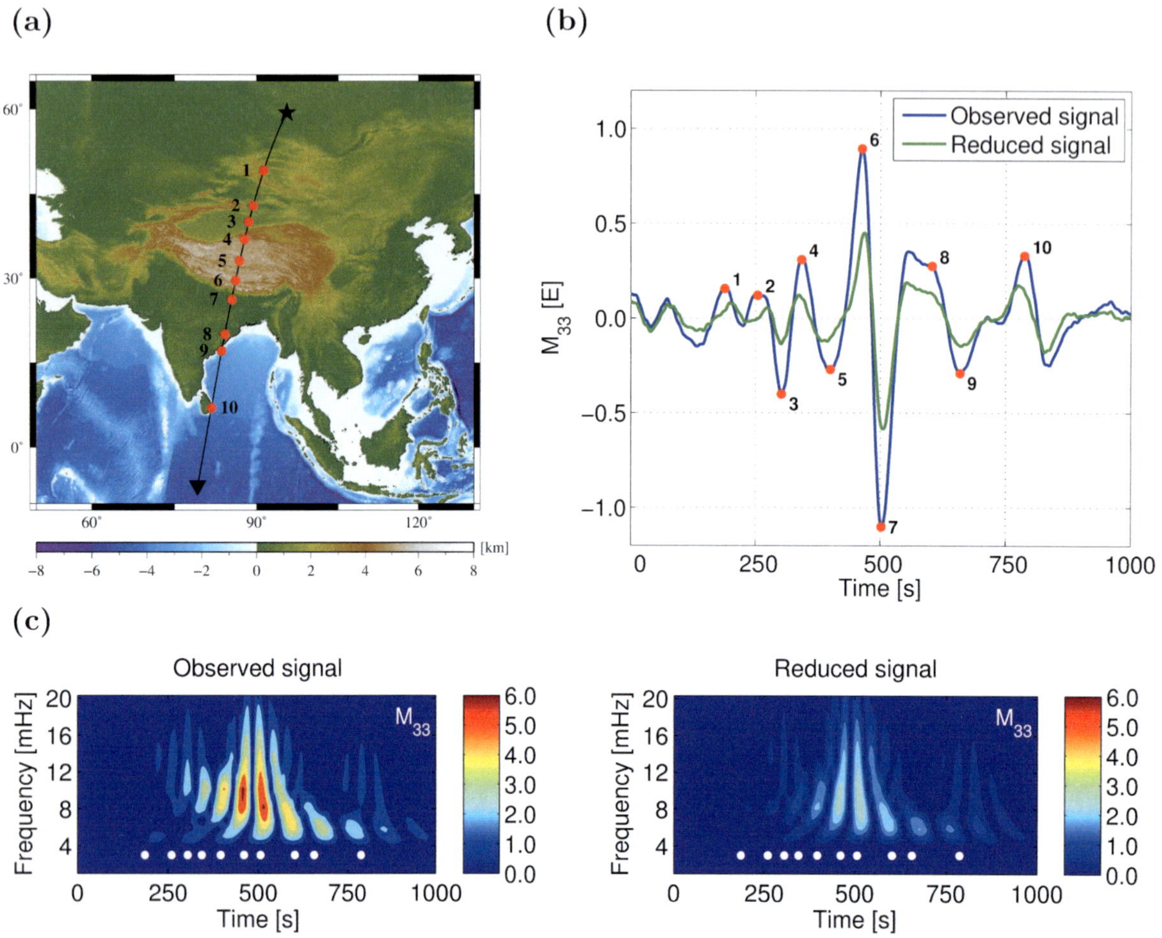

Fig. 9. Comparison of the observed M_{33} gravity gradient signal and the corresponding topographically-isostatically reduced signal along a GOCE satellite orbit that crossed the Himalayan region, as displayed in panel **(a)**. While in panel **(b)** the comparison is shown in the space domain by time series, panel **(c)** illustrates the results in the frequency domain by means of Morlet wavelet scalograms (Grombein et al., 2014a). In both cases, the signal content is band-pass filtered to the measurement bandwidth of the GOCE gradiometer. Note that the displayed absolute wavelet coefficients are dimensionless.

datum offsets are estimated within a least squares adjustment by comparing satellite-based GGM information with those of measured GNSS/leveling benchmarks, i.e., observation points that combine ellipsoidal heights derived from global navigation satellite system (GNSS) observations and physical heights in the local vertical datum. However, due to the limited spectral resolution of GOCE GGMs, their information needs to be augmented by additional data, accounting for the mid- to high-frequencies that are not captured by the satellite-based model. For a spectral extension, GOCE GGMs are usually combined with the high-resolution geopotential model EGM2008 (Pavlis et al., 2012). While this combination has been optimized by using a Hanning window (Blackman and Tukey, 1958),

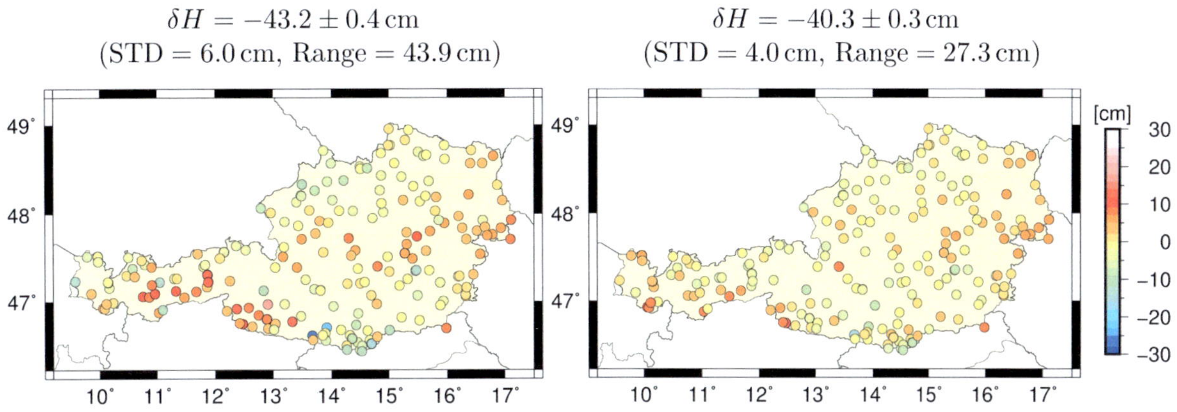

Fig. 10. Least squares adjusted residuals at 198 Austrian GNSS/leveling benchmarks with and without the consideration of topography-implied gravity signals (*right* and *left column*, respectively), cf. Grombein et al. (2017).

the investigations in Grombein et al. (2017) have mainly concentrated on the remaining high-frequency signals above the spectral resolution of EGM2008 that are assumed to be influenced by the Earth's topographic masses.

For the calculation of suitable high-frequency topography-implied gravity signals, a novel (residual) gravity forward modeling approach has been developed as an alternative to the widely-used RTM technique. In the first step of this new approach, gravity forward modeling based on tesseroid mass bodies is performed according to the RWI approach, using the input data of Release 2015 (cf. Table 3). In a second step, the resulting full spectral RWI-based topographic potential values are reduced by the effect of the topographic gravity field model RWI_TOPO_2015, thus, removing the long to medium wavelengths. Hence, in contrast to RTM, the required high-pass filtering is performed directly in the gravity domain, which avoids the doubtful assumption regarding the spectral consistency between topographic heights and implied gravity. While in principle this procedure is computationally more expensive, the availability of suitable topographic gravity field models, such as RWI_TOPO_2015, made this approach feasible.

For the analysis of the benefit and contribution of topography-implied gravity signals on the estimation of height datum offsets, three representative study areas in Germany, Austria, and Brazil have been selected. They differ strongly in size, geographical latitude, topographic properties, as well as the accuracy of their geodetic measurements. Furthermore, different GOCE GGMs of the latest (fifth) generation have been considered: TIM R5 (Brockmann et al., 2014), DIR R5 (Bruinsma et al., 2014), and GOCO05s (Mayer-Gürr et al., 2015).

To assess the performance of topography-implied gravity signals, an estimation of height datum offsets has been performed with and without the consideration of high-frequency topographic signals. Besides considerable changes in the estimated height datum offset δH of a few centimeters, the performed analyses show that significant improvements of 10–40 % can be achieved in terms of a reduced standard deviation (STD) and range of the least squares residuals, see Fig. 10 for the obtained results in Austria. While this holds true for the German and Austrian data sets, this is not the case for Brazil, where a large error level in the GNSS/leveling data seems to over-modulate the effect of topography-implied gravity signals. In summary, the analyses emphasized the importance of high-frequency topography-implied gravity signals for an accurate estimation of height datum offsets, particularly in regions with highly variable topography, where the remaining omission error of EGM2008 can still amount to a significant cm–dm level.

In addition to these investigations, the publication Grombein et al. (2016b) provides an outlook to further research work in the context of height system unification. To achieve a unification of high precision, the additional use of local terrestrial gravity data is required. For this purpose, the solution of the GBVP is used to estimate height datum offsets in a least squares approach (e.g., Rummel and Teunissen, 1988; Heck and Rummel, 1990). While previous studies have been mainly concentrated on the use of the scalar-free GBVP (e.g., Xu, 1992; Gerlach and Rummel, 2013; Amjadiparvar et al., 2016), the perspectives and benefits of the alternative use of the fixed GBVP approach have been discussed in Grombein et al. (2016b). To this end, the theory of the fixed GBVP approach has been extended for height system unification. Moreover, by conducting a closed loop simulation, it has been shown that mm-accuracy of the estimated height datum offsets can be achieved in principle. While the presented formulas were restricted to the use of terrestrial gravity data, for a practical realization of this approach, a combination with a GGM and topographic information is required in order to reduce systematic errors.

5. Publication overview

This thesis includes the following five peer-reviewed publications (four original articles in international journals and one proceedings contribution to an IAG Scientific Assembly):

1) Grombein, T., Seitz, K., and Heck, B. (2013): Optimized formulas for the gravitational field of a tesseroid. *Journal of Geodesy* 87(7):645–660. DOI: 10.1007/s00190-013-0636-1.

   Author's contributions. The first author derived the optimized tesseroid formulas, designed and carried out the numerical investigations, and wrote the manuscript for the article. The second author helped to numerically verify the formulas due to independent, comparative calculations. All authors discussed the results and contributed to the improvement of the manuscript by suitable comments and corrections.

2) Grombein, T., Luo, X., Seitz, K., and Heck, B. (2014a): A wavelet-based assessment of topographic-isostatic reductions for GOCE gravity gradients. *Surveys in Geophysics* 35(4):959–982. DOI: 10.1007/s10712-014-9283-1.

   Author's contributions. The first and second author designed the study. The first author developed the Rock-Water-Ice approach, processed the input data, carried out the computations and analyses, and wrote most of the manuscript for the article. The second author wrote the section about the mathematical background of the continuous Morlet wavelet transform and helped to implement and perform the wavelet-based analysis. The third author considerably contributed to the executed spherical harmonic analysis. All authors discussed the results and contributed to the improvement of the manuscript by suitable comments and corrections.

3) Grombein, T., Seitz, K., and Heck, B. (2016a): The Rock-Water-Ice topographic gravity field model RWI_TOPO_2015 and its comparison to a conventional rock-equivalent version. *Surveys in Geophysics* 37(5):937–976. DOI: 10.1007/s10712-016-9376-0.

   Author's contributions. The first author elaborated the refinements of the Rock-Water-Ice approach, designed the study, processed the input data, carried out the computations and analyses, and wrote the manuscript for the article. The second author considerably contributed to the executed spherical harmonic analysis. All authors discussed the results and contributed to the improvement of the manuscript by suitable comments and corrections.

4) Grombein, T., Seitz, K., and Heck, B. (2017): On high-frequency topography-implied gravity signals for height system unification using GOCE-based global geopotential models. *Surveys in Geophysics* 38(2):443–477. DOI: 10.1007/s10712-016-9400-4.

   Author's contributions. The first author developed the new residual gravity forward modeling method, designed the study, processed the input data, carried out the computations and analyses, and wrote the manuscript for the article. All authors discussed the results and contributed to the improvement of the manuscript by suitable comments and corrections.

5) Grombein, T., Seitz, K., and Heck, B. (2016b): Height system unification based on the fixed GBVP approach. In: Rizos, C., and Willis, P. (eds.) IAG 150 years. Proceedings of the IAG Scientific Assembly, Postdam, Germany, Sept. 1–6, 2013. *International Association of Geodesy Symposia*, vol. 143. Springer Berlin Heidelberg, pp. 305–311. DOI: 10.1007/1345_2015_104.

   Author's contributions. The third author formulated the idea for the fixed GBVP approach. The first author elaborated the theory and formulas, designed and carried out the closed loop simulation, and wrote the manuscript for the article. All authors discussed the results and contributed to the improvement of the manuscript by suitable comments and corrections.

6. Conclusions and outlook

6.1. Conclusions

In the framework of this thesis, the tesseroid-based Rock-Water-Ice (RWI) approach has been developed, which provides an important contribution to state-of-the-art gravity forward modeling, needed for various applications in geodesy and geophysics. The basis of the

RWI method is a three-layer decomposition of the Earth's topography with respect to its rock, water, and ice masses. This allows a rigorous separate gravity forward modeling of these masses with layer-specific density values, thus, avoiding geometry changes and mass displacements that arise in classically applied condensation methods like the use of rock-equivalent heights.

To account for the curvature of the Earth, tesseroid bodies are used for mass discretization and are arranged on an ellipsoidal reference surface. Moreover, to consider the effect of isostatic compensation masses, the Airy-Heiskanen concept has been modified by introducing additional geophysical information in terms of seismologically derived Moho depths. To reduce the increased computation time due to the more complex modeling, optimized tesseroid formulas have been elaborated that allow an efficient numerical evaluation with a significantly decreased runtime.

Besides a discussion and evaluation of the newly developed methods, two main applications in the context of the GOCE satellite mission and height system unification have been considered in this thesis. To this end, the RWI approach has been applied to different topographic input data and was used for various gravity field functionals.

In the first application, it has been shown that RWI-based topographic-isostatic effects can be used to significantly smooth measured GOCE gravity gradients. This is particularly important for a suitable harmonic downward continuation of the gradients from satellite altitude to the Earth's surface. The strongest benefit of this procedure can be seen in areas with a highly variable topography, such as the Himalayan region.

In a second application, topographic signals obtained from the RWI approach have been used to achieve a spectral extension of satellite-based GOCE global geopotential models, aiming to reduce their omission error. For this purpose, an alternative to residual terrain modeling has been suggested, which allows a high-pass filtering directly in the gravity domain. Hence, this strategy does not necessitate any assumption on the spectral consistency between topographic heights and induced gravity signals. By applying this concept in a study for the unification of height systems based on GOCE-derived global geopotential models, it has been shown that the additional consideration of topographic signals provides crucial information for an accurate estimation of height datum offsets.

In order to make the RWI method readily applicable for other researches, several high-resolution topographic-isostatic gravity field models (RWI models) have been generated and made publicly available. These models provide spherical harmonic coefficients of the Earth's topographic-isostatic potential that can be used to efficiently calculate various gravity field functionals by using spherical harmonic synthesis.

The models of Release 2012 (RWI_TOPO_2012, RWI_ISOS_2012, and RWI_TOIS_2012) are based on the $5' \times 5'$ global topographic database DTM2006.0 and are expanded up to degree and order 1800. This release particularly concentrates on the difference due to the topographic, isostatic, and topographic-isostatic potential. In contrast, the models of Release 2015 (RWI_TOPO_2015 and REQ_TOPO_2015) use updated height information of the $1' \times 1'$ Earth2014 topography suite and are expanded up to degree and order 2190. This release focuses on the differences between the RWI approach and the conventional use of rock-equivalent heights. Additionally, separate spherical harmonic coefficients are provided for the rock, water, and ice proportions of the topographic potential, namely RWI_TOPO_2015_Rock, RWI_TOPO_2015_Water, RWI_TOPO_2015_Ice, REQ_TOPO_2015_Water, and REQ_TOPO_2015_Ice.

In summary, the benefit of the RWI approach has been shown in various analyses. For instance, differences with respect to the use of rock-equivalent heights reach significant magnitudes in the case of terrestrial, airborne and satellite-based applications. Therefore, it is strongly recommended to consider a more realistic modeling, when calculating topographic-isostatic effects based on gravity forward modeling.

6.2. Outlook

This outlook provides some perspectives to research work that is aspired for the future, classified by four general scopes:

Refinements of the RWI approach. In principle, the RWI approach can easily be extended to the use of additional layers, possibly accounting for different sediments in the Earth's upper crust. This requires further information about crustal structures and corresponding density values. However, in contrast to global high-resolution information of the Earth's topography, current global crustal models still have a relatively low spatial resolution of $1^\circ \times 1^\circ$ (Laske et al., 2013) or $0.5^\circ \times 0.5^\circ$ (Reguzzoni and Sampietro, 2015). Therefore, such models are only applicable to model larger structures and mass anomalies.

In the case of water and ice masses, it is well-known that their actual density depends on pressure and temperature. This phenomenon can be properly described by using a radially varying density function $\rho = \rho(r)$, see Kuhn (2000, Chapter 5) for first attempts in this context. Although the present formulas are restricted to the use of a homogeneous density for each tesseroid, it should be straightforward to integrate a continuous radial variability of the density.

Generation of additional topographic-isostatic gravity field models. As a complement to the topographic gravity field model RWI_TOPO_2015, it is planned to also generate consistent isostatic and topographic-isostatic models, comparable to Release 2012. For this purpose, it is reasonable to update the used Moho depths and refine their spatial resolution (e.g., Reguzzoni et al., 2013).

Generally, the spatial resolution of current global digital terrain models would allow to perform a spherical harmonic analysis up to much higher degrees. As has been demonstrated in the publication Grombein et al. (2016a), the $4' \times 4'$ global ellipsoidal grid of RWI-based topographic potential values can be used to derive spherical harmonic coefficients up to degree and order 2699. By adapting the spatial resolution of the GFM-derived calculation grid to the $1' \times 1'$ resolution of Earth2014 model, spherical harmonic coefficients could theoretically even be expanded up to degree and order 10 800. However, when using a representation in spherical harmonics up to such ultra-high degrees, several problems arise:

1) Due to the term $(R/r)^n$ occurring in the formulas for spherical harmonic synthesis in Eq. (2), small errors in the spherical harmonic coefficients are strongly amplified for $r < R$ and large n. This is particularly critical in the case of ultra-high degrees and for an evaluation toward the poles (e.g., Holmes and Pavlis, 2007).

2) The ellipsoidally arranged topographic masses essentially generate an ellipsoidal harmonic spectrum. When mapping this spectrum to spherical harmonics, the resulting coefficients of higher degrees are correlated with those of lower degrees (cf. Jekeli, 1988). As a consequence, for an accurate spherical harmonic synthesis, it is necessary to evaluate the spherical harmonic coefficients up to their maximum degree, i.e., they should not be truncated. Considering a spherical harmonic expansion up to ultra-high degrees, this might not be feasible for all users and additionally seems not to be reasonable for all applications.

3) Another principal issue arises due to the numerical instability of calculating the associated Legendre functions of higher degrees (cf. Wittwer et al., 2008). In this context, suitable algorithms for ultra-high degrees are needed (e.g., Fukushima, 2012; Rexer and Hirt, 2015).

All these three aspects show the current limitations of using spherical harmonics for ultra-high degrees. Thus, in the future, it might be inevitable to find a more suitable spectral representation for high-resolution gravity field data, e.g., space localizing base functions such as harmonic splines or radial base functions (cf. Schmidt et al., 2007).

Investigation of residual GFM approaches. The results achieved with the novel (residual) GFM in the publication Grombein et al. (2017) are already very promising. To establish this approach as an appropriate alternative to the widely-used RTM techniques, it would be advantageous to conduct additional investigations. This includes (i) an adaption of this method to topographic data of higher spatial resolutions than the $1' \times 1'$ Earth2014 model, (ii) a direct and detailed comparison to the RTM method, (iii) an employment to other applications, such as a regional (quasi-)geoid determination.

Height system unification. To make the proposed fixed GBVP approach ready for practical use in height system unification, it is necessary to adapt the method with regard to the restricted availability of the global terrestrial gravity data. For this purpose, a suitable modification of Hotine's integral kernel is needed in order to restrict the global integration area (cf. Featherstone, 2013). Furthermore, to reduce systematic errors, it seems to be beneficial to combine the terrestrial data with a global geopotential model and topographic information, using a remove-compute-restore framework.

Moreover, the impact of the used spherically approximated GBVP solution on the estimation of height datum offsets needs to be analyzed, and correction terms for linearization errors as well as ellipsoidal and topographic effects are to be taken into account, where appropriate. First investigations with respect to the fixed GBVP are presented by Heck and Seitz (2014), Müßle et al. (2014), and Heck et al. (2015).

References

Abd-Elmotaal, H. (1995): Theoretical background of the Vening Meinesz isostatic model. In: Sünkel, H., and Marson, I. (eds.) Gravity and geoid, Joint symposium of the International Gravity Commission and the International Geoid Commission. Graz, Austria, Sept. 11–17, 1994. *International Association of Geodesy Symposia*, vol. 113. Springer Berlin Heidelberg, pp. 268–277. DOI: 10.1007/978-3-642-79721-7_28.

Abd-Elmotaal, H., Seitz, K., Abd-Elbaky, M., and Heck, B. (2014): Comparison among three harmonic analysis techniques on the sphere and the ellipsoid. *Journal of Applied Geodesy* 8(1):1–19. DOI: 10.1515/jag-2013-0008.

Abd-Elmotaal, H., Seitz, K., Abd-Elbaky, M., and Heck, B. (2016): Tailored reference geopotential model for Africa. In: Rizos, C., and Willis, P. (eds.) IAG 150 years. Proceedings of the IAG Scientific Assembly, Postdam, Germany, Sept. 1–6, 2013. *International Association of Geodesy Symposia*, vol. 143. Springer Berlin Heidelberg, pp. 383–390. DOI: 10.1007/1345_2015_84.

Álvarez, O., Nacif, S., Gimenez, M., Folguera, A., and Braitenberg, C. (2014): GOCE derived vertical gravity gradient delineates great earthquake rupture zones along the Chilean margin. *Tectonophysics* 622:198–215. DOI: 10.1016/j.tecto.2014.03.011.

Amjadiparvar, B., Rangelova, E., and Sideris, M. G. (2016): The GBVP approach for vertical datum unification: recent results in North America. *Journal of Geodesy* 90(1):45–63. DOI: 10.1007/s00190-015-0855-8.

Anderson, E. G. (1976): The effect of topography on solutions of Stokes' problem. Unisurv S-14, Report. School of Surveying, University of New South Wales, Australia.

Balmino, G., Vales, N., Bonvalot, S., and Briais, A. (2012): Spherical harmonic modelling to ultra-high degree of Bouguer and isostatic anomalies. *Journal of Geodesy* 86(7):499–520. DOI: 10.1007/s00190-011-0533-4.

Baran, I., Kuhn, M., Claessens, S. J., Featherstone, W. E., Holmes, S. A., and Vaníček, P. (2006): A synthetic Earth gravity model designed specifically for testing regional gravimetric geoid determination algorithms. *Journal of Geodesy* 80(1):1–16. DOI: 10.1007/s00190-005-0002-z.

Barthelmes, F. (2013): Definition of functionals of the geopotential and their calculation from spherical harmonic models. Scientific Technical Report STR09/02. German Research Centre for Geosciences (GFZ), Potsdam, Germany. URL: http://icgem.gfz-potsdam.de/ICGEM/theory/theory.html.

Bassin, C., Laske, G., and Masters, G. (2000): The current limits of resolution for surface wave tomography in North America. *Eos, Transactions, American Geophysical Union* 81:F897. URL: http://igppweb.ucsd.edu/~gabi/crust2.html.

Blackman, R. B. and Tukey, J. W. (1958): The measurement of power spectra from the point of view of communications engineering – Part I. *The Bell System Technical Journal* 37(1):185–282. DOI: 10.1002/j.1538-7305.1958.tb03874.x.

Bouman, J., Ebbing, J., Fuchs, M., Sebera, J., Lieb, V., Szwillus, W., Haagmans, R., and Novák, P. (2016): Satellite gravity gradient grids for geophysics. *Nature Scientific Reports* 6:21050. DOI: 10.1038/srep21050.

Bouman, J., Ebbing, J., Meekes, S., Fattah, R. A., Fuchs, M., Gradmann, S., Haagmans, R., Lieb, V., Schmidt, M., Dettmering, D., and Bosch, W. (2015): GOCE gravity gradient data for lithospheric modeling. *International Journal of Applied Earth Observation and Geoinformation*, Special issue on GOCE Earth science applications and models 35(Part A):16–30. DOI: 10.1016/j.jag.2013.11.001.

Brockmann, J. M., Zehentner, N., Höck, E., Pail, R., Loth, I., Mayer-Gürr, T., and Schuh, W.-D. (2014): EGM_TIM_RL05: an independent geoid with centimeter accuracy purely based on the GOCE mission. *Geophysical Research Letters* 41(22):8089–8099. DOI: 10.1002/2014GL061904.

Bruinsma, S. L., Förste, C., Abrikosov, O., Lemoine, J.-M., Marty, J.-C., Mulet, S., Rio, M.-H., and Bonvalot, S. (2014): ESA's satellite-only gravity field model via the direct approach based on all GOCE data. *Geophysical Research Letters* 41(21):7508–7514. DOI: 10.1002/2014GL062045.

Claessens, S. J. and Hirt, C. (2013): Ellipsoidal topographic potential: new solutions for spectral forward gravity modeling of topography with respect to a reference ellipsoid. *Journal of Geophysical Research* 118(11):5991–6002. DOI: 10.1002/2013JB010457.

Deng, X.-L., Grombein, T., Shen, W.-B., Heck, B., and Seitz, K. (2016): Corrections to "A comparison of the tesseroid, prism and point-mass approaches for mass reductions in gravity field modelling" (Heck and Seitz, 2007) and "Optimized formulas for the gravitational field of a tesseroid" (Grombein et al., 2013). *Journal of Geodesy* 90(6):585–587. DOI: 10.1007/s00190-016-0907-8.

Du, J., Chen, C., Lesur, V., Lane, R., and Wang, H. (2015): Magnetic potential, vector and gradient tensor fields of a tesseroid in a geocentric spherical coordinate system. *Geophysical Journal International* 201(3):1977–2007. DOI: 10.1093/gji/ggv123.

D'Urso, M. G. (2013): On the evaluation of the gravity effects of polyhedral bodies and a consistent treatment of related singularities. *Journal of Geodesy* 87(3):239–252. DOI: 10.1007/s00190-012-0592-1.

Ebbing, J., Braitenberg, C., and Götze, H.-J. (2001): Forward and inverse modelling of gravity revealing insight into crustal structures of the Eastern Alps. *Tectonophysics* 337(3–4):191–208. DOI: 10.1016/S0040-1951(01)00119-6.

ESA (1999): The four candidate Earth explorer core missions – Gravity field and steady-state ocean circulation. In: Battrick, B. (ed.) Reports for mission selection, *ESA Special Publication*, vol. 1233(1). ESA Publications Division, ESTEC, Noordwijk, The Netherlands.

Featherstone, W. E. (2013): Deterministic, stochastic, hybrid and band-limited modifications of Hotine's integral. *Journal of Geodesy* 87(5):487–500. DOI: 10.1007/s00190-013-0612-9.

Fecher, T., Pail, R., and Gruber, T. (2015): Global gravity field modeling based on GOCE and complementary gravity data. *International Journal of Applied Earth Observation and Geoinformation*, Special issue on GOCE Earth science applications and models 35(Part A):120–127. DOI: 10.1016/j.jag.2013.10.005.

Fecher, T., Pail, R., Gruber, T., and the GOCO project team (2016): The combined gravity field model GOCO05c. In: General Assembly of the European Geosciences

Union 2016. Vienna, Austria, Apr. 17–22, 2016. *Geophysical Research Abstracts*, vol. 18. EGU2016-7696.

Forsberg, R. (1984): A study of terrain reductions, density anomalies and geophysical inversion methods in gravity field modelling. Report 355. Department of Geodetic Science and Surveying, The Ohio State University, Columbus, USA.

Forsberg, R. and Tscherning, C. C. (1997): Topographic effects in gravity field modelling for BVP. In: Sansò, F., and Rummel, R. (eds.) Geodetic boundary value problems in view of the one centimeter geoid, *Lecture Notes in Earth Sciences*, vol. 65. Springer Berlin Heidelberg, pp. 239–272. DOI: 10.1007/BFb0011707.

Forsberg, R. and Tscherning, C. C. (2008): An overview manual for the GRAVSOFT geodetic gravity field modelling programs. 2nd ed. Contract report for JUPEM. URL: http://cct.gfy.ku.dk/publ_cct/cct1792.pdf.

Fuchs, M. J., Hooper, A., Broerse, T., and Bouman, J. (2016): Distributed fault slip model for the 2011 Tohoku-Oki earthquake from GNSS and GRACE/GOCE satellite gravimetry. *Journal of Geophysical Research* 121(2):1114–1130. DOI: 10.1002/2015JB012165.

Fukushima, T. (2012): Numerical computation of spherical harmonics of arbitrary degree and order by extending exponent of floating point numbers. *Journal of Geodesy* 86(4):271–285. DOI: 10.1007/s00190-011-0519-2.

Gatti, A., Reguzzoni, M., and Venuti, G. (2013): The height datum problem and the role of satellite gravity models. *Journal of Geodesy* 87(1):15–22. DOI: 10.1007/s00190-012-0574-3.

Gerlach, C. and Rummel, R. (2013): Global height system unification with GOCE: a simulation study on the indirect bias term in the GBVP approach. *Journal of Geodesy* 87(1):57–67. DOI: 10.1007/s00190-012-0579-y.

Göttl, F. and Rummel, R. (2009): A geodetic view on isostatic models. *Pure and Applied Geophysics* 166(8):1247–1260. DOI: 10.1007/s00024-004-0489-x.

Grombein, T., Luo, X., Seitz, K., and Heck, B. (2014a): A wavelet-based assessment of topographic-isostatic reductions for GOCE gravity gradients. *Surveys in Geophysics* 35(4):959–982. DOI: 10.1007/s10712-014-9283-1.

Grombein, T., Seitz, K., and Heck, B. (2010a): Untersuchung zur effizienten Berechnung topographischer Effekte auf den Gradiententensor am Fallbeispiel der Satellitengradiometriemission GOCE. *KIT Scientific Reports*, no. 7547. KIT Scientific Publishing, Karlsruhe, Germany. DOI: 10.5445/KSP/1000017531.

Grombein, T., Seitz, K., and Heck, B. (2010b): Modelling topographic effects in GOCE gravity gradients. In: Münch, U., and Dransch, W. (eds.) Observation of the system Earth from space, *GEOTECHNOLOGIEN Science Report*, vol. 17, pp. 84–93. DOI: 10.2312/GFZ.gt.17.13.

Grombein, T., Seitz, K., and Heck, B. (2011): Smoothing GOCE gravity gradients by means of topographic-isostatic reductions. In: Ouwehand, L. (ed.) Proceedings of the 4th International GOCE User Workshop, Munich, Germany, Mar. 31–Apr. 1, 2011. *ESA Special Publication*, vol. 696. ESA Publications Division, ESTEC, Noordwijk, The Netherlands.

Grombein, T., Seitz, K., and Heck, B. (2013): Optimized formulas for the gravitational field of a tesseroid. *Journal of Geodesy* 87(7):645–660. DOI: 10.1007/s00190-013-0636-1.

Grombein, T., Seitz, K., and Heck, B. (2014b): Topographic-isostatic reduction of GOCE gravity gradients. In: Rizos, C., and Willis, P. (eds.) Earth on the edge: science for a sustainable planet. Proceedings of the IAG General Assembly, Melbourne, Australia, June 28–July 2, 2011. *International Association of Geodesy Symposia*, vol. 139. Springer Berlin Heidelberg, pp. 349–356. DOI: 10.1007/978-3-642-37222-3_46.

Grombein, T., Seitz, K., and Heck, B. (2014c): Incorporating topographic-isostatic information into GOCE gravity gradient processing. In: Flechtner, F., Sneeuw, N., and Schuh, W.-D. (eds.) Observation of the system Earth from space – CHAMP, GRACE, GOCE and future missions, *Advanced Technologies in Earth Sciences, GEOTECHNOLOGIEN Science Report*, vol. 20. Springer Berlin Heidelberg, pp. 95–101. DOI: 10.1007/978-3-642-32135-1_12.

Grombein, T., Seitz, K., and Heck, B. (2016a): The Rock-Water-Ice topographic gravity field model RWI_TOPO_2015 and its comparison to a conventional rock-equivalent version. *Surveys in Geophysics* 37(5):937–976. DOI: 10.1007/s10712-016-9376-0.

Grombein, T., Seitz, K., and Heck, B. (2016b): Height system unification based on the fixed GBVP approach. In: Rizos, C., and Willis, P. (eds.) IAG 150 years. Proceedings of the IAG Scientific Assembly, Postdam, Germany, Sept. 1–6, 2013. *International Association of Geodesy Symposia*, vol. 143. Springer Berlin Heidelberg, pp. 305–311. DOI: 10.1007/1345_2015_104.

Grombein, T., Seitz, K., and Heck, B. (2017): On high-frequency topography-implied gravity signals for height system unification using GOCE-based global geopotential models. *Surveys in Geophysics* 38(2):443–477. DOI: 10.1007/s10712-016-9400-4.

Guerri, M., Cammarano, F., and Connolly, J. A. D. (2015): Effects of chemical composition, water and temperature on physical properties of continental crust. *Geochemistry, Geophysics, Geosystems* 16(7):2431–2449. DOI: 10.1002/2015GC005819.

Gutknecht, B. D., Götze, H.-J., Jahr, T., Jentzsch, G., Mahatsente, R., and Zeumann, S. (2014): Structure and state of stress of the Chilean subduction zone from terrestrial and satellite-derived gravity and gravity gradient data. *Surveys in Geophysics* 35(6):1417–1440. DOI: 10.1007/s10712-014-9296-9.

Heck, B. (1990): An evaluation of some systematic error sources affecting terrestrial gravity anomalies. *Bulletin Géodésique* 64(1):88–108. DOI: 10.1007/BF02530617.

Heck, B. and Rummel, R. (1990): Strategies for solving the vertical datum problem using terrestrial and satellite geodetic data. In: Sünkel, H., and Baker, T. (eds.) Sea surface topography and the geoid. Proceedings of the IAG General Assembly, Edinburgh, Scotland, Aug. 10–11, 1989. *International Association of Geodesy Symposia*, vol. 104. Springer Berlin Heidelberg, pp. 116–128. DOI: 10.1007/978-1-4684-7098-7_14.

Heck, B. and Seitz, K. (2007): A comparison of the tesseroid, prism and point-mass approaches for mass reductions in gravity field modelling. *Journal of Geodesy* 81(2):121–136. DOI: 10.1007/s00190-006-0094-0.

Heck, B. and Seitz, K. (2014): Ellipsoidal effects in the fixed geodetic boundary value problem. In: General Assembly of the European Geosciences Union 2014. Vienna, Austria, Apr. 27–May 2, 2014. *Geophysical Research Abstracts*, vol. 16. EGU2014-12951.

Heck, B., Seitz, K., and Grombein, T. (2015): A numerical study of up- and downward continuation effects in the solution of the geodetic boundary value problem. In: 26th General Assembly of the International Union of Geodesy and Geophysics 2015. Prague, Czech Republic, June 22–July 2, 2015.

Heiskanen, W. A. and Moritz, H. (1967): Physical geodesy. W. H. Freeman & Co., San Francisco, USA.

Hirt, C. (2013): RTM gravity forward-modeling using topography/bathymetry data to improve high-degree global geopotential models in the coastal zone. *Marine Geodesy* 36(2):183–202. DOI: 10.1080/01490419.2013.779334.

Hirt, C., Claessens, S., Fecher, T., Kuhn, M., Pail, R., and Rexer, M. (2013): New ultrahigh-resolution picture of Earth's gravity field. *Geophysical Research Letters* 40(16):4279–4283. DOI: 10.1002/grl.50838.

Hirt, C., Featherstone, W. E., and Marti, U. (2010): Combining EGM2008 and SRTM/DTM-2006.0 residual terrain model data to improve quasigeoid computations in mountainous areas devoid of gravity data. *Journal of Geodesy* 84(9):557–567. DOI: 10.1007/s00190-010-0395-1.

Hirt, C. and Kuhn, M. (2012): Evaluation of high-degree series expansions of the topographic potential to higher-order powers. *Journal of Geophysical Research* 117:B12407. DOI: 10.1029/2012JB009492.

Hirt, C. and Kuhn, M. (2014): Band-limited topographic mass distribution generates full-spectrum gravity field: gravity forward modeling in the spectral and spatial domains revisited. *Journal of Geophysical Research* 119(4):3646–3661. DOI: 10.1002/2013JB010900.

Hirt, C., Kuhn, M., Featherstone, W. E., and Göttl, F. (2012): Topographic/isostatic evaluation of new-generation GOCE gravity field models. *Journal of Geophysical Research* 117:B05407. DOI: 10.1029/2011JB008878.

Hirt, C. and Rexer, M. (2015): Earth2014: 1 arc-min shape, topography, bedrock and ice-sheet models – available as gridded data and degree-10,800 spherical harmonics. *International Journal of Applied Earth Observation and Geoinformation* 39:103–112. DOI: 10.1016/j.jag.2015.03.001. URL: http://ddfe.curtin.edu.au/models/Earth2014.

Hirt, C., Rexer, M., and Claessens, S. (2015): Topographic evaluation of fifth-generation GOCE gravity field models – globally and regionally. In: Huang, J., Reguzzoni, M., and Gruber, T. (eds.) Assessment of GOCE geopotential models, *Newton's Bulletin*, no. 5. International Association of Geodesy and International Gravity Field Service, pp. 163–186. URL: http://www.isgeoid.polimi.it/Newton/Newton_5/12_Hirt_163_186.html.

Hirt, C., Rexer, M., Scheinert, M., Pail, R., Claessens, S., and Holmes, S. (2016): A new degree-2190 (10 km resolution) gravity field model for Antarctica developed from GRACE, GOCE and Bedmap2 data. *Journal of Geodesy* 90(2):105–127. DOI: 10.1007/s00190-015-0857-6.

Holmes, S. A. and Featherstone, W. E. (2002): A unified approach to the Clenshaw summation and the recursive computation of very high degree and order normalised associated Legendre functions. *Journal of Geodesy* 76(5):279–299. DOI: 10.1007/s00190-002-0216-2.

Holmes, S. A. and Pavlis, N. K. (2007): Some aspects of harmonic analysis of data gridded on the ellipsoid. In: Kiliçoglu, A., and Forsberg, R. (eds.) Proceedings of the 1st International Symposium of the International Gravity Field Service (IGFS), Istanbul, Turkey, Aug. 28–Sept. 1, 2006. Harita Dergisi, Special Issue 18, pp. 151–156.

Ihde, J. and Sánchez, L. (2005): A unified global height reference system as a basis for IGGOS. *Journal of Geodynamics* 40(4–5):400–413. DOI: 10.1016/j.jog.2005.06.015.

Janák, J., Pitoňák, M., and Minarechová, Z. (2014): Regional quasigeoid from GOCE and terrestrial measurements. *Studia Geophysica et Geodætica* 58(4):626–649. DOI: 10.1007/s11200-013-0543-1.

Janák, J. and Wild-Pfeiffer, F. (2010): Comparison of various topographic-isostatic effects in terms of smoothing gradiometric observations. In: Mertikas, S. P. (ed.) Gravity, geoid and Earth observation, IAG Commission 2: Gravity field. Chania, Crete, Greece, June 23–27, 2008. *International Association of Geodesy Symposia*, vol. 135. Springer Berlin Heidelberg, pp. 377–381. DOI: 10.1007/978-3-642-10634-7_50.

Janák, J., Wild-Pfeiffer, F., and Heck, B. (2012): Smoothing the gradiometric observations using different topographic-isostatic models: a regional case study. In: Sneeuw, N., Novák, P., Crespi, M., and Sansò, F. (eds.) VII Hotine-Marussi Symposium on Mathematical Geodesy. Proceedings of the symposium in Rome, June 6–10, 2009. *International Association of Geodesy Symposia*, vol. 137. Springer Berlin Heidelberg, pp. 245–250. DOI: 10.1007/978-3-642-22078-4_37.

Jekeli, C. (1988): The exact transformation between ellipsoidal and spherical harmonic expansions. *Manuscripta Geodaetica* 13(2):106–113.

Kaban, M. K., Schwintzer, P., and Reigber, C. (2004): A new isostatic model of the lithosphere and gravity field. *Journal of Geodesy* 78(6):368–385. DOI: 10.1007/s00190-004-0401-6.

Kuhn, M. (2000): Geoidbestimmung unter Verwendung verschiedener Dichtehypothesen. *Deutsche Geodätische Kommission, Reihe C*, no. 520. Verlag der Bayerischen Akademie der Wissenschaften in Kommission beim Verlag C. H. Beck, Munich, Germany.

Kuhn, M. and Featherstone, W. E. (2005): Construction of a synthetic Earth gravity model by forward gravity modelling. In: Sansò, F. (ed.) A window on the future of geodesy. Proceedings of the IAG General Assembly, Sapporo, Japan, June 30–July 11, 2003. *International Association of Geodesy Symposia*, vol. 128. Springer Berlin Heidelberg, pp. 350–355. DOI: 10.1007/3-540-27432-4_60.

Kuhn, M. and Seitz, K. (2005): Comparison of Newton's integral in the space and frequency domains. In: Sansò, F. (ed.) A window on the future of geodesy. Proceedings of the IAG General Assembly, Sapporo, Japan, June 30–July 11, 2003. *International Association of Geodesy Symposia*, vol. 128. Springer Berlin Heidelberg, pp. 386–391. DOI: 10.1007/3-540-27432-4_66.

Laske, G., Masters, G., Ma, Z., and Pasyanos, M. (2013): Update on CRUST1.0 – a 1-degree global model of Earth's crust. In: General Assembly of the European Geosciences Union 2013. Vienna, Austria, Apr. 7–12, 2013. *Geophysical Research Abstracts*, vol. 15. EGU2013-2658. URL: http://igppweb.ucsd.edu/~gabi/crust1.html.

Mader, K. (1951): Das Newtonsche Raumpotential prismatischer Körper und seine Ableitungen bis zur dritten Ordnung. *Österreichische Zeitschrift für Vermessungswesen, Sonderheft* 11.

Makhloof, A. A. and Ilk, K. (2008): Effects of topographic-isostatic masses on gravitational functionals at the Earth's surface and at airborne and satellite altitudes. *Journal of Geodesy* 82(2):93–111. DOI: 10.1007/s00190-007-0159-8.

Mayer-Gürr, T., Pail, R., Gruber, T., Fecher, T., Rexer, M., Schuh, W.-D., Kusche, J., Brockmann, J.-M., Rieser, D., Zehentner, N., Kvas, A., Klinger, B., O., B., Höck, E., Krauss, S., and Jäggi, A. (2015): The combined satellite gravity field model GOCO05s. In: General Assembly of the European Geosciences Union 2015. Vienna, Austria, Apr. 12–17, 2015. *Geophysical Research Abstracts*, vol. 17. EGU2015-12364.

Mayer-Gürr, T., Rieser, D., Höck, E., Brockmann, J. M., Schuh, W.-D., Krasbutter, I., Kusche, J., Maier, A., Krauss, S., Hausleitner, W., Baur, O., Jäggi, A., Meyer, U., Prange, L., Pail, R., Fecher, T., and Gruber, T. (2012): The new combined satellite only model GOCO03s. In: International Symposium on Gravity, Geoid and Height Systems 2012. Venice, Italy, Oct. 9–12, 2012.

Moritz, H. (1980): Geodetic Reference System 1980. *Bulletin Géodésique* 54(3):395–405. DOI: 10.1007/BF02521480.

Moritz, H. (1990): The figure of the Earth. Theoretical geodesy and the Earth's interior. Wichmann-Verlag, Karlsruhe, Germany.

Müßle, M., Heck, B., Seitz, K., and Grombein, T. (2014): On the effect of planar approximation in the geodetic boundary value problem. *Studia Geophysica et Geodætica* 58(4):536–555. DOI: 10.1007/s11200-013-0249-4.

Nagy, D., Papp, G., and Benedek, J. (2000): The gravitational potential and its derivatives for the prism. *Journal of Geodesy* 74(7–8):552–560. DOI: 10.1007/s001900000116.

Novák, P. and Grafarend, E. W. (2005): Ellipsoidal representation of the topographical potential and its vertical gradient. *Journal of Geodesy* 78(11):691–706. DOI: 10.1007/s00190-005-0435-4.

Novák, P., Kern, M., Schwarz, K.-P., and Heck, B. (2003): Evaluation of band-limited topographical effects in airborne gravimetry. *Journal of Geodesy* 76(11–12):597–604. DOI: 10.1007/s00190-002-0282-5.

Novák, P. and Tenzer, R. (2013): Gravitational gradients at satellite altitudes in global geophysical studies. *Surveys in Geophysics* 34(5):653–673. DOI: 10.1007/s10712-013-9243-1.

Omang, D. O. C. and Forsberg, R. (2000): How to handle topography in practical geoid determination: three examples. *Journal of Geodesy* 74(6):458–466. DOI: 10.1007/s001900000107.

Pavlis, N. K., Factor, J. K., and Holmes, S. A. (2007): Terrain-related gravimetric quantities computed for the next EGM. In: Kiliçoglu, A., and Forsberg, R. (eds.) Proceedings of the 1st International Symposium of the International Gravity Field Service (IGFS), Istanbul, Turkey, Aug. 28–Sept. 1, 2006. Harita Dergisi, Special Issue 18, pp. 318–323.

Pavlis, N. K., Holmes, S. A., Kenyon, S. C., and Factor, J. K. (2012): The development and evaluation of the Earth Gravitational Model 2008. *Journal of Geophysical Research* 117:B04406. DOI: 10.1029/2011JB008916.

Pitonák, M., Šprlák, M., Hamácková, E., and Novák, P. (2016): Regional recovery of the disturbing gravitational potential by inverting satellite gravitational gradients. *Geophysical Journal International* 205(1):89–98. DOI: 10.1093/gji/ggw008.

Reguzzoni, M. and Sampietro, D. (2015): GEMMA: an earth crustal model based on GOCE satellite data. *International Journal of Applied Earth Observation and Geoinformation*, Special issue on GOCE Earth science applications and models 35(Part A):31–43. DOI: 10.1016/j.jag.2014.04.002.

Reguzzoni, M., Sampietro, D., and Sansò, F. (2013): Global Moho from the combination of the CRUST2.0 model and GOCE data. *Geophysical Journal International* 195(1):222–237. DOI: 10.1093/gji/ggt247.

Rexer, M. and Hirt, C. (2015): Ultra-high-degree surface spherical harmonic analysis using the Gauss-Legendre and the Driscoll/Healy quadrature theorem and application to planetary topography models of Earth, Mars and Moon. *Surveys in Geophysics* 36(6):803–830. DOI: 10.1007/s10712-015-9345-z.

Rummel, R. (2002): Global unification of height systems and GOCE. In: Sideris, M. G. (ed.) Gravity, geoid and geodynamics 2000, GGG2000 IAG International Symposium. Banff, Alberta, Canada, July 31–Aug. 4, 2000. *International Association of Geodesy Symposia*, vol. 123. Springer Berlin Heidelberg, pp. 13–20. DOI: 10.1007/978-3-662-04827-6_3.

Rummel, R., Balmino, G., Johannessen, J., Visser, P., and Woodworth, P. (2002): Dedicated gravity field missions – principles and aims. *Journal of Geodynamics* 33(1–2):3–20. DOI: 10.1016/S0264-3707(01)00050-3.

Rummel, R. and Colombo, O. L. (1985): Gravity field determination from satellite gradiometry. *Bulletin Géodésique* 59(3):233–246. DOI: 10.1007/BF02520329.

Rummel, R., Rapp, R. H., Sünkel, H., and Tscherning, C. C. (1988): Comparisons of global topographic/isostatic models to the Earth's observed gravity field. Report 388. Department of Geodetic Science and Surveying, The Ohio State University, Columbus, USA.

Rummel, R. and Teunissen, P. (1988): Height datum definition, height datum connection and the role of the geodetic boundary value problem. *Bulletin Géodésique* 62(4):477–498. DOI: 10.1007/BF02520239.

Rummel, R., Yi, W., and Stummer, C. (2011): GOCE gravitational gradiometry. *Journal of Geodesy* 85(11):777–790. DOI: 10.1007/s00190-011-0500-0.

Sánchez, L. (2015): Ein einheitliches vertikales Referenzsystem für Südamerika im Rahmen eines globalen Höhensystems. *Deutsche Geodätische Kommission, Reihe C*, no. 748. Verlag der Bayerischen Akademie der Wissenschaften in Kommission beim Verlag C. H. Beck, Munich, Germany. URL: https://www.dgk.badw.de.devweb.mwn.de/fileadmin/docs/c-748.pdf.

Schmidt, M., Fengler, M., Mayer-Gürr, T., Eicker, A., Kusche, J., Sánchez, L., and Han, S.-C. (2007): Regional gravity modeling in terms of spherical base functions. *Journal of Geodesy* 81(1):17–38. DOI: 10.1007/s00190-006-0101-5.

Seitz, K. and Heck, B. (1991): Harmonic analysis on the sphere. Internal report. Geodetic Institute, University of Karlsruhe, Germany.

Smith, D. A., Robertson, D. S., and Milbert, D. G. (2001): Gravitational attraction of local crustal masses in spherical coordinates. *Journal of Geodesy* 74(11–12):783–795. DOI: 10.1007/s001900000142.

Šprlák, M., Gerlach, C., and Pettersen, B. R. (2015): Validation of GOCE global gravitational field models in Norway. In: Huang, J., Reguzzoni, M., and Gruber, T. (eds.) Assessment of GOCE geopotential models, *Newton's Bulletin*, no. 5. International Association of Geodesy and International Gravity Field Service, pp. 13–24. URL: http://www.isgeoid.polimi.it/Newton/Newton_5/03_Sprlak_13_24.html.

Tenzer, R., Gladkikh, V., Novák, P., and Vajda, P. (2012): Spatial and spectral analysis of refined gravity data for modelling the crust-mantle interface and mantle-lithosphere structure. *Surveys in Geophysics* 33(5):817–839. DOI: 10.1007/s10712-012-9173-3.

Thong, N. C. (1989): Simulation of gradiometry using the spheroidal harmonic model of the gravitational field. *Manuscripta Geodaetica* 14(6):404–417.

Torge, W. and Müller, J. (2012): Geodesy. 4th ed. Walter de Gruyter, Berlin, Germany.

Tscherning, C. C. (1976): Computation of the second-order derivatives of the normal potential based on the representation by a Legendre series. *Manuscripta Geodaetica* 1:71–92.

Tsoulis, D. (2012): Analytical computation of the full gravity tensor of a homogeneous arbitrarily shaped polyhedral source using line integrals. *Geophysics* 77(2):F1–F11. DOI: 10.1190/geo2010-0334.1.

Tsoulis, D. and Kuhn, M. (2007): Recent developments in synthetic Earth gravity models in view of the availability of digital terrain and crustal databases of global coverage and increased resolution. In: Kiliçoglu, A., and Forsberg, R. (eds.) Proceedings of the 1st International Symposium of the International Gravity Field Service (IGFS), Istanbul, Turkey, Aug. 28–Sept. 1, 2006. Harita Dergisi, Special Issue 18, pp. 354–359.

Tsoulis, D. and Patlakis, K. (2013): A spectral assessment review of current satellite-only and combined Earth gravity models. *Reviews of Geophysics* 51(2):186–243. DOI: 10.1002/rog.20012.

Tsoulis, D., Wziontek, H., and Petrović, S. (2003): A bilinear approximation of the surface relief in terrain correction computations. *Journal of Geodesy* 77(5–6):338–344. DOI: 10.1007/s00190-003-0332-7.

Uieda, L., Bomfim, E. P., Braitenberg, C., and Molina, E. (2011): Optimal forward calculation method of the Marussi tensor due to a geologic structure at GOCE height. In: Ouwehand, L. (ed.) Proceedings of the 4th International GOCE User Workshop, Munich, Germany, Mar. 31–Apr. 1, 2011. *ESA Special Publication*, vol. 696. ESA Publications Division, ESTEC, Noordwijk, The Netherlands.

Wieczorek, M. A. (2007): Gravity and topography of the terrestrial planets. *Treatise on Geophysics*, Planets and Moons 10(5):165–206. DOI: 10.1016/B978-044452748-6.00156-5.

Wild, F. and Heck, B. (2005): A comparison of different isostatic models applied to satellite gravity gradiometry. In: Jekeli, C., Bastos, L., and Fernandes, J. (eds.) Gravity, geoid and space missions, GGSM 2004 IAG International Symposium. Porto, Portugal, Aug. 30–Sept. 3, 2004. *International Association of Geodesy Symposia*, vol. 129. Springer Berlin Heidelberg, pp. 230–235. DOI: 10.1007/3-540-26932-0_40.

Wild, F. and Heck, B. (2008): Topographic and isostatic reductions for use in satellite gravity gradiometry. In: Xu, P., Liu, J., and Dermanis, A. (eds.) VI Hotine-Marussi Symposium on Theoretical and Computational Geodesy, IAG Symposium. Wuhan, China, May 29–June 2, 2006. *International Association of Geodesy Symposia*, vol. 132. Springer Berlin Heidelberg, pp. 49–55. DOI: 10.1007/978-3-540-74584-6_8.

Wild-Pfeiffer, F. (2007): Auswirkungen topographisch-isostatischer Massen auf die Satellitengradiometrie. *Deutsche Geodätische Kommission, Reihe C*, no. 604. Verlag der Bayerischen Akademie der Wissenschaften in Kommission beim Verlag C. H. Beck, Munich, Germany. URL: https://www.dgk.badw.de.devweb.mwn.de/fileadmin/docs/c-604.pdf.

Wild-Pfeiffer, F. (2008): A comparison of different mass elements for use in gravity gradiometry. *Journal of Geodesy* 82(10):637–653. DOI: 10.1007/s00190-008-0219-8.

Wittwer, T., Klees, R., Seitz, K., and Heck, B. (2008): Ultra-high degree spherical harmonic analysis and synthesis using extended-range arithmetic. *Journal of Geodesy* 82(4):223–229. DOI: 10.1007/s00190-007-0172-y.

Xu, P. (1992): A quality investigation of global vertical datum connection. *Geophysical Journal International* 110(2):361–370. DOI: 10.1111/j.1365-246X.1992.tb00880.x.

Chapter II.

Optimized formulas for the gravitational field of a tesseroid

Thomas Grombein, Kurt Seitz, Bernhard Heck
Journal of Geodesy 87(7):645–660, 2013. DOI: 10.1007/s00190-013-0636-1.

Submitted: 17 February 2012 / Accepted: 15 April 2013 / Published online: 18 May 2013

Abstract. Various tasks in geodesy, geophysics, and related geosciences require precise information on the impact of mass distributions on gravity-field-related quantities, such as the gravitational potential and its partial derivatives. Using forward modeling based on Newton's integral, mass distributions are generally decomposed into regular elementary bodies. In classical approaches, prisms or point mass approximations are mostly utilized. Considering the effect of the sphericity of the Earth, alternative mass modeling methods based on tesseroid bodies (spherical prisms) should be taken into account, particularly in regional and global applications. Expressions for the gravitational field of a point mass are relatively simple when formulated in Cartesian coordinates. In the case of integrating over a tesseroid volume bounded by geocentric spherical coordinates, it will be shown that it is also beneficial to represent the integral kernel in terms of Cartesian coordinates. This considerably simplifies the determination of the tesseroid's potential derivatives in comparison to previously published methodologies that make use of integral kernels expressed in spherical coordinates. Based on this idea, optimized formulas for the gravitational potential of a homogeneous tesseroid and its derivatives up to second-order are elaborated in this paper. These new formulas do not suffer from the polar singularity of the spherical coordinate system and can, therefore, be evaluated for any position on the globe. Since integrals over tesseroid volumes cannot be solved analytically, the numerical evaluation is achieved by means of expanding the integral kernel in a Taylor series with fourth-order error in the spatial coordinates of the integration point. As the structure of the Cartesian integral kernel is substantially simplified, Taylor coefficients can be represented in a compact and computationally attractive form. Thus, the use of the optimized tesseroid formulas particularly benefits from a significant decrease in computation time by about 45 % compared to previously used algorithms. In order to show the computational efficiency and to validate the mathematical derivations, the new tesseroid formulas are applied to two realistic numerical experiments and are compared to previously published tesseroid methods and the conventional prism approach.

Keywords Forward modeling · Tesseroids · Gravitational field · Newton's integral

1. Introduction

Modeling the impact of mass distributions on the gravitational potential and its derivatives is a central issue in geodesy and geophysics. In physical geodesy, the most important application is the determination of mass reductions for gravity field observations, e.g., topographic reductions obtained from digital terrain models. In the classical Stokes problem, topographic (and isostatic) reductions are used to get mass-free boundary values at geoid level (Heiskanen and Moritz, 1967, Chapter 3). Moreover, these mass reductions can also be integrated into modern methods of gravity field modeling, such as the remove–compute–restore technique (Forsberg, 1984, Sect. 4) or residual terrain modeling (Forsberg and Tscherning, 1997; Hirt et al., 2010). In this context, the main purpose is to smooth the signal content of gravity data in order to improve the numerical stability for interpolation or prediction tasks as well as field transformations. Such a procedure has been proposed for different kinds of quantities like terrestrial, airborne or satellite-based gravity field observations (e.g., Novák et al., 2003; Makhloof and Ilk, 2008; Wild and Heck, 2008; Janák et al., 2012; Grombein et al., 2013). Furthermore, there are lots of interdisciplinary applications in solid earth research implicating the problem of mass modeling. For instance, this includes the construction of a synthetic Earth model (Kuhn and Featherstone, 2005), investigations of structures and density anomalies in the Earth's crust (Braitenberg and Ebbing, 2009; Álvarez et al., 2012), as well as detection of ice mass loss (Baur and Sneeuw, 2011) or hydrological mass variations (Heck and Seitz, 2008; Grombein et al., 2012).

All of the described applications can be considered in the context of forward (or inverse) modeling which is based on the evaluation of Newton's integral for the gravitational potential of a solid body $\Omega \subset \mathbb{R}^3$ (Heiskanen and Moritz, 1967, p. 3):

$$V(x_1, x_2, x_3) = G \iiint\limits_{\Omega} \frac{\rho(x_1', x_2', x_3')}{\ell} \, \mathrm{d}\Omega, \tag{1}$$

where G denotes Newton's gravitational constant, ρ the location-dependent density, and ℓ the Euclidean distance between the (attracted) computation point $P(x_1, x_2, x_3) \in \mathbb{R}^3$ and the running integration (mass) point $Q(x_1', x_2', x_3') \in \Omega$. In principle, the volume integral in Eq. (1) can either be evaluated in the space domain by direct integration methods or in the frequency domain by spherical harmonic approaches (cf. Kuhn and Seitz, 2005; Wild-Pfeiffer and Heck, 2007). In both cases, information on the geometry of the mass distribution as well as the density function $\rho(x'_1, x'_2, x'_3)$ inside the masses is required. However, in most practical applications the geometry and the density are only available in discrete form, e.g., represented by a grid with a specific resolution. To this end, numerical evaluations of Eq. (1) in the space domain rely on a mass discretization, where the integration domain Ω is decomposed into elementary geometrical bodies $\Omega_i \subset \Omega$ assuming a constant density value ρ_i. By applying the superposition principle, the effect of the whole mass distribution can be approximated by the sum of the impact over all individual mass bodies:

$$V(x_1, x_2, x_3) \approx \sum_i G\rho_i \iiint\limits_{\Omega_i} \frac{1}{\ell}\, \mathrm{d}\Omega, \quad \Omega = \bigcup_i \Omega_i. \tag{2}$$

According to the requirements of the specific application, the decomposition can be carried out using different types of mass bodies. In most of the classical approaches, prismatic bodies with rectangular form are utilized. Although the respective volume integrals in Eq. (1) can be solved analytically in the case of prisms (Mader, 1951; Nagy et al., 2000, 2002), the calculation is very time consuming due to several logarithmic and arctan functions that have to be evaluated. A significant reduction of computation time can be achieved by applying fast Fourier transform techniques (cf. Forsberg, 1985; Schwarz et al., 1990; Klose and Ilk, 1993). The price to be paid is a decreasing accuracy when the bounding surface is too rough. An alternative are approximate solutions of the prism integrals, such as MacMillan's formulas based on a Taylor series expansion of the integral kernel (MacMillan, 1930), or the use of Gauss–Legendre cubature (e.g., Ku, 1977; von Frese et al., 1981).

Using prisms is especially beneficial for local applications, where height information is generally related to planar Cartesian coordinates referenced to a map projection. For larger application areas the curvature of the Earth has to be taken into account by a vertical shift of the prisms as implemented in the widely-used TC software (Forsberg, 1984, p. 111), or additional coordinate transformations (e.g., Grombein et al., 2010, p. 30) which increases the computation time even more. Furthermore, the use of prisms with curved surfaces, e.g., topped by a bilinear surface (Smith et al., 2001; Tsoulis et al., 2003), as well as general polyhedral bodies (Petrović, 1996; Tsoulis, 2012; D'Urso, 2013) have been proposed to improve the rough approximation of classical rectangular prisms.

Particularly in regional and global applications, it is advantageous to apply mass bodies that are directly linked to the curvature of the Earth. According to Anderson (1976, p. 48ff.), mass elements bounded by geographical grid lines (B, L) and surfaces of constant ellipsoidal heights h are called tesseroids. Usually, these ellipsoidal bodies are approximated by corresponding spherical ones, which are bounded by geocentric spherical coordinates (r, φ, λ). Even though spherical tesseroids are considered in the following, the Earth's ellipticity can be taken into account by fixing these bodies on an ellipsoidal reference surface with a latitude-dependent Earth radius (cf. Heck and Seitz, 2007).

In contrast to prisms, Newton's integral in Eq. (1) cannot be solved analytically in the case of tesseroids (e.g., Grüninger, 1990, p. 76). Instead of this, approximate solutions have to be applied. Analogous to the idea of MacMillan's formulas for the prism, Heck and Seitz (2007) derived a third-order approximation which applies a Taylor series expansion of the integral kernels. While Heck and Seitz (2007) originally derived formulas for the tesseroid potential and the first radial derivative, Wild-Pfeiffer (2007, 2008) extended the approach to all components of first- and second-order derivatives. Furthermore, also Gauss–Legendre cubature can be applied as proposed by Asgharzadeh et al. (2007) and Wild-Pfeiffer (2007, 2008). For global computations, an alternative consists in analytically solving the one-dimensional integral with respect to the geocentric distance r and calculating the remaining two-dimensional surface integral numerically (cf. Martinec, 1998; Heck and Seitz, 2007).

Detailed comparisons of using different mass discretizations and evaluation techniques in forward modeling are provided by Heck and Seitz (2007), Wild-Pfeiffer (2008), and Grombein et al. (2010, Chapter 7). In terms of precision and computation time, these investigations have verified the numerical efficiency when using tesseroids instead of conventional prisms. However, practical computations often rely on a combination of different methods by subdividing the total integration domain into a near and a far zone with respect to the distance from the computation point. Since the impact of distant masses on the gravitational potential diminishes with increasing distance, the calculation procedure for remote bodies in the far zone can be substantially simplified. Assuming local mass conservation, suitable approximations are based on mass layers, mass lines, or point mass approximations (e.g., Grüninger, 1990; Tsoulis, 1999; Wild-Pfeiffer, 2008). As forward and inverse modeling is generally a time consuming task, this paper focuses on the development of optimized formulas for the gravitational field of a homogeneous tesseroid. In contrast to previously published tesseroid formulas that rely on the use of spherical integral kernels, alternative expressions based on the Cartesian integral kernels are derived. Using these new formulas allows an efficient numerical evaluation with a significantly reduced runtime.

The paper is organized as follows: in Sect. 2, theoretical basics are presented that are needed further on in this paper. Section 3 focuses on the mathematical elaboration of the optimized tesseroid formulas based on Cartesian integral kernels, whereas in Sect. 4 the consistency with formulas based on spherical integral kernels is shown. Numerical evaluation rules for an efficient implementation of the new tesseroid formulas are presented in Sect. 5. Furthermore, remarks on the validity and accuracy of the derived formulas, as well as limitations for the very near zone are indicated. The benefit of the optimized tesseroid formulas is shown by two numerical examples presented in Sect. 6. Finally, conclusions as well as an outlook to ongoing and further research work are provided in Sect. 7.

2. Theoretical preliminaries

In Sect. 2.1, the tesseroid potential is introduced, while Sect. 2.2 focuses on the definition of the partial derivatives as used in this paper.

2.1. Gravitational potential of a tesseroid

Tesseroids are bounded by three pairs of surfaces: A pair of concentric spheres (r_1 = const., r_2 = const.), a pair of meridional planes (λ_1 = const., λ_2 = const.), and a pair of coaxial circular cones, defined by the parallels φ_1 = const., φ_2 = const. (see Fig. 1). On the basis of Newton's integral in Eq. (1), the gravitational potential V^* of a tesseroid with a constant mass density ρ can be specified by introducing the integration domain $\Omega^* := [r_1, r_2] \times [\varphi_1, \varphi_2] \times [\lambda_1, \lambda_2] \subset \mathbb{R}^3$ and the spherical volume element $\mathrm{d}\Omega = r'^2 \cos\varphi' \,\mathrm{d}r' \,\mathrm{d}\varphi' \,\mathrm{d}\lambda'$:

$$V^*(r,\varphi,\lambda) = G\rho \iiint\limits_{\Omega^*} \frac{1}{\ell}\,\mathrm{d}\Omega = G\rho \int\limits_{\lambda_1}^{\lambda_2}\int\limits_{\varphi_1}^{\varphi_2}\int\limits_{r_1}^{r_2} \frac{r'^2\cos\varphi'}{\ell}\,\mathrm{d}r'\,\mathrm{d}\varphi'\,\mathrm{d}\lambda', \tag{3}$$

where

$$\ell = \ell(P,Q) = \sqrt{r^2 + r'^2 - 2rr'\cos\psi} \tag{4}$$

denotes the Euclidean distance function between the computation point $P(r,\varphi,\lambda) \in \mathbb{R}^3$ and the running integration point $Q(r',\varphi',\lambda') \in \Omega^*$. The spherical distance ψ between the position vectors of P and Q is defined by

$$\cos\psi(P,Q) := \sin\varphi\sin\varphi' + \cos\varphi\cos\varphi'\cos(\lambda' - \lambda). \tag{5}$$

As the integration with respect to λ' and φ' comprises elliptical integrals (cf. Bronstein et al., 2008, p. 430ff.), Eq. (3) cannot be solved in closed analytical form. In order to calculate the tesseroid potential, methods of numerical analysis based on the evaluation of the integral kernel

$$K(P,Q) := \frac{r'^2\cos\varphi'}{\sqrt{r^2 + r'^2 - 2rr'\cos\psi}} \tag{6}$$

are applied, which will be discussed in Sect. 5.

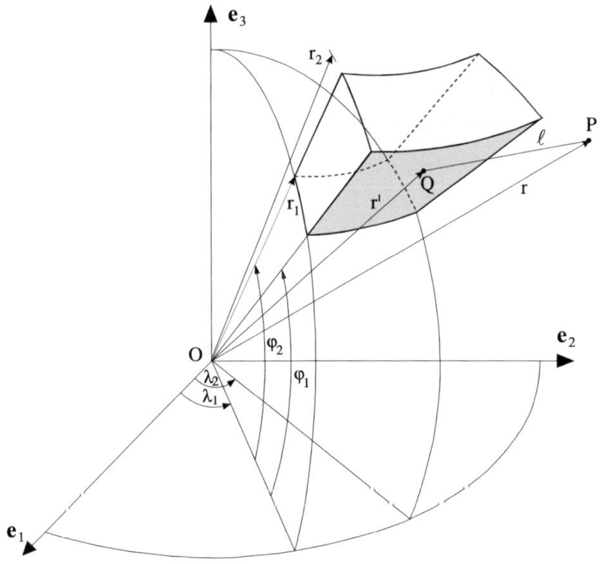

Fig. 1. Geometry of a spherical tesseroid (Heck and Seitz, 2007); the spherical coordinates are referred to the geocentric Earth-fixed equatorial reference system defined by the base vectors $\boldsymbol{e_1}$, $\boldsymbol{e_2}$, $\boldsymbol{e_3}$.

As useful for most practical applications, the computation point P is restricted to be situated outside the tesseroid domain, i.e., $P \notin \Omega^*$. It will be shown later on in Sect. 5.3 that this restriction should be attenuated in some cases by $P \notin \Omega^* \backslash \partial\Omega^*$, where $\partial\Omega^*$ denotes the boundary surface of the tesseroid.

2.2. Definition of partial derivatives

Besides the gravitational potential of a tesseroid the first- and second-order derivatives are considered in this paper. These derivatives are usually defined in a local topocentric Cartesian coordinate system $(\boldsymbol{f_1}, \boldsymbol{f_2}, \boldsymbol{f_3})$ with respect to the computation point P. The $\boldsymbol{f_1}$-axis of the left-handed system as used in the following points north, the $\boldsymbol{f_2}$-axis points east, and the $\boldsymbol{f_3}$-axis points upwards in the geocentric radial direction. The position vector of an arbitrary point in this local Cartesian coordinate system is denoted as

$$\boldsymbol{x} = x_1 \cdot \boldsymbol{f}_1 + x_2 \cdot \boldsymbol{f}_2 + x_3 \cdot \boldsymbol{f}_3 = x_i \cdot \boldsymbol{f}_i, \tag{7}$$

where $\boldsymbol{f}_i$ are the unit vectors in the direction of the coordinate axes as described above. In Eq. (7) and further on, the Einstein summation convention is used.

The first-order partial derivatives of the gravitational potential V of any massive body compose the vector field

$$\boldsymbol{a} = \operatorname{grad} V = \nabla V = a_i \cdot \boldsymbol{f}_i, \tag{8}$$

where

$$a_i := \frac{\partial V(x_1, x_2, x_3)}{\partial x_i} \tag{9}$$

are the physical components of the gravitational acceleration with respect to the basis $\boldsymbol{f}_i$.

The second-order derivatives, known as gravitational gradients, compose a rank two tensor field that is arranged in the so-called Marussi tensor

$$\mathbf{M} = (\nabla \otimes \nabla)V = M_{ij} \cdot (\boldsymbol{f}_i \otimes \boldsymbol{f}_j), \tag{10}$$

where the operator $\otimes$ denotes the tensor product and

$$M_{ij} := \frac{\partial^2 V(x_1, x_2, x_3)}{\partial x_i \, \partial x_j} \tag{11}$$

are the physical components of this tensor with respect to the basis $\boldsymbol{f}_i$. Since the gravitational field is irrotational, i.e., $\nabla \times (\nabla V) = \mathbf{0}$, and the potential V is a harmonic function outside the masses, the Marussi tensor $\mathbf{M}$ is symmetric and trace-free due to the Laplace equation, i.e.

$$\Delta V = \sum_{k=1}^{3} M_{kk} = 0. \tag{12}$$

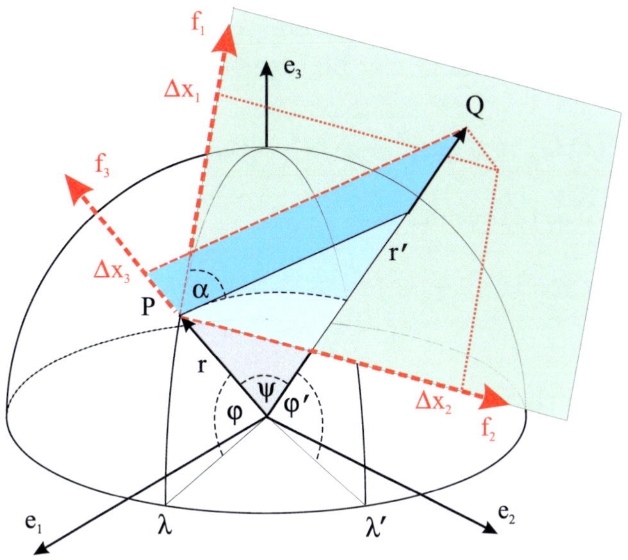

Fig. 2. Visualization of the coordinate differences Δx_1, Δx_2, and Δx_3 between the computation point P and the running integration point Q. The geocentric Cartesian and related spherical coordinate systems are indicated by the *black axes* $\boldsymbol{e}_i$, the topocentric local Cartesian coordinate system is represented by the *red axes* $\boldsymbol{f}_i$.

3. Optimized tesseroid formulas based on Cartesian integral kernels

As it can be considered from Fig. 2, the Cartesian coordinate differences between the computation point P and the running integration point Q can be expressed by

$$\begin{aligned}\Delta x_1 &= r' \sin\psi \cos\alpha,\\ \Delta x_2 &= r' \sin\psi \sin\alpha,\\ \Delta x_3 &= r' \cos\psi - r,\end{aligned} \tag{13}$$

where α denotes the azimuth angle at P in direction to Q. Note that $(\Delta x_1, \Delta x_2, \Delta x_3)$ in Eq. (13) can also be interpreted as the Cartesian coordinates of Q with respect to the topocentric system at the computation point P. Inserting well-known relations of spherical trigonometry (cf. Heiskanen and Moritz, 1967, p. 113)

$$\begin{aligned}\sin\psi\cos\alpha &= \cos\varphi\sin\varphi' - \sin\varphi\cos\varphi'\cos\delta\lambda,\\ \sin\psi\sin\alpha &= \cos\varphi'\sin\delta\lambda\end{aligned} \tag{14}$$

into Eq. (13) results in

$$\begin{aligned}\Delta x_1 &= r'\big(\cos\varphi\sin\varphi' - \sin\varphi\cos\varphi'\cos\delta\lambda\big),\\ \Delta x_2 &= r'\cos\varphi'\sin\delta\lambda,\\ \Delta x_3 &= r'\cos\psi - r,\end{aligned} \tag{15}$$

where $\delta\lambda := \lambda' - \lambda$. By making use of the derived functional relations in Eq. (15) the Euclidean distance between P and Q can be expressed in the local Cartesian system by

$$\ell(P,Q) = \sqrt{{\Delta x_1}^2 + {\Delta x_2}^2 + {\Delta x_3}^2} = \sqrt{\Delta x_i \Delta x_i}. \tag{16}$$

Inserting Eq. (16) into Eq. (3) results in

$$V^*(r,\varphi,\lambda) = G\rho \iiint\limits_{\Omega^*} \frac{1}{\sqrt{{\Delta x_1}^2 + {\Delta x_2}^2 + {\Delta x_3}^2}} \, \mathrm{d}\Omega \tag{17}$$

for the tesseroid potential that is now based on an integral kernel in Cartesian coordinates.

Since the integral kernel in Eq. (17) is an analytical function if $P \notin \Omega^*$, differentiation and integration can be interchanged according to the Leibniz integral rule (e.g., Kellogg, 1929, p. 152). For example, this implies

$$\frac{\partial V^*}{\partial x_1} = G\rho \frac{\partial}{\partial x_1} \iiint\limits_{\Omega^*} \frac{1}{\ell} \, \mathrm{d}\Omega = G\rho \iiint\limits_{\Omega^*} \frac{\partial}{\partial x_1} \frac{1}{\ell} \, \mathrm{d}\Omega. \tag{18}$$

Analogous to expressions for the prism or the point mass approximation (e.g., Heck and Seitz, 2007; Wild-Pfeiffer, 2008), the partial derivatives of the Cartesian integral kernel result in

$$\frac{\partial}{\partial x_i}\left(\frac{1}{\ell}\right) = \frac{\Delta x_i}{\ell^3}, \tag{19}$$

$$\frac{\partial^2}{\partial x_i \partial x_j}\left(\frac{1}{\ell}\right) = \left(\frac{3\Delta x_i \Delta x_j}{\ell^5} - \frac{\delta_{ij}}{\ell^3}\right), \tag{20}$$

where $i, j \in \{1,2,3\}$ and δ_{ij} denotes the Kronecker delta, i.e. $\delta_{ij} = 1$ if $i = j$, and $\delta_{ij} = 0$ otherwise.

The gravitational potential, the gravitational acceleration, and the Marussi tensor caused by a tesseroid with a homogeneous mass density ρ can therefore be described by using the general formula

$$\begin{Bmatrix} V^*(r,\varphi,\lambda) \\ a_i^*(r,\varphi,\lambda) \\ M_{ij}^*(r,\varphi,\lambda) \end{Bmatrix} = G\rho \iiint\limits_{\Omega^*} \frac{1}{\ell^3} \begin{Bmatrix} \Delta x_i \Delta x_i \\ \Delta x_i \\ \frac{3\Delta x_i \Delta x_j}{\ell^2} - \delta_{ij} \end{Bmatrix} \mathrm{d}\Omega, \tag{21}$$

where $\Delta x_i \Delta x_i = \ell^2$.

Although the integral kernels in Eq. (21) are expressed in Cartesian coordinates, it is worthwhile to mention that the integration domain is still bounded by spherical coordinates extending over the domain of a tesseroid, where $\Omega^* = [r_1, r_2] \times [\varphi_1, \varphi_2] \times [\lambda_1, \lambda_2]$ and $\mathrm{d}\Omega = r'^2 \cos\varphi' \, \mathrm{d}r' \, \mathrm{d}\varphi' \, \mathrm{d}\lambda'$. Thus, also these formulas comprise elliptic integrals that have to be evaluated numerically (see Sect. 5).

4. Comparison to tesseroid formulas based on spherical integral kernels

Considering previously published tesseroid methods based on integral kernels expressed in spherical coordinates, the consistency with the elaborated optimized formulas is demonstrated in this section. Moreover, this comparison outlines some numerical advantages of the new formulas.

Following Wild-Pfeiffer (2007, 2008), the basic idea of tesseroid formulas based on spherical integral kernels is a conversion of spherical to Cartesian derivatives. To this end, the first- and second-order spherical derivatives of the tesseroid potential are determined and converted into the moving triad of the local Cartesian coordinate system by making use of the functional relationships (e.g., Tscherning, 1976)

$$a_1^* = \frac{1}{r}\frac{\partial V^*}{\partial \varphi}, \quad a_2^* = \frac{1}{r\cos\varphi}\frac{\partial V^*}{\partial \lambda}, \quad a_3^* = \frac{\partial V^*}{\partial r}, \tag{22}$$

and

$$\begin{aligned}
M_{11}^* &= \frac{1}{r^2}\left(\frac{\partial^2 V^*}{\partial \varphi^2} + r\,\frac{\partial V^*}{\partial r}\right),\\
M_{22}^* &= \frac{1}{r^2\cos^2\varphi}\left(\frac{\partial^2 V^*}{\partial \lambda^2} - \cos\varphi\sin\varphi\,\frac{\partial V^*}{\partial \varphi} + r\cos^2\varphi\,\frac{\partial V^*}{\partial r}\right),\\
M_{33}^* &= \frac{\partial^2 V^*}{\partial r^2},\\
M_{12}^* &= \frac{1}{r^2\cos\varphi}\left(\frac{\partial^2 V^*}{\partial \varphi\,\partial \lambda} + \tan\varphi\,\frac{\partial V^*}{\partial \lambda}\right) = M_{21}^*,\\
M_{13}^* &= \frac{1}{r}\left(\frac{\partial^2 V^*}{\partial \varphi\,\partial r} - \frac{1}{r}\,\frac{\partial V^*}{\partial \varphi}\right) = M_{31}^*,\\
M_{23}^* &= \frac{1}{r\cos\varphi}\left(\frac{\partial^2 V^*}{\partial \lambda\,\partial r} - \frac{1}{r}\,\frac{\partial V^*}{\partial \lambda}\right) = M_{32}^*.
\end{aligned} \tag{23}$$

In this case, the elements a_i^* and M_{ij}^* can be considered as the physical coordinates of the tensors of first- and second-orders (covariant derivatives) represented in the topocentric reference frame with the orthonormal base vectors

$$\boldsymbol{f}_i = \frac{1}{|\boldsymbol{g}_i|}\cdot \boldsymbol{g}_i, \tag{24}$$

where

$$\boldsymbol{g}_1 := \frac{\partial \boldsymbol{x}}{\partial \varphi}, \quad \boldsymbol{g}_2 := \frac{\partial \boldsymbol{x}}{\partial \lambda}, \quad \boldsymbol{g}_3 := \frac{\partial \boldsymbol{x}}{\partial r}, \tag{25}$$

are the unnormalized vectors pointing in the direction of the increasing spherical coordinates.

By applying the Leibniz integral rule to the expression for the tesseroid potential in Eq. (3), the required first- and second-order spherical derivatives result in

$$
\begin{aligned}
\frac{\partial V^*}{\partial \varphi} &= G\rho \iiint\limits_{\Omega^*} \frac{rr'C_\varphi}{\ell^3}\,\mathrm{d}\Omega, \\
\frac{\partial V^*}{\partial \lambda} &= G\rho \iiint\limits_{\Omega^*} \frac{rr'C_\lambda}{\ell^3}\,\mathrm{d}\Omega, \\
\frac{\partial V^*}{\partial r} &= G\rho \iiint\limits_{\Omega^*} \frac{(r'\cos\psi - r)}{\ell^3}\,\mathrm{d}\Omega,
\end{aligned} \tag{26}
$$

and

$$
\begin{aligned}
\frac{\partial^2 V^*}{\partial \varphi^2} &= G\rho \iiint\limits_{\Omega^*} \frac{rr'}{\ell^3}\left(\frac{3rr'C_\varphi^2}{\ell^2} - \cos\psi\right)\mathrm{d}\Omega, \\
\frac{\partial^2 V^*}{\partial \lambda^2} &= G\rho \iiint\limits_{\Omega^*} \frac{rr'}{\ell^3}\left(\frac{3rr'C_\lambda^2}{\ell^2} - \cos\varphi'\cos\varphi\cos\delta\lambda\right)\mathrm{d}\Omega, \\
\frac{\partial^2 V^*}{\partial r^2} &= G\rho \iiint\limits_{\Omega^*} \frac{1}{\ell^3}\left[\frac{3(r'\cos\psi - r)^2}{\ell^2} - 1\right]\mathrm{d}\Omega, \\
\frac{\partial^2 V^*}{\partial \varphi \partial \lambda} &= G\rho \iiint\limits_{\Omega^*} \frac{rr'\cos\varphi'\sin\delta\lambda}{\ell^3}\left(\frac{3rr'C_\varphi\cos\varphi}{\ell^2} - \sin\varphi\right)\mathrm{d}\Omega, \\
\frac{\partial^2 V^*}{\partial \varphi \partial r} &= G\rho \iiint\limits_{\Omega^*} \frac{r'C_\varphi}{\ell^3}\left[\frac{3r(r'\cos\psi - r)}{\ell^2} + 1\right]\mathrm{d}\Omega, \\
\frac{\partial^2 V^*}{\partial \lambda \partial r} &= G\rho \iiint\limits_{\Omega^*} \frac{r'C_\lambda}{\ell^3}\left[\frac{3r(r'\cos\psi - r)}{\ell^2} + 1\right]\mathrm{d}\Omega,
\end{aligned} \tag{27}
$$

where

$$
\begin{aligned}
C_\varphi &:= \frac{\partial \cos\psi}{\partial \varphi} = \cos\varphi\sin\varphi' - \sin\varphi\cos\varphi'\cos\delta\lambda, \\
C_\lambda &:= \frac{\partial \cos\psi}{\partial \lambda} = \cos\varphi\cos\varphi'\sin\delta\lambda.
\end{aligned} \tag{28}
$$

As the volume integrals in Eqs. (26) and (27) cannot be solved analytically, Wild-Pfeiffer (2007, 2008) suggested a formalism based on a two-step sequence:

1. Numerical evaluation of the spherical derivatives in Eqs. (26) and (27),

2. Conversion of the resulting numerical values into the local frame by applying the functional relations according to Eqs. (22) and (23), respectively.

Due to the explicit evaluation of Eqs. (22) and (23), this procedure suffers from the polar singularity which is induced by the spherical coordinate system. As a consequence, this approach causes indeterminate values for the first- and second-order derivatives in Eqs. (9) and (11) in the case of the computation point P being located on the polar axis, i.e., $\varphi = \pm 90°$.

In contrast, the elaborated optimized tesseroid formulas in Eq. (21) provide direct expressions for the potential derivatives in the local Cartesian system that avoid an additional transformation and can therefore be used for any position on the globe. However, the mathematical consistency of both tesseroid approaches can be shown. To this end, the analytical expressions for the spherical derivatives in Eqs. (26) and (27) are inserted into the functional relationships in Eqs. (22) and (23). As the volume integrals extend over the same domain Ω^* they remain purely additive and can be combined. After elementary operations, elaborated exemplarily for the M_{23}^* component in the Appendix, the components of the first- and second-order derivatives of the tesseroid potential can be expressed by

$$\begin{aligned}
a_1^* &= G\rho \iiint\limits_{\Omega^*} \frac{r' C_\varphi}{\ell^3}\,\mathrm{d}\Omega,\\
a_2^* &= G\rho \iiint\limits_{\Omega^*} \frac{r' \cos\varphi' \sin\delta\lambda}{\ell^3}\,\mathrm{d}\Omega,\\
a_3^* &= G\rho \iiint\limits_{\Omega^*} \frac{r' \cos\psi - r}{\ell^3}\,\mathrm{d}\Omega,
\end{aligned} \tag{29}$$

and

$$\begin{aligned}
M_{11}^* &= G\rho \iiint\limits_{\Omega^*} \left(\frac{3r'^2 C_\varphi^2}{\ell^5} - \frac{1}{\ell^3}\right)\mathrm{d}\Omega,\\
M_{22}^* &= G\rho \iiint\limits_{\Omega^*} \left[\frac{3(r' \cos\varphi' \sin\delta\lambda)^2}{\ell^5} - \frac{1}{\ell^3}\right]\mathrm{d}\Omega,\\
M_{33}^* &= G\rho \iiint\limits_{\Omega^*} \left[\frac{3(r' \cos\psi - r)^2}{\ell^5} - \frac{1}{\ell^3}\right]\mathrm{d}\Omega,\\
M_{12}^* &= G\rho \iiint\limits_{\Omega^*} \frac{3r'^2 C_\varphi \cos\varphi' \sin\delta\lambda}{\ell^5}\,\mathrm{d}\Omega,\\
M_{13}^* &= G\rho \iiint\limits_{\Omega^*} \frac{3r' C_\varphi (r' \cos\psi - r)}{\ell^5}\,\mathrm{d}\Omega,\\
M_{23}^* &= G\rho \iiint\limits_{\Omega^*} \frac{3r' \cos\varphi' \sin\delta\lambda (r' \cos\psi - r)}{\ell^5}\,\mathrm{d}\Omega.
\end{aligned} \tag{30}$$

By applying the substitutions $\Delta x_1, \Delta x_2$ and Δx_3 in Eq. (13) it becomes obvious that Eqs. (29) and (30) are consistent with the optimized tesseroid formulas in Eq. (21).

5. Aspects of numerical evaluation

While there are quite few approaches for the numerical evaluation of elliptic volume integrals, as aforementioned in Sect. 1, this paper applies a Taylor series approach as presented in Heck and Seitz (2007). In this approach, the integration is achieved by means of expanding the integral kernel in a Taylor series with fourth-order error in the spatial coordinates of the integration point. In Sect. 5.1, the main idea of this method is recapitulated and then adapted to the optimized tesseroid formulas in Sect. 5.2. Remarks concerning the validity and the accuracy of the evaluation rules as well as limitations are indicated in Sect. 5.3.

5.1. Idea of the Taylor series approach

Following Heck and Seitz (2007), the Taylor expansion of the spherical integral kernel in Eq. (6) at the point $Q_0(r_0, \varphi_0, \lambda_0)$ can be expressed by[1]

$$K(P,Q) = \sum_{i,j,k} \frac{K_{ijk}(P,Q_0)}{i!\; j!\; k!} (r' - r_0)^i (\varphi' - \varphi_0)^j (\lambda' - \lambda_0)^k, \tag{31}$$

where

$$K_{ijk}(P,Q_0) := \left. \frac{\partial^{i+j+k} K(P,Q)}{\partial r'^i \partial \varphi'^j \partial \lambda'^k} \right|_{\substack{r'=r_0 \\ \varphi'=\varphi_0 \\ \lambda'=\lambda_0}} \tag{32}$$

are the partial derivatives of the integral kernel K evaluated at Q_0. When integrating the Taylor series in Eq. (31), the maximum efficiency is achieved by fixing the Taylor point Q_0 at a point of symmetry, e.g., at the geometrical center of the tesseroid

$$r_0 = \frac{r_1 + r_2}{2}, \quad \varphi_0 = \frac{\varphi_1 + \varphi_2}{2}, \quad \lambda_0 = \frac{\lambda_1 + \lambda_2}{2}. \tag{33}$$

By inserting Eq. (31) into Eq. (3), the integration with respect to each coordinate (r', φ', λ') can be performed by a substitution (see Heck and Seitz, 2007), e.g., in the case of r' follows

$$\int_{r_1}^{r_2} (r' - r_0)^i \,\mathrm{d}r' = \int_{-\Delta r/2}^{+\Delta r/2} (r^*)^i \,\mathrm{d}r^* = \frac{1 - (-1)^{i+1}}{(i+1)2^{i+1}} (\Delta r)^{i+1} = \begin{cases} 0, & \text{if } i \text{ odd}, \\ \frac{(\Delta r)^{i+1}}{(i+1)2^i}, & \text{if } i \text{ even}, \end{cases} \tag{34}$$

where

$$\Delta r = r_2 - r_1, \quad \Delta\varphi = \varphi_2 - \varphi_1, \quad \Delta\lambda = \lambda_2 - \lambda_1 \tag{35}$$

[1] Note that the original publication contains a minor formal misprint in the general expression of the Taylor series expansion in Eq. (31), which has been corrected in this version. There, the term in the denominator reads $(i+j+k)!$. However, this mistake has no consequences on the further derived formulas used in the article, which remain correct and unchanged. For reference, see the comment Deng, X.-L., Grombein, T., Shen, W.-B., Heck, B., and Seitz, K. (2016): Corrections to "A comparison of the tesseroid, prism and point-mass approaches for mass reductions in gravity field modelling" (Heck and Seitz, 2007) and "Optimized formulas for the gravitational field of a tesseroid" (Grombein et al., 2013). *Journal of Geodesy* 90(6):585–587, DOI: 10.1007/s00190-016-0907-8.

denote the dimensions of the tesseroid. Consequently, terms with odd-order i, j, or k in the Taylor series in Eq. (31) cancel out after a subsequent integration and only terms with even-order remain. Thus, the gravitational potential of a tesseroid can be approximated by

$$V^*(r,\varphi,\lambda) = G\rho\Delta r\Delta\varphi\Delta\lambda\Big[K_{000} + \frac{1}{24}\Big(K_{200}\Delta r^2 + K_{020}\Delta\varphi^2 + K_{002}\Delta\lambda^2\Big) + O(\Delta^4/\ell_0^5)\Big], \tag{36}$$

where the Landau symbol $O(\Delta^4/\ell_0^5)$ indicates that the omitted Taylor residual has a magnitude of fourth-order, scaled with a negative power of ℓ_0 which is the Euclidean distance between P and Q_0. In this case, Δ should be considered symbolically, e.g., in metrical units it would be $\Delta r = r_2 - r_1$, $\Delta\varphi \sim r_0(\varphi_2 - \varphi_1)$, and $\Delta\lambda \sim r_0 \cos\varphi_0(\lambda_2 - \lambda_1)$. Furthermore, it is worthwhile mentioning that $\ell_0 > 0$ is strictly finite if an exterior computation point P is considered, as presumed in Sect. 2.1, i.e., $P \notin \Omega^*$ implies $\ell_0 > \Delta/2$.

The zero-order term in Eq. (36) corresponds to the potential of a point mass that concentrates the mass

$$m_0 = \rho r_0^2 \cos\varphi_0 \Delta r\Delta\varphi\Delta\lambda \tag{37}$$

at the Taylor point Q_0. This is consistent with the zero-order approximation of the total tesseroid mass (e.g., Grüninger, 1990, p. 79):

$$m^* = \rho \int_{\lambda_1}^{\lambda_2}\int_{\varphi_1}^{\varphi_2}\int_{r_1}^{r_2} r'^2 \cos\varphi' \,\mathrm{d}r'\,\mathrm{d}\varphi'\,\mathrm{d}\lambda' = \frac{1}{3}\rho\Big(r_2^3 - r_1^3\Big)(\sin\varphi_2 - \sin\varphi_1)(\lambda_2 - \lambda_1) \approx m_0. \tag{38}$$

For the evaluation of Eq. (36) the zero- and second-order coefficients according to Eq. (32) are explicitly provided in Heck and Seitz (2007). Analogously, Wild-Pfeiffer (2008) presented corresponding zero- and second-order coefficients for the numerical evaluation of the first- and second-order spherical derivatives of the tesseroid potential. They can be utilized in combination with the relationships in Eqs. (22) and (23) in order to calculate the components of the gravitational acceleration and the Marussi tensor.

5.2. Adaption of the Taylor series approach

Analogously to the procedure in Eq. (36), a numerical solution of the optimized tesseroid formulas in Eq. (21) can be supplied by subsequently integrating Taylor series expansions of the Cartesian integral kernels at the tesseroid's geometrical center point $Q_0(r_0, \varphi_0, \lambda_0)$. To supply a more general notation that allows an efficient computational implementation, the substitutions

$$(\xi_1', \xi_2', \xi_3') := (r', \varphi', \lambda') \tag{39}$$

for the coordinates of the integration point Q and

$$(\Delta\xi_1, \Delta\xi_2, \Delta\xi_3) := (\Delta r, \Delta\varphi, \Delta\lambda) \tag{40}$$

for the dimensions of the tesseroid are used in the following. Furthermore, based on Eq. (39) the differential operators are defined as

$$\partial_k := \frac{\partial}{\partial \xi'_k}, \quad \partial_k^2 := \frac{\partial^2}{\partial {\xi'_k}^2}, \quad k \in \{1,2,3\}. \tag{41}$$

A suitable numerical solution of Eq. (21) with a fourth-order error in the spatial coordinates of the integration point is then provided by

$$\begin{Bmatrix} V^*(r,\varphi,\lambda) \\ a_i^*(r,\varphi,\lambda) \\ M_{ij}^*(r,\varphi,\lambda) \end{Bmatrix} = \omega \begin{Bmatrix} K(P,Q) \\ L_i(P,Q) \\ N_{ij}(P,Q) \end{Bmatrix} + \frac{\omega}{24} \sum_{k=1}^{3} {\Delta\xi_k}^2 \begin{Bmatrix} \partial_k^2 K(P,Q) \\ \partial_k^2 L_i(P,Q) \\ \partial_k^2 N_{ij}(P,Q) \end{Bmatrix} \Bigg|_{\substack{r'=r_0 \\ \varphi'=\varphi_0 \\ \lambda'=\lambda_0}} + \begin{Bmatrix} O(\Delta^4/\ell_0^5) \\ O(\Delta^4/\ell_0^6) \\ O(\Delta^4/\ell_0^7) \end{Bmatrix}, \tag{42}$$

where

$$\omega := G\rho \prod_{k=1}^{3} \Delta\xi_k. \tag{43}$$

In Eq. (42), the Cartesian integral kernels of Eq. (21) are denoted by

$$\begin{Bmatrix} K(P,Q) \\ L_i(P,Q) \\ N_{ij}(P,Q) \end{Bmatrix} := \frac{1}{\ell^3} \begin{Bmatrix} \Delta x_i \Delta x_i \\ \Delta x_i \\ \frac{3\Delta x_i \Delta x_j}{\ell^2} - \delta_{ij} \end{Bmatrix} \tau, \tag{44}$$

in which

$$\tau := {r'}^2 \cos\varphi' \tag{45}$$

factors the spherical volume element $\mathrm{d}\Omega$. The Cartesian coordinate differences Δx_i and the Euclidean distance ℓ can be calculated using the previous Eqs. (13) and (16), respectively. The required second-order derivatives in Eq. (44) can be represented in the general form

$$\begin{Bmatrix} \partial_k^2 K \\ \partial_k^2 L_i \\ \partial_k^2 N_{ij} \end{Bmatrix} = \frac{1}{\ell^3} \left[\partial_k^2(\tau) \begin{Bmatrix} \alpha \\ \alpha_i \\ \alpha_{ij} \end{Bmatrix} + \partial_k(\tau) \begin{Bmatrix} \beta_k \\ \beta_{ik} \\ \beta_{ijk} \end{Bmatrix} + \tau \begin{Bmatrix} \gamma_k \\ \gamma_{ik} \\ \gamma_{ijk} \end{Bmatrix} \right], \tag{46}$$

where

$$\alpha := \Delta x_i \Delta x_i, \quad \alpha_i := \Delta x_i, \quad \alpha_{ij} := \frac{3\Delta x_i \Delta x_j}{\ell^2} - \delta_{ij}, \tag{47}$$

$$\beta_k := 2\partial_k(\bar{\ell}),$$

$$\beta_{ik} := 2\left[\partial_k(\Delta x_i) + \frac{3\Delta x_i \partial_k(\bar{\ell})}{\ell^2}\right],$$

$$\beta_{ijk} := \frac{6}{\ell^2}\left[\Delta x_i \partial_k(\Delta x_j) + \Delta x_j \partial_k(\Delta x_i) + \frac{5\Delta x_i \Delta x_j \partial_k(\bar{\ell})}{\ell^2} - \delta_{ij}\partial_k(\bar{\ell})\right], \tag{48}$$

$$\gamma_k := \partial_k^2(\bar{\ell}) + \frac{3(\partial_k(\bar{\ell}))^2}{\ell^2},$$

$$\gamma_{ik} := \partial_k^2(\Delta x_i) + \frac{3}{\ell^2}\left\{2\partial_k(\Delta x_i)\partial_k(\bar{\ell}) + \Delta x_i\left[\partial_k^2(\bar{\ell}) + \frac{5(\partial_k(\bar{\ell}))^2}{\ell^2}\right]\right\},$$

$$\begin{aligned}\gamma_{ijk} := \frac{3}{\ell^2}\Bigg\{&\Delta x_i \partial_k^2(\Delta x_j) + \Delta x_j \partial_k^2(\Delta x_i) + 2\partial_k(\Delta x_i)\partial_k(\Delta x_j)\\ &+\frac{5}{\ell^2}\left[2\partial_k(\bar{\ell})\Big(\Delta x_i \partial_k(\Delta x_j) + \partial_k(\Delta x_i)\Delta x_j\Big) + \Delta x_i \Delta x_j\left(\partial_k^2(\bar{\ell}) + \frac{7(\partial_k(\bar{\ell}))^2}{\ell^2}\right)\right]\\ &-\delta_{ij}\left[\partial_k^2(\bar{\ell}) + \frac{5(\partial_k(\bar{\ell}))^2}{\ell^2}\right]\Bigg\},\end{aligned} \tag{49}$$

and

$$\bar{\ell} := -\frac{1}{2}\ell^2 = -\frac{1}{2}\Delta x_i \Delta x_i. \tag{50}$$

Particularly, for $\xi_1' = r'$ holds:

$$\begin{aligned}\partial_1(\tau) &= 2r'\cos\varphi',\\ \partial_1^2(\tau) &= 2\cos\varphi',\\ \partial_1(\bar{\ell}) &= r\cos\psi - r',\\ \partial_1^2(\bar{\ell}) &= -1,\\ \partial_1(\Delta x_1) &= C_\varphi,\\ \partial_1(\Delta x_2) &= \cos\varphi'\sin\delta\lambda,\\ \partial_1(\Delta x_3) &= \cos\psi,\\ \partial_1^2(\Delta x_1) &= 0,\\ \partial_1^2(\Delta x_2) &= 0,\\ \partial_1^2(\Delta x_3) &= 0.\end{aligned} \tag{51}$$

Particularly, for $\xi_2' = \varphi'$ holds:

$$\begin{aligned}\partial_2(\tau) &= -r'^2\sin\varphi',\\ \partial_2^2(\tau) &= -r'^2\cos\varphi',\\ \partial_2(\bar{\ell}) &= r\partial_2(\Delta x_3),\\ \partial_2^2(\bar{\ell}) &= r\partial_2^2(\Delta x_3),\\ \partial_2(\Delta x_1) &= r'\big[\cos\varphi\cos\varphi' + \sin\varphi\sin\varphi'\cos\delta\lambda\big],\end{aligned}$$

$$
\begin{aligned}
\partial_2(\Delta x_2) &= -r' \sin\varphi' \sin\delta\lambda, \\
\partial_2(\Delta x_3) &= r' \big[\sin\varphi \cos\varphi' - \cos\varphi \sin\varphi' \cos\delta\lambda\big], \\
\partial_2^2(\Delta x_1) &= -\Delta x_1, \\
\partial_2^2(\Delta x_2) &= -\Delta x_2, \\
\partial_2^2(\Delta x_3) &= -r' \cos\psi.
\end{aligned} \tag{52}
$$

Particularly, for $\xi_3' = \lambda'$ holds:

$$
\begin{aligned}
\partial_3(\tau) &= 0, \\
\partial_3^2(\tau) &= 0, \\
\partial_3(\bar{\ell}) &= r\partial_3(\Delta x_3), \\
\partial_3^2(\bar{\ell}) &= r\partial_3^2(\Delta x_3), \\
\partial_3(\Delta x_1) &= \Delta x_2 \sin\varphi, \\
\partial_3(\Delta x_2) &= r' \cos\varphi' \cos\delta\lambda, \\
\partial_3(\Delta x_3) &= -\Delta x_2 \cos\varphi, \\
\partial_3^2(\Delta x_1) &= \partial_3(\Delta x_2) \sin\varphi, \\
\partial_3^2(\Delta x_2) &= -\Delta x_2, \\
\partial_3^2(\Delta x_3) &= -\partial_3(\Delta x_2) \cos\varphi.
\end{aligned} \tag{53}
$$

Note that according to Eq. (42), the analytical expressions of the Cartesian integral kernels in Eq. (44) and its second-order derivatives in Eq. (46) have to be evaluated at the Taylor point Q_0, i.e., the primed coordinates of the running integration point (r', φ', λ') have to be replaced by the coordinates of the Taylor point $(r_0, \varphi_0, \lambda_0)$.

5.3. Remarks on the validity and accuracy

From potential theory it is known that the gravitational potential of a voluminous mass and its first-order derivatives are defined and continuous in the whole $\mathbb{R}^3$, even if the computation point P is located at the boundary surface or inside a field generating mass distribution of continuous density (e.g., Kellogg, 1929, p. 151). The singularity of Newton's integral when the computation point P coincides with the integration point Q, i.e., $\ell = 0$, is weak and therefore removable. If the density function satisfies a Hölder condition, the second-order derivatives are also defined for interior computation points P, but generally they are not defined for points on the boundary surface as it represents an interface of density discontinuity (cf. Kellogg, 1929, p. 156).

In the context of a homogeneous tesseroid this implies that values for V^*, a_i^*, and M_{ij}^* in Eq. (21) are well-defined for interior computation points P, even though the presented Taylor series approach in Eq. (42) is limited to exterior points due to the required condition $P \notin \Omega^*$. Considering the limiting behavior of Eq. (21) the restriction can be attenuated to $P \notin \Omega^* \backslash \partial\Omega^*$ in the case of the gravitational potential and the elements of the gravitational acceleration. Thus, V^* and a_i^* can also be evaluated in the case of a computation point P placed on the tesseroid surface, which is particularly useful for terrestrial applications.

Instead of an analytical error analysis of Eq. (42), the approximation error of the method is estimated by a realistic numerical experiment in the following section. Generally, as mentioned in Heck and Seitz (2007) and Grombein et al. (2010, p. 34), the accuracy of the presented approach is strongly sensitive to the geometrical shape of the tesseroid, particularly for small distances between the computation point P and the Taylor point Q_0. Besides the tesseroid dimensions Δr, $\Delta\varphi$, and $\Delta\lambda$ that can imply for example a flat or columnar shape, the position of P relative to the tesseroid also impacts the computation. Due to the meridional convergence of the spherical coordinate system, the geometrical shape of a tesseroid changes with respect to the latitude of its position. For instance, the ground surface of a tesseroid located near the equator is almost quadratic whereas its surface degenerates to a triangular shape in the polar region.

As noted in Heck and Seitz (2007), special care should be taken when applying the Taylor series approach for tesseroids in the near zone around the computation point P which is particularly the case for terrestrial applications. Since Heck and Seitz (2007) found unacceptably large errors, they recommended to replace tesseroids by equivalent prisms in the direct vicinity of the computation point. Another possibility that will be shown in the following section is the horizontal respectively vertical subdivision of the tesseroids in the near zone. Although numerical investigations on these two possibilities provide satisfactory results, the origin of the occurring numerical problems of tesseroids in the very near zone is the subject of ongoing investigations.

6. Numerical investigations

In order to validate the derived evaluation rules and to show the computational efficiency, two realistic numerical experiments are presented in this section. The elaborated optimized tesseroid formulas based on Cartesian integral kernels (cf. Sect. 3) are compared to the previously published tesseroid approaches based on spherical integral kernels (cf. Sect. 4), and conventional rectangular prism formulas (Nagy et al., 2000, 2002). In the following, these approaches are denoted as tesseroid (Cartesian), tesseroid (spherical), and prism, respectively. The software implementation is performed in the programming language C++. The source code may be made available upon request. Section 6.1 focuses on the required computation time, while in Sect. 6.2 the approximation error induced by the different methods is analyzed.

6.1. Comparison of the computation time

In a first experiment, the required computation time of forward modeling based on the different approaches is compared. As a realistic application, topographic reductions are estimated by utilizing the digital terrain model DTM2006.0 (Pavlis et al., 2007) that provides a $5' \times 5'$ global grid of heights h_i^{DTM}. The topographic information of this model can be represented by 9 931 200 individual tesseroid bodies Ω_i^* with the constant horizontal dimensions $\Delta_{\mathrm{hor}} = \Delta\varphi = \Delta\lambda = 5'$ and the variable vertical dimensions $\Delta r_i = h_i^{\mathrm{DTM}}$.

According to Eq. (2) the total topographic effect is approximated by the sum over the impact of all individual tesseroids, i.e.

$$\begin{Bmatrix} V \\ a_i \\ M_{ij} \end{Bmatrix} \approx \begin{Bmatrix} \sum V^* \\ \sum a_i^* \\ \sum M_{ij}^* \end{Bmatrix}. \tag{54}$$

Note that in the case of the prism formulas the tesseroid dimensions are converted to equivalent rectangular prisms under the assumption of the same mass density (cf. Heck and Seitz, 2007; Grombein et al., 2010, p. 29ff.). Furthermore, as mentioned in Sect. 1, an additional coordinate transformation has to be performed in the case of prisms in order to take the curvature of the Earth into account (cf. Grombein et al., 2010, p. 30).

In Table 1, the required relative computation time for calculating the components V, a_i, and M_{ij} is given with respect to the applied method. A significant speed-up of computation time by using the optimized (Cartesian) tesseroid formulas instead of the spherical ones can be recognized, which increases with the order of the derivatives. Generally, this can be considered as a consequence of a reduced number of occurring arithmetic operations which are specified in Table 2. In the case of the elements of the Marussi tensor, the reduction is mainly caused by the fact that only six volume integrals have to be evaluated instead of nine as in former representations.

In total, when computing V, a_i, and M_{ij} together, only 55 % of the run time of the spherical tesseroid methods is needed for the optimized (Cartesian) approach. In comparison to the conventional prism approach, the advantage of using tesseroids for global applications is impressively shown.

Table 1. Comparison of computation times t [%] using tesseroid methods with spherical and optimized Cartesian integral kernels as well as prisms to compute the gravitational potential V, the components of the gravitational acceleration a_i and the Marussi tensor M_{ij} of topographic masses. All values are specified in percentage relative to the computation time of the spherical tesseroid method.

	Tesseroid (spherical)	Tesseroid (Cartesian)	Prism
$t(V)$	100	80	1265
$t(a_i)$	100	72	620
$t(M_{ij})$	100	44	125
$t(V, a_i, M_{ij})$	100	55	402

Table 2. Comparison of the optimized (Cartesian) tesseroid formulas with respect to previously published spherical methods in terms of the number of occurring arithmetic operations n. Note that these are approximate values as they are strongly dependent on the actual computational implementation.

	Tesseroid (spherical)		Tesseroid (Cartesian)	
Operations	$+ / -$	$\cdot / \div$	$+ / -$	$\cdot / \div$
$n(V^*)$	25	80	20	65
$n(a_i^*)$	70	250	60	200
$n\left(M_{ij}^*\right)$	250	890	160	550

6.2. Comparison of the approximation error

In a second experiment, the approximation error induced by the different methods is analyzed and compared. In order to obtain reference values, an analytical solution is needed. Therefore, a spherical shell approximation is considered as a simple synthetic Earth model, where topographic masses with a constant thickness h' and a homogeneous mass density ρ are fixed on a sphere with a mean Earth radius R. Thus, the inner radius of the spherical shell is set to $R_1 = R$, the outer radius to $R_2 = R + h'$.

Analytical solution of a spherical shell. The gravitational potential in the external domain of a spherical shell Ω^s with constant density ρ can be determined analytically from the potential difference between two solid spheres with radii R_1 and R_2 (e.g., Vaníček et al., 2001):

$$V^s(r) := \frac{G(m_2 - m_1)}{r}, \tag{55}$$

where

$$m_1 = \rho \frac{4}{3} \pi R_1^3, \tag{56}$$

$$m_2 = \rho \frac{4}{3} \pi R_2^3, \tag{57}$$

$$r = \sqrt{x_1^2 + x_2^2 + x_3^2}. \tag{58}$$

By determining the partial derivatives of V^s in Eq. (55) the elements of the gravitational acceleration and the Marussi tensor result in

$$a_i^s(r) = G(m_2 - m_1) \frac{x_i}{r^3}, \tag{59}$$

$$M_{ij}^s(r) = G(m_2 - m_1) \left(\frac{3 x_i x_j}{r^5} - \frac{\delta_{ij}}{r^3} \right). \tag{60}$$

Table 3. Parameters defining the spherical shell.

$R_1 = R$	$6378.137\,\mathrm{km}$
$R_2 = R + h'$	$6379.137\,\mathrm{km}$
G	$6.672 \cdot 10^{-11}\,\mathrm{m^3\,kg^{-1}\,s^{-2}}$
ρ	$2670\,\mathrm{kg\,m^{-3}}$

Due to the isotropy of the spherical shell, a computation point P on the polar axis can be considered without loss of generality, i.e., $x_1 = 0,\ x_2 = 0,\ x_3 = r \geq R_2$. This implies $a_1^s = a_2^s = M_{12}^s = M_{13}^s = M_{23}^s = 0$ and

$$a_3^s(r) = \frac{V^s(r)}{r}, \tag{61}$$

$$M_{11}^s(r) = M_{22}^s(r) = -\frac{V^s(r)}{r^2}, \tag{62}$$

$$M_{33}^s(r) = \frac{2V^s(r)}{r^2}. \tag{63}$$

Setting of the experiment. In order to quantify realistic approximation errors, the absolute differences between the exact values (V^s, a_i^s, M_{ij}^s) of the analytical solution and the actual values when applying forward modeling based on a mass discretization with tesseroids are analyzed:

$$\begin{Bmatrix} \delta V \\ \delta a_i \\ \delta M_{ij} \end{Bmatrix} := \left| \begin{Bmatrix} V^s \\ a_i^s \\ M_{ij}^s \end{Bmatrix} - \begin{Bmatrix} \sum V^* \\ \sum a_i^* \\ \sum M_{ij}^* \end{Bmatrix} \right|. \tag{64}$$

In the experiment, a spherical shell with a constant thickness of $h' = 1\,\mathrm{km}$ is chosen which approximately corresponds to a mean topographic height. Furthermore, the spherical shell is defined by the parameters as specified in Table 3. In the case of the gravitational potential and acceleration, mostly related to terrestrial applications, the computation point is placed on the surface of the spherical shell. For the second-order derivatives of the Marussi tensor, the satellite altitude of the gravity gradiometry field mission GOCE (Gravity field and steady-state Ocean Circulation Explorer) is utilized. In Table 4, the settings of both applications and the resulting reference values for the spherical shell are indicated, where $1\,\mathrm{mGal} = 10^{-5}\,\mathrm{m\,s^{-2}}$ and $1\,\mathrm{mE} = 10^{-12}\,\mathrm{s^{-2}}$.

To get the actual values for the developed method the spherical shell Ω^s is decomposed into individual tesseroid bodies Ω_i^* bounded by spherical grid lines. The horizontal dimensions of the tesseroids are again set to $\Delta_{\mathrm{hor}} = \Delta\varphi = \Delta\lambda = 5'$. According to the thickness of the spherical shell, the vertical dimension is fixed to $\Delta r = h' = 1\,\mathrm{km}$.

Table 4. Setting of the terrestrial and satellite application and resulting reference values according to Eq. (55) and Eqs. (61) – (63).

Application	Terrestrial	Satellite
h	1 km	260 km
$r = R + h$	6379.137 km	6638.137 km
V^s	14 278.119 $\mathrm{m^2\,s^{-2}}$	–
a_3^s	223.825 mGal	–
$M_{11}^s = M_{22}^s$	–	−311.383 mE
M_{33}^s	–	622.765 mE

In order to improve the accuracy of the tesseroid approaches in the case of terrestrial applications, a subdivision of the mass elements located in the near zone around the computation point P is performed as proposed in Sect. 5.3. To this end, for tesseroids, whose geometrical center Q_0 is located inside a spherical distance ψ_c with respect to the computation point, a 100×100 horizontal subdivision is performed. Based on the $5' \times 5'$ resolution of the original tesseroid bodies this implies an increase of the grid resolution to $3'' \times 3'' \approx 100\,\mathrm{m} \times 100\,\mathrm{m}$ which is consistent with current high-resolution global DTMs like the SRTM3 model (Farr et al., 2007). Different extensions of the near zone are analyzed, where the spherical distance ψ_c is set to integer multiples of the horizontal tesseroid dimension, i.e.

$$\psi_c = \kappa \cdot \Delta_{\mathrm{hor}}, \quad \kappa \in \mathbb{N}_0. \tag{65}$$

For the cases $\kappa = \{1, \ldots, 5\}$, Fig. 3 illustrates the numbers of mass elements in the near zone that are subdivided. Due to the meridional convergence these numbers are dependent on the latitude φ of the computation point P showing a strong increase toward the pole. Note again that instead of a subdivision it is also possible to utilize equivalent prisms which has been shown in the numerical investigations in Heck and Seitz (2007) and will, therefore, not be presented in this paper. In the case of satellite applications, no subdivision is performed as there is a large distance between the computation points and the surface of the spherical shell.

As mentioned in Sect. 5.3 the geographical position of the tesseroid impacts the accuracy due to a changing geometry according to the meridional convergence. Concerning the strong influence of the near zone (cf. Heck and Seitz, 2007), it can be assumed that the total approximation error may also depend on the latitude φ of the computation point P. The approximation error in Eq. (64) is therefore evaluated for different positions on the globe. Due to the spherical symmetry, this can be restricted to a computation point P running along an arbitrary but fixed meridian on the northern hemisphere. All other cases provide analogous results.

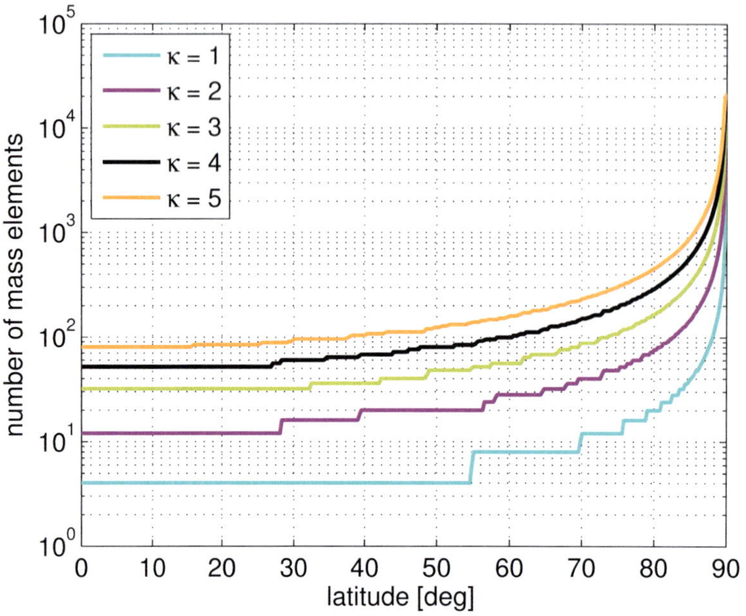

Fig. 3. Visualization of the number of mass elements in the near zone as a function of the latitude φ of the computation point. According to Eq. (65) the near zone is bounded by a spherical distance $\psi_c = \kappa \cdot \Delta_{\text{hor}}$ with respect to the position of the computation point. All values in this figure refer to a horizontal dimension of $\Delta_{\text{hor}} = 5'$.

Results for a terrestrial application. In Fig. 4, the estimated approximation error is presented for the gravitational potential δV on the left panel and the radial component of the gravitational acceleration δa_3 on the right panel. As indicated in Table 4 the computation point P is located on top of the spherical shell. Both tesseroid approaches (green and overlaid red dashed curve) show the same, nearly constant behavior with respect to the latitude φ. The approximation error δV is in a range of about $10^0 - 10^{-1}\,\text{m}^2\,\text{s}^{-2}$, while the order of magnitude for δa_3 is about $10^2\,\text{mGal}$. This error behavior clearly demonstrates the above indicated numerical problems of the tesseroid approach when the computation point P is located in the direct vicinity of the particular mass bodies. Compared with the conventional prism approach (blue curve) the approximation error for tesseroids is inferior by about three orders of magnitude in the case of δV and two orders of magnitude in the case of δa_3.

Furthermore, it can be seen that the approximation errors for the prism approach show significant dependencies on the latitude φ, particularly in the case of the potential. In the polar region, where there is the largest difference in the geometrical shape between a tesseroid and a rectangular prism, the tesseroid approaches supply slightly better, but still bad results.

When performing the intended subdivision in the near zone, it can clearly be seen in Fig. 4 that the approximation error for the optimized (Cartesian) tesseroid approach in both

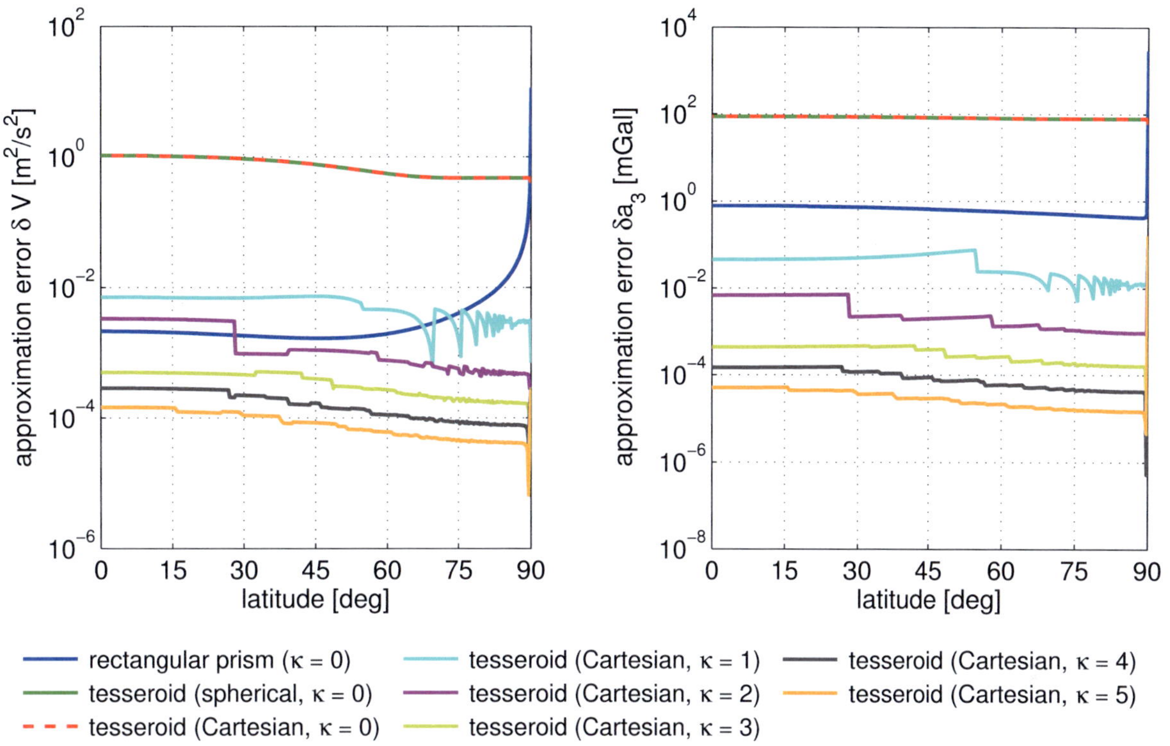

Fig. 4. Visualization of the estimated approximation error δV (*left panel*) and δa_3 (*right panel*) as a function of the latitude φ of the computation point that is located on top of the respective spherical shell, i.e., $h = h' = 1\,\text{km}$. The *blue curve* is obtained by applying conventional prism formulas. The *green curve* represents the use of the spherical tesseroid approach and is completely overlaid by the *red dashed curve* that indicates the utilization of the optimized (Cartesian) tesseroid formulas. These three cases are calculated without a special consideration of the near zone, i.e., $\kappa = 0$. The *remaining curves* represent the use of the optimized (Cartesian) tesseroid formulas by performing an additional 100×100 horizontal subdivision of mass elements located in the near zone. According to Eq. (65) the cases of $\kappa = \{1, \ldots, 5\}$ are displayed.

cases δV and δa_3 can be largely reduced. For increasing values of κ respectively ψ_c, the approximation error is rapidly decreasing. Occurring discontinuities in the approximation errors can be associated with a changing number of mass elements in the near zone (cf. Fig. 3). In the case of $\kappa = 3$, the approximation error δV is below $10^{-3}\,\text{m}^2\,\text{s}^{-2}$, which is consistent with a sub-millimeter error in derived geoid heights. Similarly, δa_3 is below $1\,\mu\text{Gal}$, which corresponds to the accuracy of actual gravimeters. Note that a comparable behavior is provided if the subdivision is applied to the tesseroid method based on spherical integral kernels. The corresponding cases are therefore not illustrated in Fig. 4.

To summarize, the achieved accuracy will be sufficient for most practical terrestrial applications if a subdivision is performed in the near zone extended by a spherical distance of $\psi_c \geq 3 \cdot \Delta_{\text{hor}}$ with respect to the computation point. However, it should be mentioned that the computation time is increased due to the densification in the very near zone, but it is still considerably smaller in comparison to conventional prism formulas (see Table 5).

Table 5. Comparison of computation time t [%] using optimized (Cartesian) tesseroid methods and conventional prism formulas to compute the gravitational potential V, the components of the gravitational acceleration a_i and the Marussi tensor M_{ij}. The values for the optimized tesseroid method are indicated in relation to the extension of a near zone, in which a 100×100 horizontal subdivision is performed. The near zone is bounded by a spherical distance of $\psi_c = \kappa \cdot \Delta_{\text{hor}}$ around each computation point. All values are specified in percentage relative to the computation time of the spherical tesseroid method with $\kappa = 0$.

	Tesseroid (spherical)	Tesseroid (Cartesian)						Prism
κ	0	0	1	2	3	4	5	0
t	100	55	60	67	72	79	88	402

Results for a satellite application. According to Table 4 the approximation errors in the case of the Marussi tensor are estimated in the context of the satellite gravity gradiometry mission GOCE, i.e., the computation point P is fixed at a height of $h = 260\,\text{km}$ above the sphere of radius $R = R_1$. In Fig. 5, the approximation errors according to Eq. (64) for δM_{11} (upper left panel), δM_{22} (upper right panel), and δM_{33} (lower left panel) are visualized.

For all three components nearly the same behavior is visible showing a considerable dependency on the latitude φ of the computation point. Generally, the approximation error rises with increasing latitude, while a rapid increase can be seen in the polar region at $\varphi > 85°$. Due to the logarithmic scale a change of sign from a positive to a negative approximation error induces a behavior as visible in δM_{11} and δM_{33} at $\varphi \approx 25°$ in the case of the prism approach.

The approximation error of the conventional prism approach is in a range of about $10^0 - 10^{-5}\,\text{mE}$, while the tesseroid approaches comprise significant smaller errors of about $10^{-5} - 10^{-13}\,\text{mE}$. Due to the large distance of the computation points to the tesseroid bodies, it is not necessary to take special care for the near zone, i.e., $\kappa = 0$ can be fixed without any problems. Again the green curve for the spherical tesseroid approach is mostly overlaid by the red curve of the optimized (Cartesian) tesseroid approach showing that both variants provide the same approximation errors. Some small oscillations can be detected near the equator indicating the limitation of the numerical stability. Furthermore, it is worth mentioning that in the case of δM_{22} a large difference between the two tesseroid approaches can be detected at the pole point. This effect clearly illustrates the polar singularity problem of the spherical tesseroid approach.

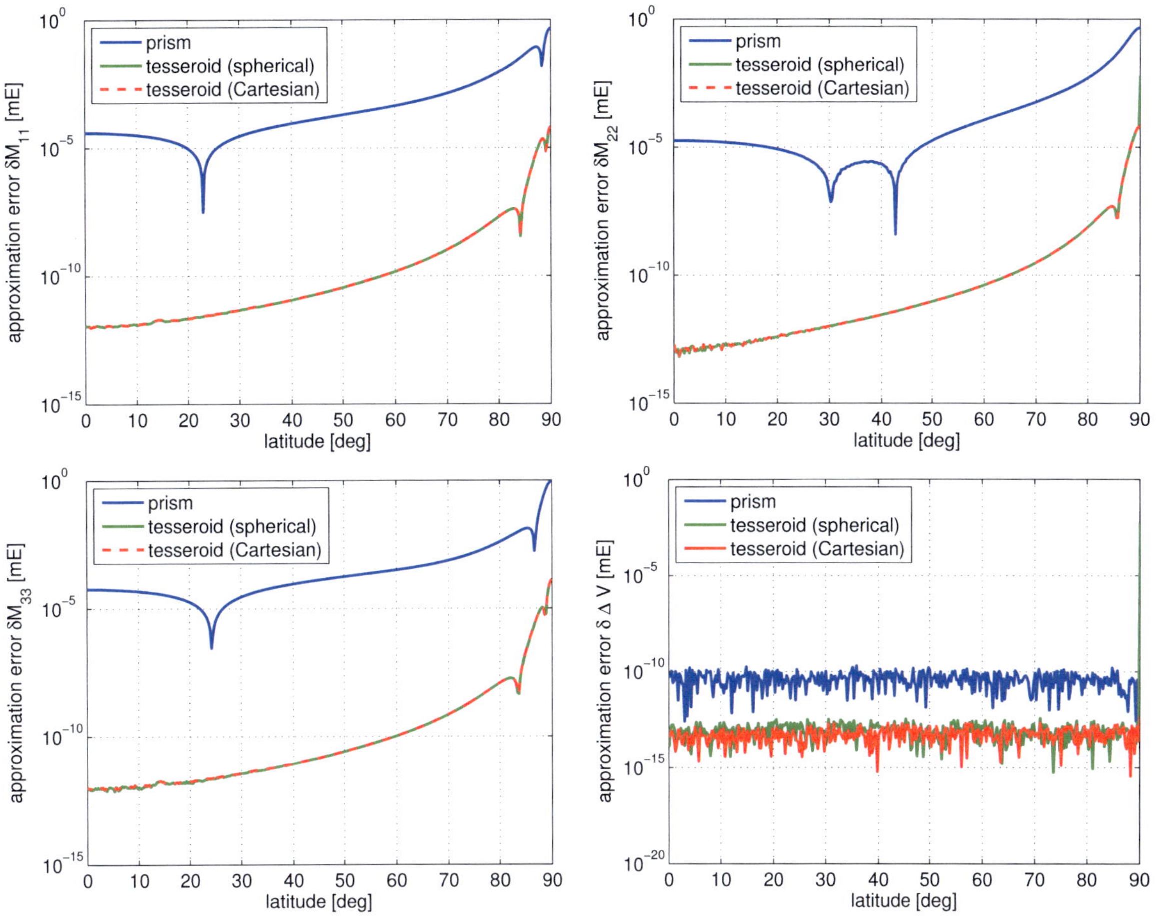

Fig. 5. Visualization of the estimated approximation error δM_{11} (*upper left*), δM_{22} (*upper right*), and δM_{33} (*lower left*) as a function of the latitude φ. The *lower right* illustrates the Laplace condition $\delta\Delta V$ according to Eq. (66). The computation point P is located at a satellite height of $h = 260\,\text{km}$. The *blue curve* is obtained by using prisms, the *red dashed curve* by applying the optimized (Cartesian) tesseroid formulas. The *green curve* represents the spherical tesseroid approach and is overlaid by the *red dashed curve* in most cases.

As an additional quality characteristic the discrepancy in the Laplace equation

$$\delta\,\Delta\,V := \sum\left(\sum_{k=1}^{3} M_{kk}^{*}\right) \tag{66}$$

is displayed on the lower right panel of Fig. 5 supporting the findings indicated above.

7. Conclusions and outlook

When using forward (or inverse) modeling based on Newton's integral, tesseroid bodies are the natural mass discretization when dealing with data parameterized in geodetic

D'Urso, M. G. (2013): On the evaluation of the gravity effects of polyhedral bodies and a consistent treatment of related singularities. *Journal of Geodesy* 87(3):239–252. DOI: 10.1007/s00190-012-0592-1.

Farr, T. G., Rosen, P. A., Caro, E., Crippen, R., Duren, R., Hensley, S., Kobrick, M., Paller, M., Rodriguez, E., Roth, L., Seal, D., Shaffer, S., Shimada, J., Umland, J., Werner, M., Oskin, M., Burbank, E., and Alsdorf, D. (2007): The shuttle radar topography mission. *Reviews of Geophysics* 45(2):RG2004. DOI: 10.1029/2005RG000183.

Forsberg, R. (1984): A study of terrain reductions, density anomalies and geophysical inversion methods in gravity field modelling. Report 355. Department of Geodetic Science and Surveying, The Ohio State University, Columbus, USA.

Forsberg, R. (1985): Gravity field terrain effect computations by FFT. *Bulletin Géodésique* 59(4):342–360. DOI: 10.1007/BF02521068.

Forsberg, R. and Tscherning, C. C. (1997): Topographic effects in gravity field modelling for BVP. In: Sansò, F., and Rummel, R. (eds.) Geodetic boundary value problems in view of the one centimeter geoid, *Lecture Notes in Earth Sciences*, vol. 65. Springer Berlin Heidelberg, pp. 239–272. DOI: 10.1007/BFb0011707.

Grombein, T., Seitz, K., Awange, J. L., and Heck, B. (2012): Detection of hydrological mass variations by means of an inverse tesseroid approach. In: General Assembly of the European Geosciences Union 2012. Vienna, Austria, Apr. 22–27, 2012. *Geophysical Research Abstracts*, vol. 14. EGU2012-7548.

Grombein, T., Seitz, K., and Heck, B. (2010): Untersuchung zur effizienten Berechnung topographischer Effekte auf den Gradiententensor am Fallbeispiel der Satellitengradiometriemission GOCE. *KIT Scientific Reports*, no. 7547. KIT Scientific Publishing, Karlsruhe, Germany. DOI: 10.5445/KSP/1000017531.

Grombein, T., Seitz, K., and Heck, B. (2013): Topographic-isostatic reduction of GOCE gravity gradients. In: Rizos, C., and Willis, P. (eds.) Earth on the edge: science for a sustainable planet. Proceedings of the IAG General Assembly, Melbourne, Australia, June 28–July 2, 2011. *International Association of Geodesy Symposia*, vol. 139. Springer Berlin Heidelberg, pp. 349–356. DOI: 10.1007/978-3-642-37222-3_46. in print.

Grüninger, W. (1990): Zur topographisch-isostatischen Reduktion der Schwere. PhD thesis. Universität Karlsruhe, Germany.

Heck, B. and Seitz, K. (2007): A comparison of the tesseroid, prism and point-mass approaches for mass reductions in gravity field modelling. *Journal of Geodesy* 81(2):121–136. DOI: 10.1007/s00190-006-0094-0.

Heck, B. and Seitz, K. (2008): Representation of the time variable gravity field due to hydrological mass variations by surface layer potentials. In: General Assembly of the European Geosciences Union 2008. Vienna, Austria, Apr. 13–18, 2008. *Geophysical Research Abstracts*, vol. 10. EGU2008-A-09355.

Heiskanen, W. A. and Moritz, H. (1967): Physical geodesy. W. H. Freeman & Co., San Francisco, USA.

Hirt, C., Featherstone, W. E., and Marti, U. (2010): Combining EGM2008 and SRTM/DTM-2006.0 residual terrain model data to improve quasigeoid computations in mountainous areas devoid of gravity data. *Journal of Geodesy* 84(9):557–567. DOI: 10.1007/s00190-010-0395-1.

Janák, J., Wild-Pfeiffer, F., and Heck, B. (2012): Smoothing the gradiometric observations using different topographic-isostatic models: a regional case study. In: Sneeuw, N., Novák, P., Crespi, M., and Sansò, F. (eds.) VII Hotine-Marussi Symposium on Mathematical Geodesy. Proceedings of the symposium in Rome, June 6–10, 2009. *International Association of Geodesy Symposia*, vol. 137. Springer Berlin Heidelberg, pp. 245–250. DOI: 10.1007/978-3-642-22078-4_37.

Kellogg, O. D. (1929): Foundations of potential theory. Springer, Berlin, Germany.

Klose, U. and Ilk, K. (1993): A solution to the singularity problem occurring in the terrain correction formula. *Manuscripta Geodaetica* 18(5):263–279.

Ku, C. C. (1977): A direct computation of gravity and magnetic anomalies caused by 2- and 3-dimensional bodies of arbitrary shape and arbitrary magnetic polarization by equivalent-point method and a simplified cubic spline. *Geophysics* 42(3):610–622. DOI: 10.1190/1.1440732.

Kuhn, M. and Featherstone, W. E. (2005): Construction of a synthetic Earth gravity model by forward gravity modelling. In: Sansò, F. (ed.) A window on the future of geodesy. Proceedings of the IAG General Assembly, Sapporo, Japan, June 30–July 11, 2003. *International Association of Geodesy Symposia*, vol. 128. Springer Berlin Heidelberg, pp. 350–355. DOI: 10.1007/3-540-27432-4_60.

Kuhn, M. and Seitz, K. (2005): Comparison of Newton's integral in the space and frequency domains. In: Sansò, F. (ed.) A window on the future of geodesy. Proceedings of the IAG General Assembly, Sapporo, Japan, June 30–July 11, 2003. *International Association of Geodesy Symposia*, vol. 128. Springer Berlin Heidelberg, pp. 386–391. DOI: 10.1007/3-540-27432-4_66.

MacMillan, W. D. (1930): Theoretical mechanics, vol. 2: the theory of the potential. McGraw-Hill, New York, USA (reprinted by Dover Publications, New York, USA 1958).

Mader, K. (1951): Das Newtonsche Raumpotential prismatischer Körper und seine Ableitungen bis zur dritten Ordnung. *Österreichische Zeitschrift für Vermessungswesen, Sonderheft* 11.

Makhloof, A. A. and Ilk, K. (2008): Effects of topographic-isostatic masses on gravitational functionals at the Earth's surface and at airborne and satellite altitudes. *Journal of Geodesy* 82(2):93–111. DOI: 10.1007/s00190-007-0159-8.

Martinec, Z. (1998): Boundary-value problems for gravimetric determination of a precise geoid. *Lecture Notes in Earth Sciences*, vol. 73. Springer, Berlin Heidelberg, Germany. DOI: 10.1007/BFb0010337.

Nagy, D., Papp, G., and Benedek, J. (2000): The gravitational potential and its derivatives for the prism. *Journal of Geodesy* 74(7–8):552–560. DOI: 10.1007/s001900000116.

Nagy, D., Papp, G., and Benedek, J. (2002): Corrections to "The gravitational potential and its derivatives for the prism". *Journal of Geodesy* 76(8):475. DOI: 10.1007/s00190-002-0264-7.

Novák, P., Kern, M., Schwarz, K.-P., and Heck, B. (2003): Evaluation of band-limited topographical effects in airborne gravimetry. *Journal of Geodesy* 76(11–12):597–604. DOI: 10.1007/s00190-002-0282-5.

Pavlis, N. K., Factor, J. K., and Holmes, S. A. (2007): Terrain-related gravimetric quantities computed for the next EGM. In: Kiliçoglu, A., and Forsberg, R. (eds.) Proceedings of the 1st International Symposium of the International Gravity Field Service (IGFS), Istanbul, Turkey, Aug. 28–Sept. 1, 2006. Harita Dergisi, Special Issue 18, pp. 318–323.

Petrović, S. (1996): Determination of the potential of homogeneous polyhedral bodies using line integrals. *Journal of Geodesy* 71(1):44–52. DOI: 10.1007/s001900050074.

Schwarz, K.-P., Sideris, M. G., and Forsberg, R. (1990): The use of FFT techniques in physical geodesy. *Geophysical Journal International* 100(3):485–514. DOI: 10.1111/j.1365-246X.1990.tb00701.x.

Smith, D. A., Robertson, D. S., and Milbert, D. G. (2001): Gravitational attraction of local crustal masses in spherical coordinates. *Journal of Geodesy* 74(11–12):783–795. DOI: 10.1007/s001900000142.

Tscherning, C. C. (1976): Computation of the second-order derivatives of the normal potential based on the representation by a Legendre series. *Manuscripta Geodaetica* 1:71–92.

Tsoulis, D. (1999): Analytical and numerical methods in gravity field modelling of ideal and real masses. *Deutsche Geodätische Kommission, Reihe C*, no. 510. Verlag der Bayerischen Akademie der Wissenschaften in Kommission beim Verlag C. H. Beck, Munich, Germany.

Tsoulis, D. (2012): Analytical computation of the full gravity tensor of a homogeneous arbitrarily shaped polyhedral source using line integrals. *Geophysics* 77(2):F1–F11. DOI: 10.1190/geo2010-0334.1.

Tsoulis, D., Wziontek, H., and Petrović, S. (2003): A bilinear approximation of the surface relief in terrain correction computations. *Journal of Geodesy* 77(5–6):338–344. DOI: 10.1007/s00190-003-0332-7.

Vaníček, P., Novák, P., and Martinec, Z. (2001): Geoid, topography, and the Bouguer plate or shell. *Journal of Geodesy* 75(4):210–215. DOI: 10.1007/s001900100165.

Von Frese, R. R. B., Hinze, W. J., Braile, L. W., and Luca, A. J. (1981): Spherical-Earth gravity and magnetic anomaly modeling by Gauss-Legendre quadrature integration. *Journal of Geophysics* 49(3):234–242. URL: http://ntrs.nasa.gov/search.jsp?R=19820016644.

Wild, F. and Heck, B. (2008): Topographic and isostatic reductions for use in satellite gravity gradiometry. In: Xu, P., Liu, J., and Dermanis, A. (eds.) VI Hotine-Marussi Symposium on Theoretical and Computational Geodesy, IAG Symposium. Wuhan, China, May 29–June 2, 2006. *International Association of Geodesy Symposia*, vol. 132. Springer Berlin Heidelberg, pp. 49–55. DOI: 10.1007/978-3-540-74584-6_8.

Wild-Pfeiffer, F. (2007): Auswirkungen topographisch-isostatischer Massen auf die Satellitengradiometrie. *Deutsche Geodätische Kommission, Reihe C*, no. 604. Verlag der Bayerischen Akademie der Wissenschaften in Kommission beim Verlag C. H. Beck, Munich, Germany. URL: https://www.dgk.badw.de.devweb.mwn.de/fileadmin/docs/c-604.pdf.

Wild-Pfeiffer, F. (2008): A comparison of different mass elements for use in gravity gradiometry. *Journal of Geodesy* 82(10):637–653. DOI: 10.1007/s00190-008-0219-8.

Wild-Pfeiffer, F. and Heck, B. (2007): Comparison of the modelling of topographic and isostatic masses in the space and the frequency domain for use in satellite gravity gradiometry. In: Kiliçoglu, A., and Forsberg, R. (eds.) Proceedings of the 1st International Symposium of the International Gravity Field Service (IGFS), Istanbul, Turkey, Aug. 28–Sept. 1, 2006. Harita Dergisi, Special Issue 18, pp. 312–317.

Chapter III.

A wavelet-based assessment of topographic-isostatic reductions for GOCE gravity gradients

Thomas Grombein, Xiaoguang Luo, Kurt Seitz, Bernhard Heck
Surveys in Geophysics 35(4):959–982, 2014. DOI: 10.1007/s10712-014-9283-1.

Submitted: 13 September 2013 / Accepted: 1 February 2014 / Published online: 19 March 2014

Abstract. Gravity gradient measurements from ESA's satellite mission GOCE (Gravity field and steady-state Ocean Circulation Explorer) contain significant high- and mid-frequency signal components, which are primarily caused by the attraction of the Earth's topographic and isostatic masses. In order to mitigate the resulting numerical instability of a harmonic downward continuation, the observed gradients can be smoothed with respect to topographic-isostatic effects using a remove-compute-restore technique. For this reason, topographic-isostatic reductions are calculated by forward modeling that employs the advanced Rock-Water-Ice methodology. The basis of this approach is a three-layer decomposition of the topography with variable density values and a modified Airy-Heiskanen isostatic concept incorporating a depth model of the Mohorovičić discontinuity. Moreover, tesseroid bodies are utilized for mass discretization and arranged on an ellipsoidal reference surface. To evaluate the degree of smoothing via topographic-isostatic reduction of GOCE gravity gradients, a wavelet-based assessment is presented in this paper and compared with statistical inferences in the space domain. Using the Morlet wavelet, continuous wavelet transforms are applied to measured GOCE gravity gradients before and after reducing topographic-isostatic signals. By analyzing a representative data set in the Himalayan region, an employment of the reductions leads to significantly smoothed gradients. In addition, smoothing effects that are invisible in the space domain can be detected in wavelet scalograms, making a wavelet-based spectral analysis a powerful tool.

Keywords GOCE · Topographic-isostatic reduction · Rock-Water-Ice (RWI) decomposition · Forward modeling · Tesseroids · Continuous Morlet wavelet transform

1. Introduction

The high- and mid-frequency constituents of gravity-field-related quantities, observed on or close to the Earth's surface, are mainly affected by the variable topography and the isostatic compensation masses. Using forward modeling, these topographic-isostatic signals can be simulated from digital terrain models (DTM) and applied to the measurements. In classical physical geodesy, topographic-isostatic mass reductions are utilized to obtain boundary values at geoid level according to the Stokes problem that requires a mass-free space outside the boundary surface (Heiskanen and Moritz, 1967, Chapter 3). By applying the remove-compute-restore technique (Forsberg, 1984, Sect. 4) or residual terrain modeling (Forsberg and Tscherning, 1997), topographic-isostatic reductions can also be used to smooth the high- and mid-frequency signal content of gravity field observations. The main benefit of such a procedure is an improved numerical stability for interpolation and prediction tasks as well as field transformations. One example is the stabilization of the harmonic downward continuation process of airborne or satellite-based gravity field observations (Novák et al., 2003; Makhloof and Ilk, 2008). While in the present paper the smoothing behavior of topographic-isostatic signals will be analyzed, there are many other geodetic and geophysical applications that require precise topographic mass reductions or even account for a compensation of deeper lying mass components. This includes, for instance, the compilation of combined high-resolution gravity field models (Fecher et al., 2013; Hirt et al., 2013) or investigations of structures in the Earth's crust (Álvarez et al., 2012; Novák and Tenzer, 2013).

The present paper focuses on the topographic-isostatic reduction of the second-order derivatives of the Earth's gravitational potential V as observed by ESA's satellite mission GOCE. These derivatives can be arranged in the symmetric Marussi tensor $\mathbf{M}$, where

$$M_{ij} = \frac{\partial^2 V(x_1, x_2, x_3)}{\partial x_i \, \partial x_j} = M_{ji}, \quad i, j \in \{1, 2, 3\} \tag{1}$$

are its physical components known as gravity gradients that are usually specified in a local Cartesian coordinate system. Based on the concept of satellite gravity gradiometry (Rummel and Colombo, 1985), GOCE measures gravity gradients at an altitude of about 250 km. Within the measurement bandwidth of the GOCE gradiometer (5 to 100 mHz), the main diagonal tensor components M_{11}, M_{22}, M_{33} and the off-diagonal component M_{13} are observed with a precision of about 1 to $2\,\mathrm{mE}/\sqrt{\mathrm{Hz}}$ ($1\,\mathrm{E} = 10^{-9}\,\mathrm{s}^{-2}$), while the other off-diagonal components M_{12} and M_{23} are less accurately measured with a precision of about $1\,\mathrm{E}/\sqrt{\mathrm{Hz}}$ (Rummel et al., 2011).

Previous investigations based on simulated GOCE gravity gradients have shown that significant smoothing effects can be achieved by considering topographic-isostatic reductions (Wild-Pfeiffer, 2007, 2008; Janák et al., 2012). Furthermore, a closed loop simulation presented in Janák and Wild-Pfeiffer (2010) verified the benefit of incorporating topographic-isostatic information into the downward continuation process of satellite gradiometry measurements.

Due to the high sensitivity of gravity gradients, a refinement of the commonly used topographic-isostatic reduction method is recommended. In comparison to conventional modeling approaches, where topographic masses are usually characterized with a constant density value, the advanced Rock-Water-Ice (RWI) method has been developed by Grombein et al. (2010, 2011). This method applies a three-layer decomposition of the topography, in which the rock, water, and ice masses are modeled individually with layer-specific density values. In terms of isostasy, the classical Airy-Heiskanen model is adapted to the RWI approach and extended by means of a seismologically derived depth model of the Mohorovičić discontinuity.

In previous studies, the smoothing impact of topographic-isostatic reductions on GOCE gravity gradients was analyzed by conducting statistical inferences in the space domain (Wild-Pfeiffer, 2007, 2008; Janák et al., 2012; Grombein et al., 2014). The degree of smoothing is thereby evaluated based on the percentage changes in the standard deviation and range of the gradients before and after reduction.

In addition to this approach, this paper uses the continuous wavelet transform to assess the performance of topographic-isostatic reductions. Analyzing gradient time series by means of wavelets has the advantage of simultaneously examining the signal structure in both time and frequency domains.

Taking advantage of their prominent time-frequency localization, wavelets have been applied for pattern recognition in geoscientific applications (Keller, 2004, Sect. 3.1), e.g., characterizing temporally variable features of polar motion (Gibert et al., 1998; Liu et al., 2007), studying bathymetric profiles and seafloor sediments (Little et al., 1993; Atallah et al., 2002), and detecting cycle slips in observations from the U.S. Global Positioning System (GPS) (Collin and Warnant, 1995; Yi et al., 2006).

Furthermore, wavelet algorithms contribute to reducing multipath effects on GPS measurements (Satirapod and Rizos, 2005; Wu et al., 2009), particularly in the form of an adaptive filter combined with the cross-validation technique (Zhong et al., 2008). In order to appropriately assess GPS phase errors, Satirapod et al. (2001) used a wavelet-based method to decompose observation residuals into unmodeled systematic biases and random measurement noise. By combining the empirical mode decomposition (EMD) and wavelet techniques, Wang et al. (2009) proposed an improved EMD-wavelet approach to mitigate systematic errors in GPS phase observations. In the context of GPS stochastic modeling, Luo (2014, Sect. 8.1.7 and 8.2.7) incorporated wavelet transforms for visual inspection and interpretation of the results from a residual-based temporal correlation analysis.

The paper is organized as follows: Section 2 describes the details and input data of the RWI topographic-isostatic reduction method. In Sect. 3, the mathematical background of the continuous Morlet wavelet transform is explained in a concise manner. Sect. 4 presents the study area and the used GOCE gravity gradient data set. The results are discussed in Sect. 5, where the degree of smoothing is analyzed and compared in both time and frequency domains. Finally, Sect. 6 provides a summary and concluding remarks.

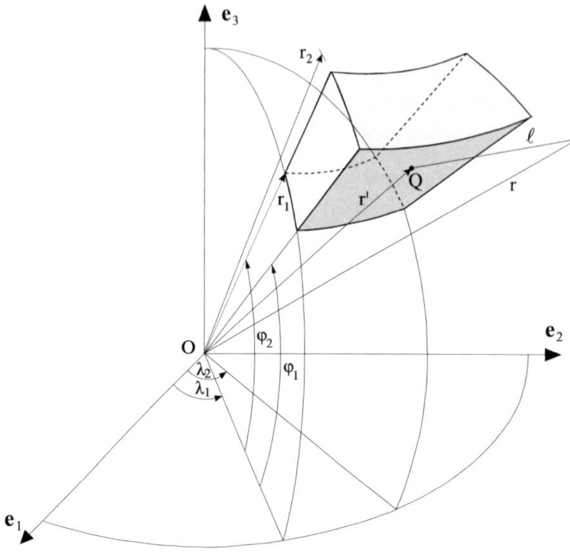

Fig. 1. Geometry of a tesseroid used for mass discretization (Heck and Seitz, 2007); the spherical coordinates (r, φ, λ) are referred to the geocentric Earth-fixed equatorial reference system defined by the base vectors $\boldsymbol{e_1}$, $\boldsymbol{e_2}$, $\boldsymbol{e_3}$.

2. RWI topographic-isostatic reduction method

Using forward modeling, the impact of topographic-isostatic masses on gravity gradients can generally be obtained by evaluating the second-order derivatives of Newton's integral (Heiskanen and Moritz, 1967, p. 3):

$$M_{ij}(P) = G \iiint\limits_{\Omega} \frac{\partial^2}{\partial x_i \, \partial x_j} \frac{\rho}{\ell} \, \mathrm{d}\Omega, \tag{2}$$

where G denotes Newton's gravitational constant, $\rho = \rho(Q)$ is the location-dependent mass density function, and $\ell = \ell(P, Q)$ is the Euclidean distance between the attracted computation point P and the running integration mass point Q. The integration domain Ω extends over the topographic and isostatic masses, requiring global information on the Earth's geometry and density. Generally, Newton's integral in Eq. (2) can either be evaluated in the space domain by direct integration methods or in the frequency domain by spectral approaches (Kuhn and Seitz, 2005; Wild-Pfeiffer and Heck, 2007; Tenzer and Novák, 2013).

According to the grid resolution of the used input data, a mass discretization of the volume integrals in Eq. (2) has to be performed. By using space domain techniques, the integration domain Ω is decomposed into elementary mass bodies Ω_i^* with constant density values, where $\Omega = \bigcup \Omega_i^*$. For applications at a global scale, a discretization using tesseroid mass bodies (spherical prisms) is particularly beneficial (Heck and Seitz, 2007; Wild-Pfeiffer, 2008). As shown in Fig. 1, tesseroids are bounded by geocentric spherical coordinate lines (r, φ, λ) and are therefore directly related to the curvature of the Earth. Representing gravity gradients in a local north-oriented frame (LNOF), where the x_1-axis points north, the x_2-axis west, and the x_3-axis upwards in the geocentric radial direction, optimized tesseroid formulas are used as proposed by Grombein et al. (2013):

$$\left\{M_{ij}^*(P)\right\}_{\text{LNOF}} = G\rho \int_{r_1}^{r_2}\int_{\varphi_1}^{\varphi_2}\int_{\lambda_1}^{\lambda_2} N_{ij}(P,Q)\,\mathrm{d}\Omega^*, \tag{3}$$

where $\mathrm{d}\Omega^* = r'^2\cos\varphi'\,\mathrm{d}r'\,\mathrm{d}\varphi'\,\mathrm{d}\lambda'$ is the spherical volume element. The integral kernel function of Eq. (3) can be formulated as

$$N_{ij}(P,Q) = \left(\frac{3\Delta x_i \Delta x_j}{\ell^5} - \frac{\delta_{ij}}{\ell^3}\right), \tag{4}$$

where

$$\begin{aligned} \Delta x_1 &= r'\left[\cos\varphi\sin\varphi' - \sin\varphi\cos\varphi'\cos(\lambda'-\lambda)\right], \\ \Delta x_2 &= -r'\cos\varphi'\sin(\lambda'-\lambda), \\ \Delta x_3 &= r'\left[\sin\varphi\sin\varphi' + \cos\varphi\cos\varphi'\cos(\lambda'-\lambda)\right] - r, \end{aligned} \tag{5}$$

denote the coordinate differences between $P(r,\varphi,\lambda)$ and $Q(r',\varphi',\lambda')$,

$$\ell = \sqrt{\Delta x_1{}^2 + \Delta x_2{}^2 + \Delta x_3{}^2}, \tag{6}$$

and δ_{ij} is the Kronecker delta, i.e., $\delta_{ij} = 1$ if $i = j$, and $\delta_{ij} = 0$ otherwise.

2.1. Topographic reduction

To characterize the topographic mass distribution, the $5' \times 5'$ global topographic database DTM2006.0 (Pavlis et al., 2007) is utilized, which provides information on surface elevation, ocean and lake depth, as well as ice thickness. Moreover, each grid element is classified into one of the following six terrain types: dry land above mean sea level (MSL), lake or pond, ocean, grounded glacier, oceanic ice shelf, dry land below MSL. As each of these terrain types can be represented by a rock, water, and ice proportion (see Fig. 2a), DTM2006.0 allows to characterize topographic masses with variable density values.

A simple and common way to account for differences in density is the concept of rock-equivalent heights, where the DTM heights are condensed such as to refer to a constant reference density (Rummel et al., 1988; Kuhn and Seitz, 2005; Hirt et al., 2012). The main drawback of this approach is that the geometry of the mass distribution changes considerably if the actual density differs strongly from the adopted constant value (Tsoulis and Kuhn, 2007). As numerically shown in Grombein et al. (2010), this deficiency has a significant impact on the topographic reduction signal at the GOCE satellite altitude.

To establish a more realistic topographic model, the RWI method is used, which enables a rigorous separate modeling of topographic masses with variable density values. For this

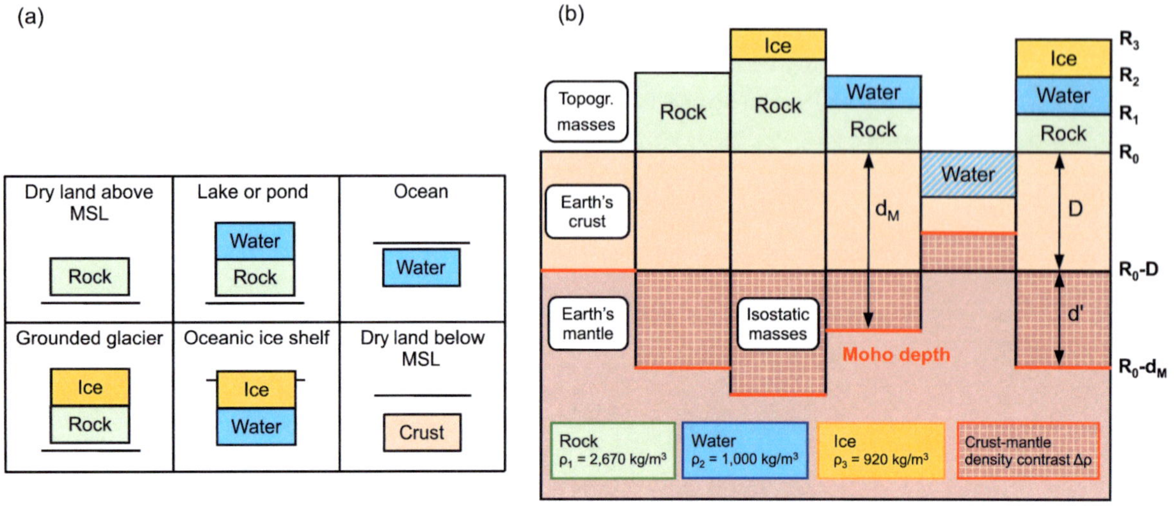

Fig. 2. Schematic representation of the Rock-Water-Ice (RWI) topographic model and the associated modified Airy-Heiskanen isostatic concept. Note that a planar illustration is used for the purpose of simplification, whereas all formulas for the RWI model correspond to a spherical (or ellipsoidal) approximation.

purpose, the information from DTM2006.0 is exploited to construct a more appropriate three-layer RWI terrain and density model parameterized by the geodetic coordinates

$$B_k = 90^\circ - (k - 1/2) \cdot \Delta B, \qquad k = 1, \ldots, n \in \mathbb{N}, \tag{7}$$

$$L_l = (l - 1/2) \cdot \Delta L, \qquad l = 1, \ldots, m \in \mathbb{N}, \tag{8}$$

where $\Delta B = \Delta L = 5'$, $n = 2160$, and $m = 4320$. Each grid element (k, l) of this model contains a rock (1), water (2), and ice (3) component with different MSL heights of the corresponding upper boundary surfaces (h_1, h_2, h_3) and layer-specific density values ($\rho_1 = 2670\,\mathrm{kg\,m^{-3}}$, $\rho_2 = 1000\,\mathrm{kg\,m^{-3}}$, $\rho_3 = 920\,\mathrm{kg\,m^{-3}}$).

To calculate the topographic reduction signal, each grid element of the RWI model is represented by three vertically arranged tesseroids with different density values. The geodetic coordinates (B_k, L_l) are thereby transformed to the corresponding geocentric spherical coordinates (φ_k, λ_l) using the formulas provided by Heck (2003, p. 69):

$$\varphi_k = \arctan\left(\frac{1}{1 + e'^2} \tan B_k\right), \qquad \lambda_l = L_l, \tag{9}$$

where the square of the second numerical eccentricity is set to $e'^2 = 0.006\,739\,496\,775\,48$ according to the parameters of the GRS80 reference ellipsoid (Moritz, 1980). Furthermore, taking the ellipticity of the Earth's shape into account, the tesseroids are fixed on the surface of the GRS80 ellipsoid by introducing a latitude-dependent Earth radius (Heck, 2003, p. 68), i.e.,

$$R_0(\varphi_k) = \frac{a}{\sqrt{1 + e'^2 \sin^2 \varphi_k}}, \tag{10}$$

where $a = 6\,378\,137\,\text{m}$ is the semi-major axis of the GRS80 ellipsoid. The geocentric radii of the upper boundary surfaces of rock, water, and ice are then approximated by $R_1 = R_0 + h_1$, $R_2 = R_0 + h_2$, and $R_3 = R_0 + h_3$, respectively (see Fig. 2b).

By applying the superposition principle, topographic reductions based on the RWI model are calculated as the sum of the impact over all individual tesseroids:

$$\left\{M_{ij}^{\text{Topo}}(P)\right\}_{\text{LNOF}} = G \sum_{k=1}^{n} \sum_{l=1}^{m} \sum_{s=1}^{3} \rho_s \int_{r_1}^{r_2} \iint_{\sigma_{kl}^*} N_{ij}(P,Q)\, \mathrm{d}\Omega^*, \tag{11}$$

where $r_1 = R_{s-1}$, $r_2 = R_s$ according to Fig. 2b, and

$$\sigma_{kl}^* = \left[\varphi_k - \frac{\Delta\varphi}{2}, \varphi_k + \frac{\Delta\varphi}{2}\right] \times \left[\lambda_l - \frac{\Delta\lambda}{2}, \lambda_l + \frac{\Delta\lambda}{2}\right] \tag{12}$$

denotes the spherical tesseroid base surface associated with the grid element (k, l). As the integration over σ_{kl}^* comprises elliptic integrals, Eq. (11) can only be evaluated approximately. Therefore, in analogy to Heck and Seitz (2007), a Taylor series expansion of the integral kernel $N_{ij}(P,Q)$ and a subsequent integration is performed. The applied evaluation rules are explicitly presented in Grombein et al. (2013).

2.2. Isostatic reduction

The isostatic masses compensate the topographic load and can be quantified by applying a mass equality condition with respect to a particular normal compensation depth D (see Fig. 2b). The classical Airy-Heiskanen isostatic concept (Heiskanen and Moritz, 1967, p. 135ff.) uses a local and column-based mass compensation that can be easily adapted to the previously described RWI decomposition method. For each mass column, the mass equality condition can be expressed by

$$m^{\text{Iso}} = \sum_{s=1}^{3} m_s, \tag{13}$$

where m^{Iso} denotes the compensating isostatic mass, and m_1, m_2, and m_3 are the masses of the topographic rock, water, and ice proportions, respectively. In the classical approach of Airy-Heiskanen, the thickness d' of an isostatic mass column varies, while its density contrast $\Delta\rho$ is kept constant. In this case, $D + d'$ corresponds to the assumed depth of the boundary surface between the Earth's crust and mantle, known as Mohorovičić discontinuity (Moho), and $\Delta\rho$ represents the crust-mantle density contrast. This strongly simplified concept has two main disadvantages: 1) It fails over deep ocean trenches, where d' may rise above the ocean bottom (Wild and Heck, 2005), and 2) it does not reflect the lateral heterogeneity in $\Delta\rho$ (Kaban et al., 2004).

To overcome these deficiencies, several attempts have been made by varying either the normal compensation depth D or the density contrast $\Delta\rho$, as well as by applying different isostatic concepts over continental and ocean areas (Wild and Heck, 2005; Hirt et al., 2012).

As such combinations do not significantly influence the smoothing behavior at satellite altitude, improvements can rather be made by utilizing additional geophysical information. Therefore, the Airy-Heiskanen concept is modified by incorporating seismic Moho depths d_M. In particular, $D + d'$ is set to d_M for each mass column, which fixes the geometry of isostatic masses (see Fig. 2b). Since the mass equality condition according to Eq. (13) is no longer satisfied for a constant density, Grombein et al. (2011, 2014) suggested using variable density values $\Delta\rho = \Delta\rho(k, l)$, which implies a location-dependent crust-mantle density contrast. Employing spherical approximation, and regarding Fig. 2b, Eq. (13) can be formulated as

$$\Delta\rho \int_{R_0-d_{\mathrm{M}}}^{R_0-D} r'^2 \,\mathrm{d}r' = \sum_{s=1}^{3} \rho_s \int_{R_{s-1}}^{R_s} r'^2 \,\mathrm{d}r'. \tag{14}$$

Accordingly, for each grid element (k, l), the crust-mantle density contrast is computed as

$$\Delta\rho = \frac{\sum_{s=1}^{3} \rho_s (R_s^3 - R_{s-1}^3)}{(R_0 - D)^3 - (R_0 - d_{\mathrm{M}})^3}. \tag{15}$$

In order to accommodate the isostatic information to the topographic model, a $5' \times 5'$ global grid of smoothed Moho depths d_M is derived from the $2° \times 2°$ global CRUST 2.0 model (Bassin et al., 2000) by harmonic analysis and synthesis (Wittwer et al., 2008; Abd-Elmotaal et al., 2014). For the normal compensation depth D, an optimal value of 31 km was derived and is applied in the RWI approach for the definition of isostatic masses.

Analogously to Eq. (11), the isostatic reduction applied to the gravity gradients is calculated by

$$\left\{M_{ij}^{\mathrm{Iso}}(P)\right\}_{\mathrm{LNOF}} = G \sum_{k=1}^{n} \sum_{l=1}^{m} \Delta\rho \int_{r_1}^{r_2} \iint_{\sigma_{kl}^*} N_{ij}(P, Q) \,\mathrm{d}\Omega^*, \tag{16}$$

where $r_1 = R_0 - d_{\mathrm{M}}$ and $r_2 = R_0 - D$ according to Fig. 2b. Again the numerical evaluation of Eq. (16) is performed by means of Taylor series expansions.

2.3. Reduction values at GOCE satellite altitude

To obtain an impression of the magnitude of the RWI-based reduction values, Fig. 3a and 3b visualize the topographic and isostatic effects on the radial-radial gravity gradient component M_{33}, evaluated on a $5' \times 5'$ global grid at the mean altitude of the GOCE satellite ($h = 254.9$ km). Both reductions reach extreme values of about ± 8 E and largely cancel out each other. Thus, the combined topographic-isostatic reductions

$$\left\{M_{ij}^{\mathrm{T/I}}(P)\right\}_{\mathrm{LNOF}} = \left\{M_{ij}^{\mathrm{Topo}}(P)\right\}_{\mathrm{LNOF}} + \left\{M_{ij}^{\mathrm{Iso}}(P)\right\}_{\mathrm{LNOF}} \tag{17}$$

are nearly one order of magnitude smaller, i.e., about ± 1 E (see Fig. 3c). A strong local correlation between $M_{33}^{\mathrm{T/I}}$ and the topography can be detected, where the maximum values are found in the regions with highly variable topography like the Andes and the Himalayas.

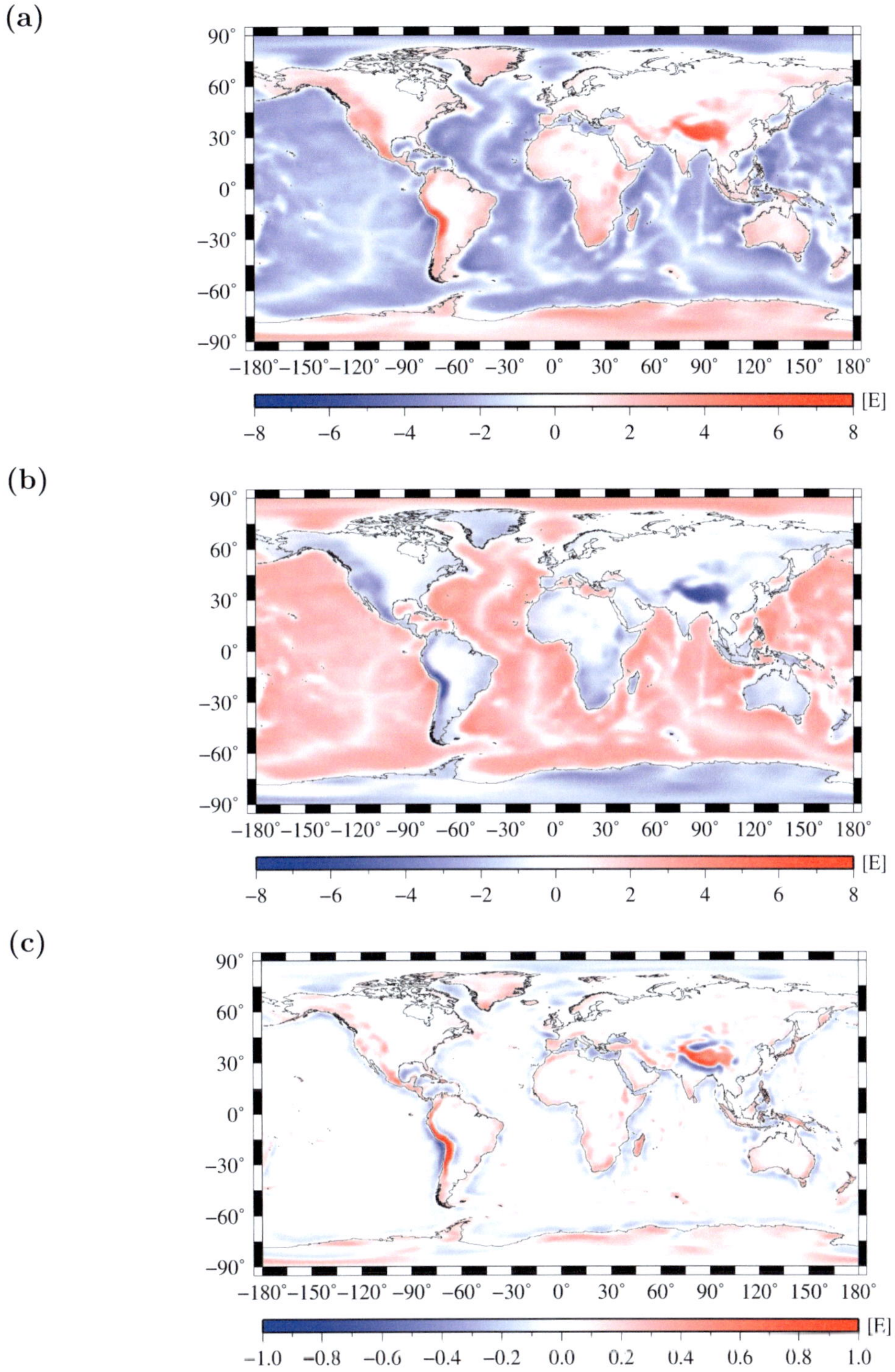

Fig. 3. Reduction values for the radial-radial gravity gradient component M_{33} evaluated on a global grid at the mean GOCE satellite altitude of $h = 254.9\,\mathrm{km}$. **(a)** topographic reductions M_{33}^{Topo} (see Eq. (11)), **(b)** isostatic reductions M_{33}^{Iso} (see Eq. (16)), **(c)** combined topographic-isostatic reductions $M_{33}^{\mathrm{T/I}}$ (see Eq. (17)). All values refer to the LNOF.

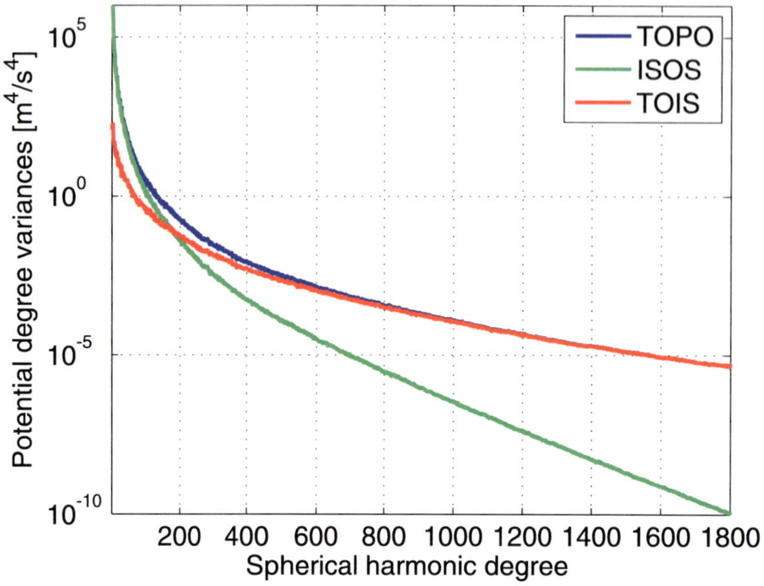

Fig. 4. Potential degree variances of the RWI model: topographic signal (TOPO, *blue curve*), isostatic signal (ISOS, *green curve*), topographic-isostatic signal (TOIS, *red curve*).

By applying harmonic analysis, gridded values of RWI topographic-isostatic effects are further used to derive a representation in spherical harmonics. The resulting global RWI topographic-isostatic gravity field model can be downloaded at http://www.gik.kit.edu/rwi_model.php in terms of spherical harmonic coefficients up to degree and order 1800. Note that there are three different versions available: RWI_TOPO_2012 (topographic signal), RWI_ISOS_2012 (isostatic signal), and RWI_TOIS_2012 (combined topographic-isostatic signal). In Fig. 4, the spectral information of each model version is displayed in terms of potential degree variances. As can be expected, the influence of the isostatic component is mainly regional and shows a decreasing effect on the combined topographic-isostatic signal with an increasing spherical harmonic degree.

3. Continuous Morlet wavelet transform

To assess the smoothing effect of topographic-isostatic reductions on GOCE gravity gradients, the continuous Morlet wavelet transform is employed in this study, which enables a signal component analysis in both time and frequency domains.

3.1. Wavelets and Morlet wavelet

Wavelets are small packages of wave-like oscillations that approach zero at both ends. Morlet et al. (1982a,b) introduced wavelets as a family of functions derived by translating and dilating a single function known as the "mother wavelet" $\psi(t)$:

$$\psi_{a,b}(t) = \frac{1}{\sqrt{a}}\psi\left(\frac{t-b}{a}\right), \tag{18}$$

where $a \in \mathbb{R}^+$ denotes the scale (or dilation) parameter measuring the degree of compression, and $b \in \mathbb{R}$ is the position (or translation) parameter providing the time location of the wavelet. For $0 < a < 1$, $\psi_{a,b}(t)$ represents the contracted version of $\psi(t)$ and corresponds to high frequencies. On the other hand, if $a > 1$, $\psi_{a,b}(t)$ has a larger time width than $\psi(t)$ and corresponds to low frequencies. The resolution of wavelets at different scales varies in the time and frequency domains according to Heisenberg's uncertainty principle (Heisenberg, 1927). For large (small) scale values, the resolution is coarse (fine) in the time domain, but fine (coarse) in the frequency domain (Debnath, 2002, p. 12).

One important property of a wavelet is the so-called admissibility condition, mathematically expressed as

$$0 < C_\psi = 2\pi \int_{-\infty}^{+\infty} \frac{\left|\hat{\psi}(\omega)\right|^2}{|\omega|} d\omega < \infty, \tag{19}$$

where $\hat{\psi}(\omega)$ denotes the Fourier transform of the mother wavelet $\psi(t)$ (Debnath, 2002, p. 14). From Eq. (19) it follows that

$$\hat{\psi}(\omega = 0) = 0 \Longleftrightarrow \frac{1}{\sqrt{2\pi}} \int_{-\infty}^{+\infty} \psi(t)dt = 0, \tag{20}$$

implying that the Fourier transform of a wavelet must be equal to zero at the zero frequency. Equivalently, in the time domain, the wavelet must be an oscillatory function with zero mean (Holschneider, 1995, p. 4).

One of the most frequently used mother wavelets in geodetic and geophysical applications is the symmetric Morlet wavelet (Goupillaud et al., 1984). Following Trauth (2007, p. 115), the Morlet wavelet can be approximated by

$$\psi_M(t) = \pi^{-1/4} e^{i\omega_0 t} e^{-t^2/2}, \tag{21}$$

where $e^{i\omega_0 t} = \cos(\omega_0 t) + i\sin(\omega_0 t)$ is Euler's formula, t is the non-dimensional time, and $\omega_0 \geq 5$ is the wave number representing the number of oscillations within the wavelet itself. The Fourier transform of $\psi_M(t)$ is

$$\hat{\psi}_M(\omega) = \pi^{-1/4} e^{-(\omega-\omega_0)^2/2}, \tag{22}$$

indicating that the Morlet wavelet can be obtained by shifting a Gaussian function in Fourier space (Torrence and Compo, 1998). By substituting $\omega = 0$, Eq. (22) becomes

$$\hat{\psi}_M(\omega = 0) = \pi^{-1/4} e^{-\omega_0^2/2}, \tag{23}$$

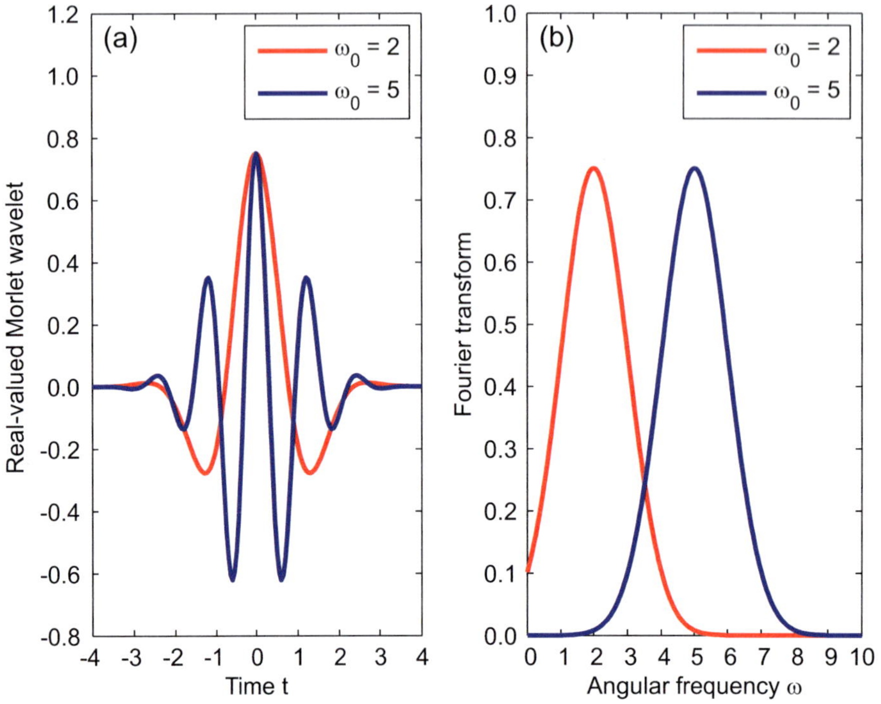

Fig. 5. Examples of the real-valued Morlet wavelet in time and frequency domains. **(a)** Morlet wavelets $\psi_M(t)$ (see Eq. (24)), **(b)** Fourier transforms $\hat{\psi}_M(\omega)$ (see Eq. (22)). Note that t is dimensionless.

which is approximately equal to zero if the wave number ω_0 is large enough, e.g., $\omega_0 \geq 5$ (Holschneider, 1995, p. 31). For different wave numbers ω_0, Fig. 5 illustrates examples of the real-valued Morlet wavelet with $e^{i\omega_0 t} = \cos(\omega_0 t)$, i.e.,

$$\psi_M(t) = \pi^{-1/4} \cos(\omega_0 t) e^{-t^2/2}, \tag{24}$$

as well as the associated Fourier transforms. In the time domain (see Fig. 5a), the Morlet wavelet with $\omega_0 = 5$ better represents an oscillatory function with zero mean. In the frequency domain (see Fig. 5b), Eq. (20) is not satisfied for $\omega_0 = 2$, while $\hat{\psi}_M(\omega = 0)$ is insignificantly different from zero for $\omega_0 = 5$. Accordingly, in this study, the real-valued Morlet wavelet with $\omega_0 = 5$ is used to perform the continuous wavelet transform.

3.2. Continuous wavelet transform

The continuous wavelet transform (CWT) decomposes a function by wavelets and operates over each possible scale a and translation b. The CWT of a function $F(t) \in L^p(\mathbb{R})$, $1 \leq p < \infty$, is written as

$$\mathscr{W}_\psi[F](a,b) = \int\limits_{-\infty}^{+\infty} F(t)\psi^*_{a,b}(t)dt = \frac{1}{\sqrt{a}} \int\limits_{-\infty}^{+\infty} F(t)\psi^*\left(\frac{t-b}{a}\right)dt, \tag{25}$$

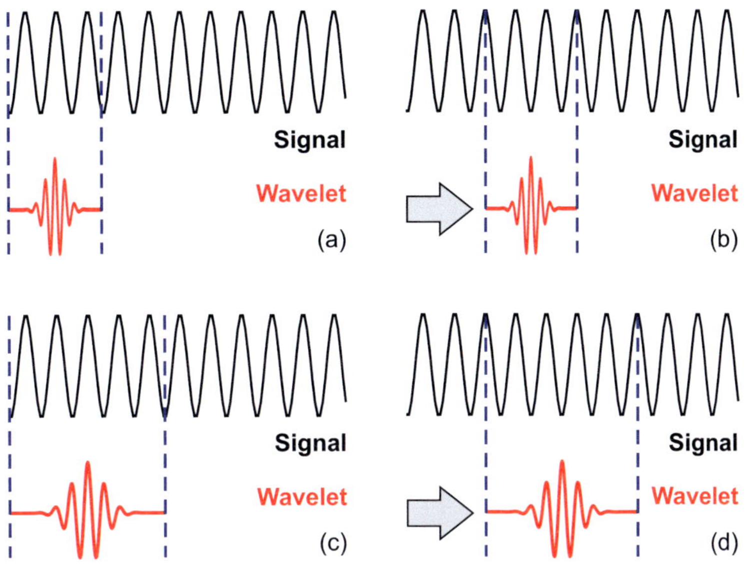

Fig. 6. Continuous wavelet transform (CWT) using the MATLAB® Wavelet Toolbox™ (MATLAB function: `cwt`).

where $L^p(\mathbb{R})$ is the vector space of all complex-valued p-th power Lebesgue integrable functions defined on $\mathbb{R}$, and ψ^* is the complex conjugate of ψ defined on the open time and scale real (b, a) half plane (Holschneider, 1995, p. 5). The numbers $\mathscr{W}_\psi[F](a,b)$ are referred to as wavelet coefficients of the function $F(t)$ with respect to the mother wavelet $\psi(t)$. Generally, the output function $\mathscr{W}_\psi[F](a,b)$ is real if the CWT is applied with a real-valued mother wavelet. Like the Fourier transform, the CWT is also linear. Differing from the Fourier transform, which is normally localized only in the frequency domain, the CWT is localized in both time and frequency domains via translations and dilations, respectively.

In this study, the CWT is carried out using the MATLAB® Wavelet Toolbox™. In this implementation, the wavelet coefficients $\mathscr{W}_\psi[F](a,b)$ are obtained by the sums of the signal $F(t)$ multiplied by the scaled and shifted versions of the mother wavelet $\psi^*_{a,b}(t)$ over all time t (see Eq. (25)). As schematically illustrated in Fig. 6, the CWT is performed within the following steps:

1. Select a mother wavelet (e.g., the real-valued Morlet wavelet) and compare it to a section at the start of the original signal (see Fig. 6a).

2. Compute the correlation between the wavelet and this signal section (see Fig. 6a), where the result depends on the shape of the chosen wavelet.

3. Shift the wavelet and repeat steps 1 and 2 until the whole signal has been covered (see Fig. 6b).

4. Scale the mother wavelet and repeat steps 1 through 3 (see Fig. 6c and 6d).

5. Repeat steps 1 through 4 for all scale values.

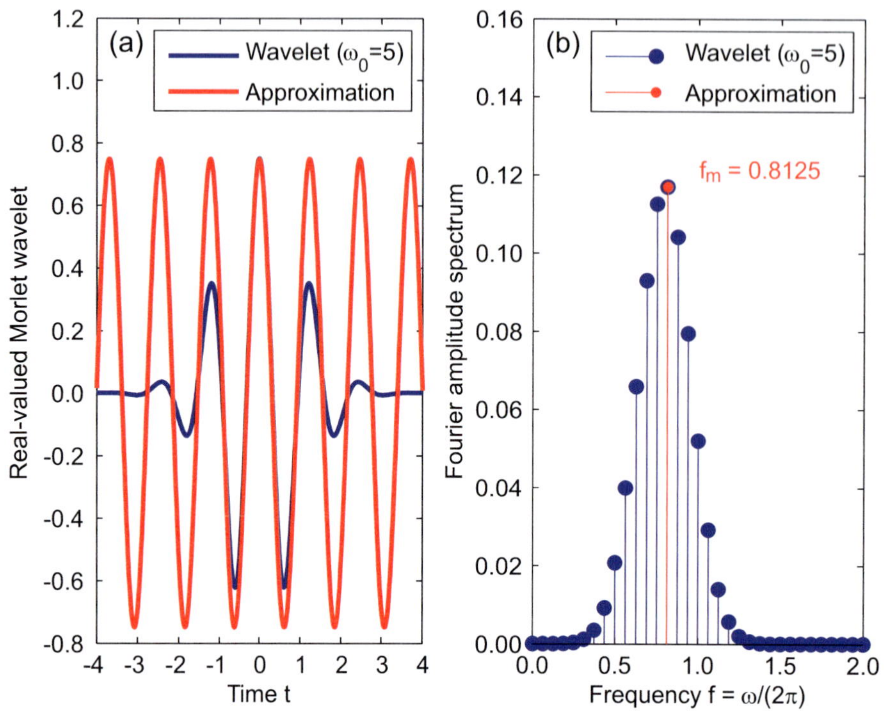

Fig. 7. Determination of the center frequency for the real-valued Morlet wavelet with a wave number of $\omega_0 = 5$ (see Eq. (24) and Fig. 5). **(a)** Wavelet approximation, **(b)** frequency determination. Note that t and f are dimensionless.

The wavelet coefficients $\mathscr{W}_\psi[F](a, b)$ obtained at different scales a and positions b constitute the results of a regression of the signal $F(t)$ on the wavelets $\psi_{a,b}(t)$. Based on the center frequency of the mother wavelet $\psi(t)$, denoted as f_m, the pseudo-frequency f_a corresponding to scale a can be computed by means of

$$f_a = \frac{f_m}{a \cdot \Delta t}, \tag{26}$$

where Δt is the sampling period of $F(t)$ (MATLAB function: `scal2frq`). The basic idea behind the center frequency is to approximate a given mother wavelet using a purely periodic signal with frequency f_m (MATLAB function: `centfrq`). A reasonable choice for f_m is the frequency at which the Fourier amplitude spectrum of the mother wavelet is as large as possible.

Taking the real-valued Morlet wavelet with $\omega_0 = 5$ as an example, Fig. 7 shows the results of wavelet approximation and center frequency determination. It can be seen from Fig. 7a that the main lobe of the Morlet wavelet is well approximated by the periodic signal with a frequency of $f_m = 0.8125$. At this frequency, the Fourier amplitude spectrum reaches its maximum, as illustrated in Fig. 7b.

To visualize the results of a CWT, the so-called scalogram is commonly used, which communicates the time-frequency localization property of wavelet transforms. The x-axis of a wavelet scalogram represents time t, the y-axis represents scale a or pseudo-frequency f_a derived from Eq. (26), and the z-axis represents absolute wavelet coefficients $|\mathscr{W}_\psi[F](a,b)|$ that can be interpreted as signal amplitudes (or energies) (Trauth, 2007, p. 115).

4. Study area and data sets

Previous studies have mainly analyzed the impact of topographic-isostatic reductions on GOCE gravity gradients on a global scale and showed significant smoothing effects in regions with highly variable topography (e.g. Grombein et al., 2014). This paper will take the Himalayas as a representative example that provides topographic-isostatic reductions on a regional scale and at different magnitudes (see Fig. 3c).

The GOCE gradiometry measurements of Level 2 are available by the two products EGG_NOM_2 and EGG_TRF_2 (Gruber et al., 2010, Sect. 5). For the intended analysis, the EGG_NOM_2 gravity gradients, related to the instrument-fixed Gradiometer Reference Frame (GRF), are more suitable, as they are calibrated and corrected, but not filtered or rotated (Bouman et al., 2011). Moreover, in contrast to the EGG_TRF_2 gravity gradients, the less accurate components M_{12} and M_{23} contain the original measurements and are not replaced by modeled gradients (Gruber et al., 2010, Sect. 5). The EGG_NOM_2 gravity gradients are represented as time series along the satellite orbit with the time tags

$$t_p = t_0 + p \cdot \Delta t, \quad p = 0, \ldots, q \in \mathbb{N}_0, \tag{27}$$

which are specified in GPS time and have a sampling period of $\Delta t = 1\,\mathrm{s}$. For the numerical investigations, a representative time series of 1001 epochs is selected, when the GOCE satellite crossed the Himalayan region on Nov. 4, 2010 ($t_0 = 972\,867\,514\,\mathrm{s}$, $q = 1000$). For each time tag t_p, the observed gravity gradients $M_{ij}(t_p)$ and the associated 3-dimensional measurement positions $P(t_p) = [r(t_p), \varphi(t_p), \lambda(t_p)]$ were obtained from the GOCE Virtual Archive at http://eo-virtual-archive1.esa.int. In Fig. 8, the black curve shows the GOCE ground track for the selected time, while Fig. 9 illustrates the corresponding topographic profile.

As presented in Table 1, the time series covers different topographic features crossed by the GOCE satellite. Starting in the lowlands of the Central Siberian Plateau, the satellite first crossed the Altai Mountains (ALT, T_1), the Tian Shan Mountains (TSM, T_2), and the Taklamakan Desert Basin (TDB, T_3). Then, it followed the rapidly increasing topography of the Kunlun Mountains (KUN, T_4) marking the northern edge of the highlands of the Tibetan Plateau (TPL, T_5). The Himalayan Mountain Range (HMR) is passed by the satellite at T_6 toward the end of the TPL, where the topography decreases strongly in the direction to the North Indian River Plain (NIR, T_7). Afterward, the satellite crossed the Indian midland with some lower mountain ranges like the Eastern Ghats (EAG, T_8), before it reached the Indian East Coast (IEC, T_9). Following the continuously increasing bathymetric depth of the Indian Ocean, the satellite striped the island of Sri Lanka (ISL, T_{10}). The selected time tags $T_1, \ldots, T_{10}$ are summarized in Table 1 and marked by red dots in Figs. 8 and 9.

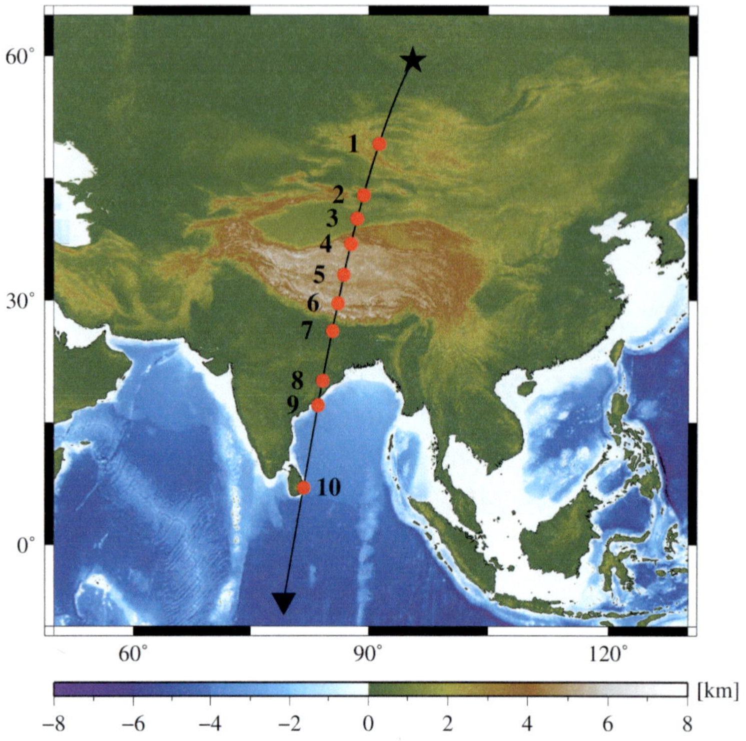

Fig. 8. DTM2006.0 topography of the Himalayan region. The *black curve* visualizes the GOCE ground track for the selected time series. The start position of the time series is indicated by a *star* and the flight direction by an *arrow*. Additionally, *red dots* mark the time tags $T_1, \ldots, T_{10}$ as specified in Table 1.

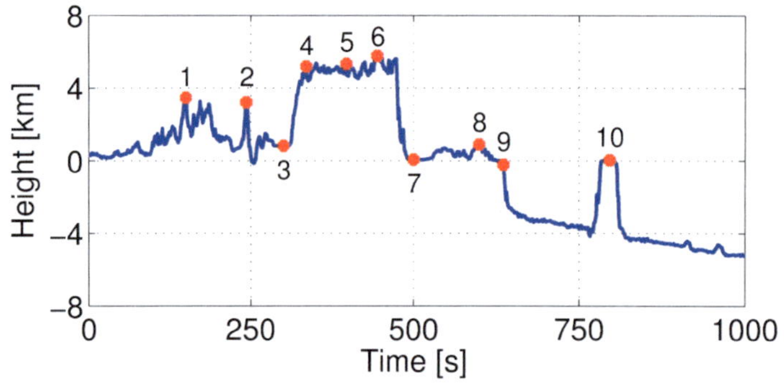

Fig. 9. Topographic/bathymetric profile along the ground track of the GOCE satellite for the selected time series. *Red dots* correspond to those shown in Fig. 8. Topographic heights are derived from DTM2006.0.

Table 1. Time tags of topographic features crossed by the GOCE satellite within the selected time series.

Time tag	Topographic feature	Abbr.
$T_1 = t_{150}$	Altai Mountains	ALT
$T_2 = t_{250}$	Tian Shan Mountains	TSM
$T_3 = t_{300}$	Taklamakan Desert Basin	TDB
$T_4 = t_{350}$	Kunlun Mountains	KUN
$T_5 = t_{400}$	Tibetan Plateau	TPL
$T_6 = t_{450}$	Himalayan Mountain Range	HMR
$T_7 = t_{500}$	North Indian River Plain	NIR
$T_8 = t_{600}$	Eastern Ghats	EAG
$T_9 = t_{650}$	Indian East Coast	IEC
$T_{10} = t_{800}$	Island of Sri Lanka	ISL

5. Assessment of topographic-isostatic reductions

Using the measurement positions $P(t_p)$ of the GOCE satellite as computation points, the RWI-based topographic-isostatic reductions $M_{ij}^{\mathrm{T/I}}(t_p)$ are calculated along the selected orbit by means of Eq. (17). As described in Sect. 2, these reductions are related to LNOF, whereas the measured EGG_NOM_2 gravity gradients $M_{ij}(t_p)$ are specified in the instrument-fixed GRF. In this orthogonal frame, the x_1-axis points in the flight direction, the x_2-axis is orthogonal to the instantaneous orbit plane, and the x_3-axis points upwards, approximately in the geocentric radial direction.

Applying epoch-wise rotation matrices $\mathbf{R}_{\mathrm{LNOF}}^{\mathrm{GRF}}(t_p)$, the computed reduction values are transformed from LNOF to GRF by evaluating the matrix operation

$$\left\{\mathbf{M}^{\mathrm{T/I}}\right\}_{\mathrm{GRF}} = \mathbf{R}_{\mathrm{LNOF}}^{\mathrm{GRF}} \cdot \left\{\mathbf{M}^{\mathrm{T/I}}\right\}_{\mathrm{LNOF}} \cdot \left(\mathbf{R}_{\mathrm{LNOF}}^{\mathrm{GRF}}\right)^{\mathrm{T}} \tag{28}$$

for every time tag t_p. A detailed description of this transformation step is provided in Gruber et al. (2010, Sect. 4.4) and Grombein et al. (2011).

In Fig. 10, the time series of the rotated RWI-based topographic-isostatic reductions $M_{ij}^{\mathrm{T/I}}(t_p)$ are displayed for each gravity gradient component. According to their magnitudes, the gradients are classified into two groups: 1) M_{11}, M_{13}, and M_{33}; 2) M_{12}, M_{22}, and M_{23}.

Generally, the first group exhibits larger signal amplitudes (see the upper row of Fig. 10) and stronger correlations with the topographic profile illustrated in Fig. 9. As the x_1-derivative is particularly sensitive to structures extending in east-west direction (e.g., most of the crossed mountain ranges), the time tags $T_1, \ldots, T_{10}$ can be identified as local maxima

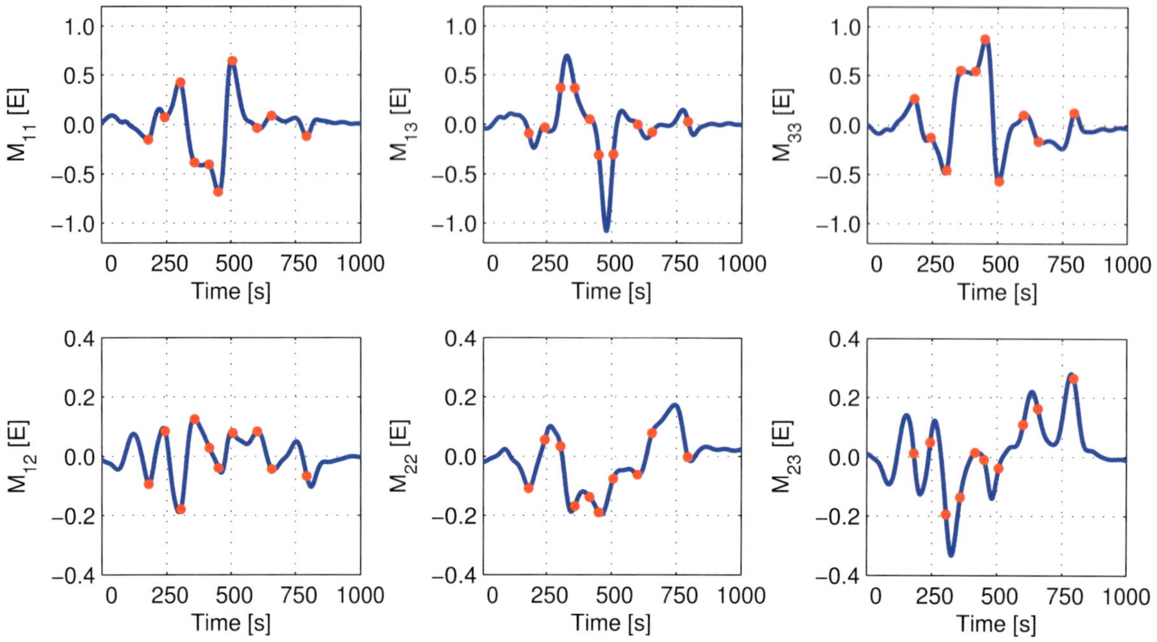

Fig. 10. Time series of the RWI-based topographic-isostatic reductions $M_{ij}^{\mathrm{T/I}}(t_p)$ calculated along the GOCE satellite orbit. *Red dots* mark the time tags $T_1, \ldots, T_{10}$ as specified in Table 1. Note the different scale of the y-axis in the *first* and *second row*.

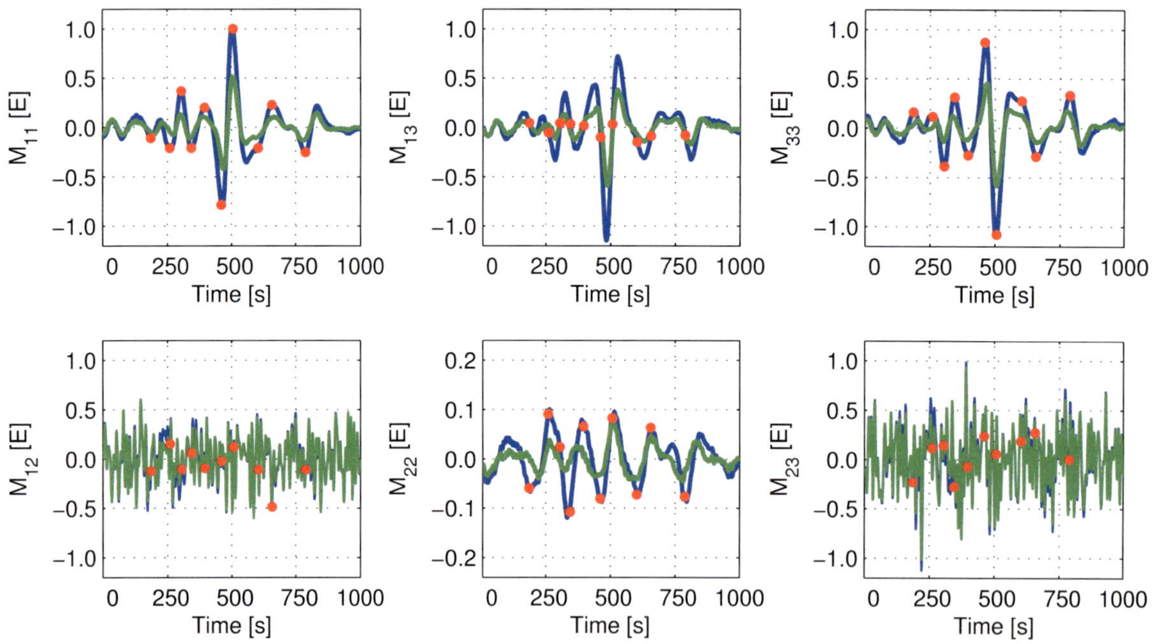

Fig. 11. Band-pass filtered time series $\overline{M}_{ij}(t_p)$ (*blue curve*) and the corresponding reduced time series $\overline{\delta M}_{ij}(t_p)$ (*green curve*). *Red dots* mark the time tags $T_1, \ldots, T_{10}$ as specified in Table 1. Note the different scale of the y-axis for the component M_{22}.

and minima of the gradient signal for M_{11}. Moreover, in the component M_{33} that reflects radial changes, these time tags can also be clearly localized, where the TPL area shows the largest amplitudes (T_3–T_7). In contrast, the extreme values of the mixed derivative M_{13} indicate strong topographic increases (T_3–T_4) or decreases (T_6–T_7).

The gradients of the second group comprise the x_2-derivative that is particularly affected by structures in north-south direction. In the absence of such topographic formations in the Himalayan region, these gradients possess smaller signal amplitudes (see the lower row of Fig. 10). Nevertheless, the most significant amplitudes can also be associated with the time tags $T_1, \ldots, T_{10}$, in particular for M_{12}.

In order to analyze the smoothing impact of the proposed RWI topographic-isostatic reduction method, the observed GOCE gravity gradients $M_{ij}(t_p)$ are compared to the reduced gradients resulting from

$$\delta M_{ij}(t_p) = M_{ij}(t_p) - \left\{ M_{ij}^{\mathrm{T/I}}(t_p) \right\}_{\mathrm{GRF}}. \tag{29}$$

To provide a deeper insight into the high- and mid-frequency signal components, this comparison will be performed within the limited measurement bandwidth of the GOCE gradiometer (MBW, 5 to 100 mHz). For this purpose, the gravity gradient time series are filtered to the MBW by applying a symmetric, non-recursive band-pass filter as presented by Schuh (2010):

$$\left\{ \begin{matrix} \overline{M}_{ij}(t_p) \\ \overline{\delta M}_{ij}(t_p) \end{matrix} \right\} = \sum_{k=-N}^{N} c_{|k|} \left\{ \begin{matrix} M_{ij}(t_{p-k}) \\ \delta M_{ij}(t_{p-k}) \end{matrix} \right\}, \tag{30}$$

where the used filter coefficients $c_{|k|}$ are adapted from Hamming (1998, p. 127ff.) and are explicitly provided in the Appendix . As proposed by Schuh (2010) the filter length is set to $N = 1000$ for this investigation. Note that in order to evaluate Eq. (30), the time series of $M_{ij}(t_p)$ and $\delta M_{ij}(t_p)$ are extended by 1000 s in both directions to overcome the warm-up of the filter.

5.1. Analysis in the space domain

In Fig. 11, the time series of the band-pass filtered GOCE measurements $\overline{M}_{ij}(t_p)$ and the corresponding reduced observations $\overline{\delta M}_{ij}(t_p)$ are illustrated. Significant smoothing effects can be detected for the gradient components M_{11}, M_{13}, and M_{33}, which is particularly observable at the selected time tags $T_1, \ldots, T_{10}$ (see the upper row of Fig. 11). As an example, signal amplitudes affected by the TPL area (T_3–T_7) can be largely reduced. From HMR (T_6) to NIR (T_7), when the satellite crossed a strongly decreasing topography, the most significant smoothing effects are present. Some smaller reductions can be seen for TSM (T_2), EAG (T_8), IEC (T_9), and ISL (T_{10}). Only insignificant changes are visible at the beginning and the end of the time series, when the satellite crossed the Central Siberian Plateau and the Indian Ocean. Regarding the gradient components M_{12}, M_{22}, and M_{23} (see the lower row of Fig. 11), smoothing effects can only be detected for M_{22}, especially between ALT (T_1) and HMR (T_6). For the less accurately measured gradients M_{12} and M_{23}, the observations are dominated by high-frequency noise, leading to no visible changes between the measured and reduced signals.

As proposed in Janák et al. (2012) and Grombein et al. (2014), the performance of topographic-isostatic reductions can be analyzed using statistical inferences. For this purpose, the degree of smoothing is quantified by percentage changes in the standard deviation (std) and range (rng) of the time series before and after reduction, i.e.,

$$P_{ij}^{\text{std}} = \left[\text{std}(\overline{M}_{ij}) - \text{std}(\overline{\delta M}_{ij})\right]/\text{std}(\overline{M}_{ij}), \tag{31}$$

$$P_{ij}^{\text{rng}} = \left[\text{rng}(\overline{M}_{ij}) - \text{rng}(\overline{\delta M}_{ij})\right]/\text{rng}(\overline{M}_{ij}). \tag{32}$$

In Table 2, these values are provided for each gradient component. Coinciding with the visual inspection, a significant impact of smoothing can be observed for M_{11}, M_{13}, M_{33}, and M_{22}. In this case, the standard deviations can be reduced by about 50 %, while the changes in range amount to about 48 %. Regarding the components M_{12} and M_{23}, only insignificant smoothing effects are detectable. This indicates that the high-frequency measurement noise makes smoothing by means of topographic-isostatic reductions not feasible for these components (Grombein et al., 2014).

Table 2. Percentage changes P_{ij} in the standard deviation (std) and range (rng) of the GOCE gravity gradients before and after applying topographic-isostatic reduction (see Eqs. (31) and (32)).

	std [%]	rng [%]
P_{11}	51.5	48.8
P_{13}	51.1	48.2
P_{33}	50.2	48.2
P_{12}	5.0	−1.2
P_{22}	48.5	46.4
P_{23}	4.7	8.0

5.2. Analysis in the frequency domain

To analyze the smoothing impact of the RWI-based topographic-isostatic reductions in the frequency domain, a wavelet-based assessment is performed. Using the real-valued Morlet wavelet as the mother wavelet, the CWT is applied to the band-pass filtered measurement signals $\overline{M}_{ij}(t_p)$ and the corresponding reduced signals $\overline{\delta M}_{ij}(t_p)$ (see Eqs. (24) and (25)). The resulting dimensionless, absolute wavelet coefficients $|\mathscr{W}_\psi|$ are visualized by means of scalograms in Fig. 12. They represent the signal structure of the gradients in both time

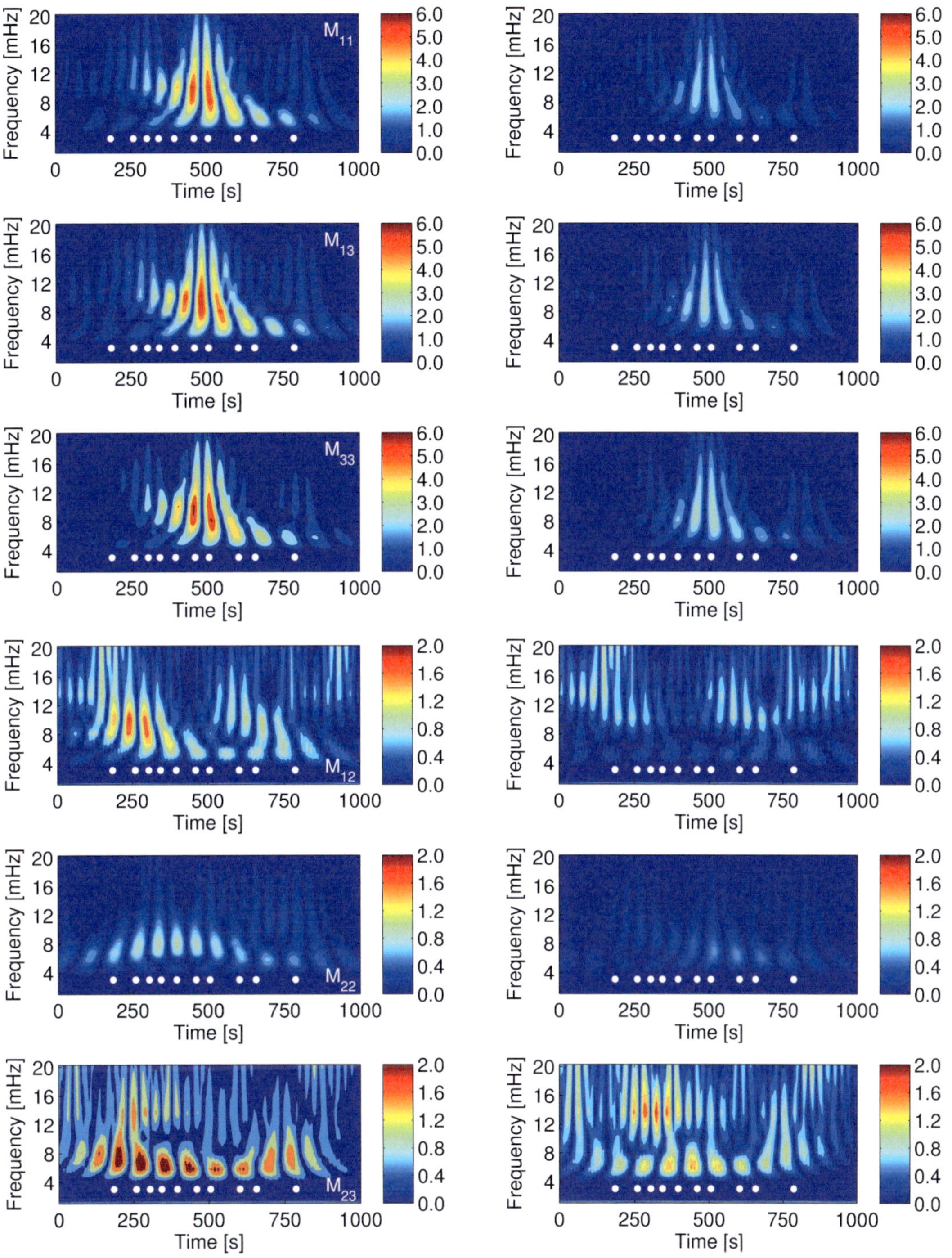

Fig. 12. Morlet wavelet scalograms of the band-pass filtered measurement signals $\overline{M}_{ij}(t_p)$ (*left panel*) and the corresponding reduced signals $\overline{\delta M}_{ij}(t_p)$ (*right panel*). *White dots* mark the time tags $T_1, \ldots, T_{10}$ as specified in Table 1. Note that the displayed absolute wavelet coefficients are dimensionless; consider the different color bar scales between (M_{11}, M_{13}, M_{33}) and (M_{12}, M_{22}, M_{23}).

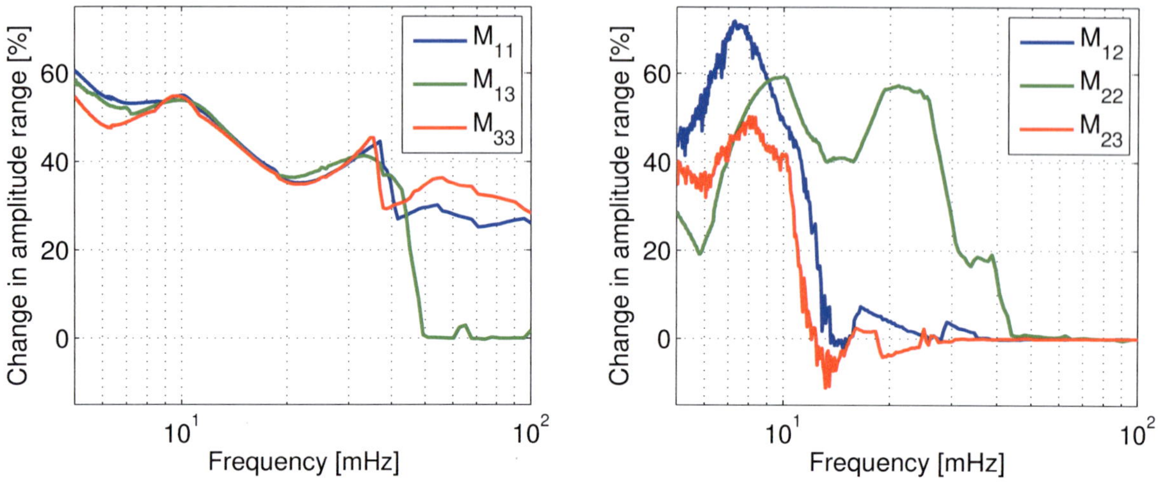

Fig. 13. Percentage changes of the amplitude range of the GOCE gravity gradients before and after applying topographic-isostatic reduction. Note that the frequency axes are plotted on a logarithmic scale.

and frequency domains, where the colors reflect the signal amplitudes (or energies). The frequency axis is limited to $f \leq 20$ mHz, where the most significant signal structures can be observed. Due to the used band-pass filter, signals with frequencies less than 5 mHz are absent.

The scalograms visualized in the upper three rows of Fig. 12 are related to the gradients M_{11}, M_{13}, and M_{33}. The measured and reduced signals are displayed in the left and right panel, respectively. All three gradients show quite similar characteristics, where most of the amplitude peaks can be associated with the time tags $T_1, \ldots, T_{10}$ marked by white dots. The largest amplitudes can be detected in the frequency range of about 8 – 12 mHz between TPL and EAG (T_5–T_8). Comparing the measured and reduced gradients with each other, the absolute wavelet coefficients are strongly reduced from about 5–6 to 1–2, indicating significant smoothing effects.

Analogously to Table 2 and Eq. (32), the degree of smoothing in the frequency domain is quantified in Fig. 13, where the percentage changes of the amplitude range are plotted against the frequency values of the MBW. In this context, the amplitude range is calculated by subtracting the minimum (absolute) wavelet coefficient from the corresponding maximum one. In the case of the gradient components M_{11}, M_{13}, and M_{33}, the amplitude range can be reduced by about 40 – 60 % for $f < 40$ mHz. For frequency values above 40 mHz, the reduction rates decrease strongly for M_{13}, while for M_{11} and M_{33} they remain at a level of about 30 %.

The gradients M_{12}, M_{22}, and M_{23}, displayed in the lower three rows of Fig. 12, exhibit much smaller signals, where the different color bar scales of the scalograms should be noted. Regarding the measured gradient M_{22}, the frequency increases slightly between TSM and NIR (T_2–T_7), which is consistent with the corresponding time series in Fig. 11. Comparing the scalograms of the measured and reduced signals, the absolute wavelet coefficients change from about 0.8 to 0.4 (50 %), which supports the findings presented in Table 2 and Fig. 13.

Considering the less accurately observed components M_{12} and M_{23}, the benefit of the wavelet-based assessment becomes more obvious. As can be expected from the associated time series in Fig. 11, the scalograms of these two components illustrate much more high-frequency signals with $f > 12$ mHz. Moreover, the scalograms also indicate significant signal structures at lower frequencies that are not clearly visible in the space domain analysis. Examining the M_{12} component, larger amplitudes are found at frequencies of about 8 – 10 mHz that can be attributed to the time tags ALT (T_1), TSM (T_2), and TDB (T_3). Although the high-frequency components still remain in the reduced signals, significant smoothing effects can be detected for $f < 10$ mHz. Such a strong smoothing impact cannot be realized in the space domain analysis, since all frequencies are superposed (see Fig. 11 and Table 2). However, it can clearly be recognized and distinguished by the frequency-dependent changing rates in Fig. 13.

6. Conclusions

In this paper, the smoothing effects of topographic-isostatic reductions on GOCE gravity gradients have been analyzed. In addition to statistical inferences used in previous investigations, a wavelet-based assessment has been proposed.

Taking the Himalayas as a suitable regional study area, topographic-isostatic reductions were calculated along a representative GOCE satellite orbit. By applying forward modeling based on mass discretization with tesseroids, these reductions were obtained using the Rock-Water-Ice (RWI) method. This approach relies upon a three-layer decomposition of the topography with variable density values and a modified Airy-Heiskanen concept that incorporates seismic Moho depths. Comparing the calculated RWI-based reductions with the topographic/bathymetric profile along the satellite ground track, representative waypoints in different topographic regions have been specified.

In order to investigate the smoothing impact of topographic-isostatic reductions on the high- and mid-frequency signals, the observed GOCE gravity gradients have been compared to the corresponding reduced ones within the measurement bandwidth of the GOCE gradiometer (5 to 100 mHz). In a first step, this comparison was performed in the space domain by utilizing statistical inferences to quantify the degree of smoothing. The results showed a significant reduction of the standard deviation and range of about 50 % for the gradients M_{11}, M_{22}, M_{33}, and M_{13}, where the smoothing effects are highly correlated with the topography. Furthermore, the less accurately observed gradients M_{12} and M_{23} seem to be insignificantly affected by topographic-isostatic reductions due to their high measurement noise. These results confirm the findings of previous studies that analyzed the smoothing effects on a global scale.

In a second step, a spectral analysis has been performed, where a continuous wavelet transform was applied to the measured and reduced GOCE gravity gradients. The resulting scalograms enable a signal component analysis in both time and frequency domains, clearly showing which frequencies are actually affected and reduced. Generally, the smoothing impact of the RWI-based topographic-isostatic signals is confirmed, supporting the reduction rates detected in the space domain analysis. The benefit of the spectral analysis is particularly impressive be shown for the less accurately observed gradients M_{12} and M_{23}. For these components, the scalograms indicate significant smoothing effects at lower frequencies of $f < 10\,\mathrm{mHz}$, which were invisible in the space domain analysis. Therefore, the wavelet-based assessment can be considered as a much more powerful tool for the evaluation of topographic-isostatic reductions than the commonly applied statistical inferences in the space domain.

Acknowledgements. This research was funded by the German Federal Ministry of Education and Research under grant number 03G0726F within the REAL GOCE project of the GEOTECHNOLOGIEN Programme. The authors would like to thank N.K. Pavlis for providing the global topographic data base DTM2006.0 and the Steinbuch Center for Computing at the Karlsruhe Institute of Technology for the allocation of computing time on the high performance parallel computer system HC3. Finally, two anonymous reviewers as well as the Editor-in-Chief are acknowledged for their valuable comments, which helped to improve the manuscript.

Appendix

For the band-pass filter in Eq. (30) the filter coefficients used

$$c_{|k|} = \begin{cases} 2\Delta t(f_h - f_l), & k = 0, \\ \frac{\sin 2\pi f \Delta t}{\pi k}\Big|_{f_l}^{f_h} \frac{\sin \pi k/N}{\pi k/N}, & k = 1, \dots, N-1, \\ -\frac{1}{2} \sum\limits_{j=-N+1}^{N-1} c_{|j|}, & k = N, \end{cases}$$

are adapted from Hamming (1998, p. 127ff.). The sampling period is denoted by Δt, f_l and f_h are the lower and upper cutoff frequencies, respectively, and N is the filter length. In this study, these parameters have been set to $\Delta t = 1\,\mathrm{s}$, $f_l = 5\,\mathrm{mHz}$, $f_h = 100\,\mathrm{mHz}$, and $N = 1000$.

References

Abd-Elmotaal, H., Seitz, K., Abd-Elbaky, M., and Heck, B. (2014): Comparison among three harmonic analysis techniques on the sphere and the ellipsoid. *Journal of Applied Geodesy* 8(1):1–19. DOI: 10.1515/jag-2013-0008.

Álvarez, O., Gimenez, M., Braitenberg, C., and Folguera, A. (2012): GOCE satellite derived gravity and gravity gradient corrected for topographic effect in the South Central Andes region. *Geophysical Journal International* 190(2):941–959. DOI: 10.1111/j.1365-246X.2012.05556.x.

Atallah, L., Smith, P. J. P., and Bates, C. R. (2002): Wavelet analysis of bathymetric sidescan sonar data for the classification of seafloor sediments in Hopvågen Bay - Norway. *Marine Geophysical Researches* 23(5–6):431–442. DOI: 10.1023/B:MARI.0000018239.07561.76.

Bassin, C., Laske, G., and Masters, G. (2000): The current limits of resolution for surface wave tomography in North America. *Eos, Transactions, American Geophysical Union* 81:F897. URL: http://igppweb.ucsd.edu/~gabi/crust2.html.

Bouman, J., Fiorot, S., Fuchs, M., Gruber, T., Schrama, E., Tscherning, C., Veicherts, M., and Visser, P. (2011): GOCE gravitational gradients along the orbit. *Journal of Geodesy* 85(11):791–805. DOI: 10.1007/s00190-011-0464-0.

Collin, F. and Warnant, R. (1995): Application of the wavelet transform for GPS cycle slip correction and comparison with Kalman filter. *Manuscripta Geodaetica* 20(3):161–172. URL: http://orbi.ulg.ac.be/handle/2268/84669.

Debnath, L. (2002): Wavelet transforms and their applications. 1st ed. Springer, New York, USA. DOI: 10.1007/978-1-4612-0097-0.

Fecher, T., Pail, R., and Gruber, T. (2013): Global gravity field modeling based on GOCE and complementary gravity data. *International Journal of Applied Earth Observation and Geoinformation* 2013. DOI: 10.1016/j.jag.2013.10.005.

Forsberg, R. (1984): A study of terrain reductions, density anomalies and geophysical inversion methods in gravity field modelling. Report 355. Department of Geodetic Science and Surveying, The Ohio State University, Columbus, USA.

Forsberg, R. and Tscherning, C. C. (1997): Topographic effects in gravity field modelling for BVP. In: Sansò, F., and Rummel, R. (eds.) Geodetic boundary value problems in view of the one centimeter geoid, *Lecture Notes in Earth Sciences*, vol. 65. Springer Berlin Heidelberg, pp. 239–272. DOI: 10.1007/BFb0011707.

Gibert, D., Holschneider, M., and Le Mouël, J.-L. (1998): Wavelet analysis of the Chandler wobble. *Journal of Geophysical Research* 103(B11):27069–27089. DOI: 10.1029/98JB02527.

Goupillaud, P., Grossman, A., and Morlet, J. (1984): Cycle-octave and related transforms in seismic signal analysis. *Geoexploration* 23(1):85–102. DOI: 10.1016/0016-7142(84)90025-5.

Grombein, T., Seitz, K., and Heck, B. (2010): Modelling topographic effects in GOCE gravity gradients. In: Münch, U., and Dransch, W. (eds.) Observation of the system Earth from space, *GEOTECHNOLOGIEN Science Report*, vol. 17, pp. 84–93. DOI: 10.2312/GFZ.gt.17.13.

Grombein, T., Seitz, K., and Heck, B. (2011): Smoothing GOCE gravity gradients by means of topographic-isostatic reductions. In: Ouwehand, L. (ed.) Proceedings of the 4th International GOCE User Workshop, Munich, Germany, Mar. 31–Apr. 1, 2011. *ESA Special Publication*, vol. 696. ESA Publications Division, ESTEC, Noordwijk, The Netherlands.

Grombein, T., Seitz, K., and Heck, B. (2013): Optimized formulas for the gravitational field of a tesseroid. *Journal of Geodesy* 87(7):645–660. DOI: 10.1007/s00190-013-0636-1.

Grombein, T., Seitz, K., and Heck, B. (2014): Topographic-isostatic reduction of GOCE gravity gradients. In: Rizos, C., and Willis, P. (eds.) Earth on the edge: science for a sustainable planet. Proceedings of the IAG General Assembly, Melbourne, Australia, June 28–July 2, 2011. *International Association of Geodesy Symposia*, vol. 139. Springer Berlin Heidelberg, pp. 349–356. DOI: 10.1007/978-3-642-37222-3_46.

Gruber, T., Rummel, R., Abrikosov, O., and van Hees, R. (2010): GOCE level 2 Product Data Handbook. no. 4.3. Document ID GO-MA-HPF-GS-0110.

Hamming, R. W. (1998): Digital filters. Dover Pubn Inc, Mineola, USA.

Heck, B. (2003): Rechenverfahren und Auswertemodelle der Landesvermessung. Klassische und moderne Methoden. 3rd ed. Wichmann-Verlag, Heidelberg, Germany.

Heck, B. and Seitz, K. (2007): A comparison of the tesseroid, prism and point-mass approaches for mass reductions in gravity field modelling. *Journal of Geodesy* 81(2):121–136. DOI: 10.1007/s00190-006-0094-0.

Heisenberg, W. K. (1927): On the perceptual content of quantum theoretical kinematics and mechanics. *Zeitschrift für Physik* 43(2–4):172–198.

Heiskanen, W. A. and Moritz, H. (1967): Physical geodesy. W. H. Freeman & Co., San Francisco, USA.

Hirt, C., Claessens, S., Fecher, T., Kuhn, M., Pail, R., and Rexer, M. (2013): New ultrahigh-resolution picture of Earth's gravity field. *Geophysical Research Letters* 40(16):4279–4283. DOI: 10.1002/grl.50838.

Hirt, C., Kuhn, M., Featherstone, W. E., and Göttl, F. (2012): Topographic/isostatic evaluation of new-generation GOCE gravity field models. *Journal of Geophysical Research* 117:B05407. DOI: 10.1029/2011JB008878.

Holschneider, M. (1995): Wavelets: an analysis tool. *Oxford Mathematical Monographs*. Oxford University Press, New York, USA.

Janák, J. and Wild-Pfeiffer, F. (2010): Comparison of various topographic-isostatic effects in terms of smoothing gradiometric observations. In: Mertikas, S. P. (ed.) Gravity, geoid and Earth observation, IAG Commission 2: Gravity field. Chania, Crete, Greece, June 23–27, 2008. *International Association of Geodesy Symposia*, vol. 135. Springer Berlin Heidelberg, pp. 377–381. DOI: 10.1007/978-3-642-10634-7_50.

Janák, J., Wild-Pfeiffer, F., and Heck, B. (2012): Smoothing the gradiometric observations using different topographic-isostatic models: a regional case study. In: Sneeuw, N., Novák, P., Crespi, M., and Sansò, F. (eds.) VII Hotine-Marussi Symposium on Mathematical Geodesy. Proceedings of the symposium in Rome, June 6–10, 2009. *International Association of Geodesy Symposia*, vol. 137. Springer Berlin Heidelberg, pp. 245–250. DOI: 10.1007/978-3-642-22078-4_37.

Kaban, M. K., Schwintzer, P., and Reigber, C. (2004): A new isostatic model of the lithosphere and gravity field. *Journal of Geodesy* 78(6):368–385. DOI: 10.1007/s00190-004-0401-6.

Keller, W. (2004): Wavelets in geodesy and geodynamics. Walter de Gruyter, Berlin, Germany. DOI: 10.1515/9783110198188.

Kuhn, M. and Seitz, K. (2005): Comparison of Newton's integral in the space and frequency domains. In: Sansò, F. (ed.) A window on the future of geodesy. Proceedings of the IAG General Assembly, Sapporo, Japan, June 30–July 11, 2003. *International*

Association of Geodesy Symposia, vol. 128. Springer Berlin Heidelberg, pp. 386–391. DOI: 10.1007/3-540-27432-4_66.

Little, S. A., Carter, P. H., and Smith, D. K. (1993): Wavelet analysis of a bathymetric profile reveals anomalous crust. *Geophysical Research Letters* 20(18):1915–1918. DOI: 10.1029/93GL01880.

Liu, L., Hsu, H., and Grafarend, E. W. (2007): Normal Morlet wavelet transform and its application to the Earth's polar motion. *Journal of Geophysical Research* 112:B08401. DOI: 10.1029/2006JB004895.

Luo, X. (2014): GPS stochastic modelling. Signal quality measures and ARMA processes. *Springer Theses: Recognizing Outstanding Ph.D. Research.* Springer, Berlin Heidelberg, Germany. DOI: 10.1007/978-3-642-34836-5.

Makhloof, A. A. and Ilk, K. (2008): Effects of topographic-isostatic masses on gravitational functionals at the Earth's surface and at airborne and satellite altitudes. *Journal of Geodesy* 82(2):93–111. DOI: 10.1007/s00190-007-0159-8.

Moritz, H. (1980): Geodetic Reference System 1980. *Bulletin Géodésique* 54(3):395–405. DOI: 10.1007/BF02521480.

Morlet, J., Arens, G., Fourgeau, E., and Giard, D. (1982a): Wave propagation and sampling theory. Part I: Complex signal and scattering in multilayered media. *Geophysics* 47(2):203–221. DOI: 10.1190/1.1441328.

Morlet, J., Arens, G., Fourgeau, E., and Giard, D. (1982b): Wave propagation and sampling theory. Part II: Sampling theory and complex waves. *Geophysics* 47(2):222–236. DOI: 10.1190/1.1441329.

Novák, P., Kern, M., Schwarz, K.-P., and Heck, B. (2003): Evaluation of band-limited topographical effects in airborne gravimetry. *Journal of Geodesy* 76(11–12):597–604. DOI: 10.1007/s00190-002-0282-5.

Novák, P. and Tenzer, R. (2013): Gravitational gradients at satellite altitudes in global geophysical studies. *Surveys in Geophysics* 34(5):653–673. DOI: 10.1007/s10712-013-9243-1.

Pavlis, N. K., Factor, J. K., and Holmes, S. A. (2007): Terrain-related gravimetric quantities computed for the next EGM. In: Kiliçoglu, A., and Forsberg, R. (eds.) Proceedings of the 1st International Symposium of the International Gravity Field Service (IGFS), Istanbul, Turkey, Aug. 28–Sept. 1, 2006. Harita Dergisi, Special Issue 18, pp. 318–323.

Rummel, R. and Colombo, O. L. (1985): Gravity field determination from satellite gradiometry. *Bulletin Géodésique* 59(3):233–246. DOI: 10.1007/BF02520329.

Rummel, R., Rapp, R. H., Sünkel, H., and Tscherning, C. C. (1988): Comparisons of global topographic/isostatic models to the Earth's observed gravity field. Report 388. Department of Geodetic Science and Surveying, The Ohio State University, Columbus, USA.

Rummel, R., Yi, W., and Stummer, C. (2011): GOCE gravitational gradiometry. *Journal of Geodesy* 85(11):777–790. DOI: 10.1007/s00190-011-0500-0.

Satirapod, C., Ogaja, C., Wang, J., and Rizos, C. (2001): An approach to GPS analysis incorporating wavelet decomposition. *Artificial Satellites* 36(2):27–35.

Satirapod, C. and Rizos, C. (2005): Multipath mitigation by wavelet analysis for GPS base station applications. *Survey Review* 38(295):2–10. DOI: 10.1179/003962605791521699.

Schuh, W.-D. (2010): Filtering of correlated data – stochastical considerations within GOCE data processing. In: GOCE Summer School. Herrsching, Germany, May 31–June 4, 2010. Lecture material.

Tenzer, R. and Novák, P. (2013): Effect of crustal density structures on GOCE gravity gradient observables. *Terrestrial, Atmospheric and Oceanic Sciences* 24(5):793–807. DOI: 10.3319/TAO.2013.05.08.01(T).

Torrence, C. and Compo, G. P. (1998): A practical guide to wavelet analysis. *Bulletin of the American Meteorological Society* 79(1):61–78. DOI: 10.1175/1520-0477(1998)079<0061:APGTWA>2.0.CO;2.

Trauth, M. H. (2007): MATLAB recipes for Earth sciences. 2nd ed. Springer, Berlin Heidelberg, Germany.

Tsoulis, D. and Kuhn, M. (2007): Recent developments in synthetic Earth gravity models in view of the availability of digital terrain and crustal databases of global coverage and increased resolution. In: Kiliçoglu, A., and Forsberg, R. (eds.) Proceedings of the 1st International Symposium of the International Gravity Field Service (IGFS), Istanbul, Turkey, Aug. 28–Sept. 1, 2006. Harita Dergisi, Special Issue 18, pp. 354–359.

Wang, J., Wang, J., and Roberts, C. (2009): Reducing GPS carrier phase errors with EMD-wavelet for precise static positioning. *Survey Review* 41(312):152–161. DOI: 10.1179/003962609X390067.

Wild, F. and Heck, B. (2005): A comparison of different isostatic models applied to satellite gravity gradiometry. In: Jekeli, C., Bastos, L., and Fernandes, J. (eds.) Gravity, geoid and space missions, GGSM 2004 IAG International Symposium. Porto, Portugal, Aug. 30–Sept. 3, 2004. *International Association of Geodesy Symposia*, vol. 129. Springer Berlin Heidelberg, pp. 230–235. DOI: 10.1007/3-540-26932-0_40.

Wild-Pfeiffer, F. (2007): Auswirkungen topographisch-isostatischer Massen auf die Satellitengradiometrie. *Deutsche Geodätische Kommission, Reihe C*, no. 604. Verlag der Bayerischen Akademie der Wissenschaften in Kommission beim Verlag C. H. Beck, Munich, Germany. URL: https://www.dgk.badw.de.devweb.mwn.de/fileadmin/docs/c-604.pdf.

Wild-Pfeiffer, F. (2008): A comparison of different mass elements for use in gravity gradiometry. *Journal of Geodesy* 82(10):637–653. DOI: 10.1007/s00190-008-0219-8.

Wild-Pfeiffer, F. and Heck, B. (2007): Comparison of the modelling of topographic and isostatic masses in the space and the frequency domain for use in satellite gravity gradiometry. In: Kiliçoglu, A., and Forsberg, R. (eds.) Proceedings of the 1st International Symposium of the International Gravity Field Service (IGFS), Istanbul, Turkey, Aug. 28–Sept. 1, 2006. Harita Dergisi, Special Issue 18, pp. 312–317.

Wittwer, T., Klees, R., Seitz, K., and Heck, B. (2008): Ultra-high degree spherical harmonic analysis and synthesis using extended-range arithmetic. *Journal of Geodesy* 82(4):223–229. DOI: 10.1007/s00190-007-0172-y.

Wu, J., Gao, J., Li, M., and Wang, Y. (2009): Wavelet transform for GPS carrier phase multipath mitigation. In: Proceedings of the 1st International Conference on Information

Science and Engineering, Nanjing, China, Dec. 26–28, 2009. IEEE, pp. 1019–1022. DOI: 10.1109/ICISE.2009.1344.

Yi, T., Li, H., and Wang, G. (2006): Cycle slip detection and correction of GPS carrier phase based on wavelet transform and neural network. In: Proceedings of the 6th International Conference on Intelligent Systems Design and Applications, Jinan, China, Oct. 16–18, 2006. IEEE, pp. 46–50. DOI: 10.1109/ISDA.2006.129.

Zhong, P., Ding, X. L., Zheng, D. W., Chen, W., and Huang, D. F. (2008): Adaptive wavelet transform based on cross-validation method and its application to GPS multipath mitigation. *GPS Solutions* 12(2):109–117. DOI: 10.1007/s10291-007-0071-y.

Chapter IV.

The Rock-Water-Ice topographic gravity field model RWI_TOPO_2015 and its comparison to a conventional rock-equivalent version

Thomas Grombein, Kurt Seitz, Bernhard Heck
Surveys in Geophysics 37(5):937–976, 2016. DOI: 10.1007/s10712-016-9376-0.

Submitted: 16 March 2016 / Accepted: 25 May 2016 / Published online: 21 June 2016

Abstract. RWI_TOPO_2015 is a new high-resolution spherical harmonic representation of the Earth's topographic gravitational potential that is based on a refined Rock-Water-Ice (RWI) approach. This method is characterized by a three-layer decomposition of the Earth's topography with respect to its rock, water, and ice masses. To allow a rigorous separate modeling of these masses with variable density values, gravity forward modeling is performed in the space domain using tesseroid mass bodies arranged on an ellipsoidal reference surface. While the predecessor model RWI_TOPO_2012 was based on the $5' \times 5'$ global topographic database DTM2006.0, the new RWI model uses updated height information of the $1' \times 1'$ Earth2014 topography suite. Moreover, in the case of RWI_TOPO_2015, the representation in spherical harmonics is extended to degree and order 2190 (formerly 1800).

Besides a presentation of the used formalism, the processing for RWI_TOPO_2015 is described in detail, and the characteristics of the resulting spherical harmonic coefficients are analyzed in the space and frequency domain. Furthermore, this paper focuses on a comparison of the RWI approach to the conventionally used rock-equivalent method. For this purpose, a consistent rock-equivalent version REQ_TOPO_2015 is generated, in which the heights of water and ice masses are condensed to the constant rock density. When evaluated on the surface of the GRS80 ellipsoid, the differences of RWI_TOPO_2015 and REQ_TOPO_2015 reach maximum amplitudes of about 1 m, 50 mGal, and 20 mE in terms of height anomaly, gravity disturbance, and the radial-radial gravity gradient, respectively. Although these differences are attenuated with increasing height above the ellipsoid, significant magnitudes can even be detected in the case of the satellite altitudes of current gravity field missions. In order to assess their performance, RWI_TOPO_2015, REQ_TOPO_2015, and RWI_TOPO_2012 are validated against independent gravity information of current global geopotential models, clearly demonstrating the attained improvements in the case of the new RWI model.

Keywords Gravity forward modeling · Topographic potential · Rock-Water-Ice (RWI) decomposition · Tesseroids

1. Introduction

Global high-resolution digital terrain models (DTMs) provide precise information on Earth's topography. By applying gravity forward modeling (GFM), this data can be used to determine the Earth's topographic potential, i.e., the gravitational potential generated by the attraction of the Earth's topographic masses. These encompass the rock, water, and ice masses of the continents, oceans, lakes, and ice sheets. With respect to a regularized Earth model, the topographic potential combines the positive effect of the mass excesses above a specified reference surface, e.g., the ellipsoid or geoid, and the negative effect of mass deficits below this surface. While the topographic potential contributes to all spectral scales of the gravity field, it particularly contains short wavelengths. This spectral characteristic makes the topographic potential and related functionals valuable for various applications in geodesy and geophysics.

On the one hand, many calculation methods in physical geodesy are embedded in a remove-compute-restore framework, where topography information is used to smooth gravity field observations, achieving a numerically stabilized interpolation and field transformation (Forsberg, 1984, Sect. 4). Such a procedure is frequently used for harmonic downward continuation of airborne and satellite-based measurements (Novák et al., 2003; Janák et al., 2014; Grombein et al., 2014a) or gravimetric (quasi-)geoid determination (Omang and Forsberg, 2000).

On the other hand, topography information can also be used for the spectral extension of band-limited measurements or global geopotential models (GGMs), e.g., using residual terrain modeling (Forsberg and Tscherning, 1997) to reduce the omission error of satellite-based GGMs. Moreover, topography-implied gravity can be utilized to refine the spatial resolution of terrestrial data sets. This is particularly important in regions where no or only limited gravity data is available, as in Africa (Abd-Elmotaal et al., 2016). Here, topography-implied gravity data is used as fill-in information for the construction of high-resolution regional and global geopotential models, like EGM2008 (Pavlis et al., 2012).

Moreover, topographic information can be used to construct a synthetic Earth gravity model (Kuhn and Featherstone, 2005; Baran et al., 2006) or to assess the performance of satellite-based and combined GGMs (Tsoulis and Patlakis, 2013; Hirt et al., 2015). Last but not least, topographic mass reductions are required for the geoid determination based on Stokes' problem (Heiskanen and Moritz, 1967, Chapter 3).

In geophysical studies, topographic corrections of gravity field observations are used to interpret crustal structures (Ebbing et al., 2001; Tenzer et al., 2012), to analyze mass anomalies in the Earth's interior (Wieczorek, 2007), or to perform a general lithospheric modeling (Bouman et al., 2015). Furthermore, global information of the topographic potential has been combined with GGMs to generate global Bouguer gravity maps (Balmino et al., 2012; Claessens and Hirt, 2013).

Generally, the Earth's topographic potential V^{Topo} can be represented by Newton's integral (Heiskanen and Moritz, 1967, p. 3):

$$V^{\text{Topo}}(P) = G \iiint\limits_{\Omega} \frac{\rho}{\ell} \, \mathrm{d}\Omega, \tag{1}$$

where $G = 6.672 \cdot 10^{-11}\,\mathrm{m}^3\,\mathrm{kg}^{-1}\,\mathrm{s}^{-2}$ denotes Newton's gravitational constant, $\rho = \rho(Q)$ is the location-dependent mass density function, and $\ell = \ell(P, Q)$ is the Euclidean distance between the attracted computation point P and the running integration mass point Q. The integration domain Ω extends over the topographic masses, requiring global information on the Earth's geometry and its mass density distribution. As V^{Topo} is harmonic outside Ω, Newton's integral in Eq. (1) can be expanded into a series of solid spherical harmonics for $P \notin \Omega$. By using geocentric spherical coordinates (r, φ, λ), this spherical harmonic expansion is expressed by Heiskanen and Moritz (1967, p. 59):

$$V^{\text{Topo}}(P) = \frac{GM}{r} \sum_{n=0}^{\infty} \left(\frac{R}{r}\right)^n \sum_{m=0}^{n} \left(\overline{C}_{nm}^{\text{Topo}}\, \overline{Y}_{nm}^{C} + \overline{S}_{nm}^{\text{Topo}}\, \overline{Y}_{nm}^{S}\right), \tag{2}$$

where GM is the geocentric gravitational constant, R is the radius of a reference sphere, $\overline{Y}_{nm}^{C}$ and $\overline{Y}_{nm}^{S}$ are the fully normalized surface spherical harmonic functions of degree n and order m, and $\overline{C}_{nm}^{\text{Topo}}$ and $\overline{S}_{nm}^{\text{Topo}}$ are the fully normalized spherical harmonic (SH) coefficients. A major advantage of the representation in terms of spherical harmonics is that the coefficients can be used to efficiently calculate various functionals of the topographic potential in different heights (e.g., Barthelmes, 2013). In the following, a set of SH coefficients ($\overline{C}_{nm}^{\text{Topo}}$, $\overline{S}_{nm}^{\text{Topo}}$) up to a maximum degree $N_{\max}$ is called a topographic gravity field model (or topographic potential model).

According to the representations in Eqs. (1) and (2), different GFM methods have been proposed that can be classified by space and frequency domain techniques (e.g., Kuhn and Seitz, 2005). In Fig. 1, a general processing scheme for GFM in both domains is presented and is briefly described in the following.

By using space domain GFM, Newton's integral in Eq. (1) is evaluated by a mass discretization. In the first step, the integration domain Ω is decomposed into elementary mass bodies according to the grid resolution of the used DTM. Based on the specific application, various mass bodies like point-masses, prisms or tesseroids can be used (e.g., Nagy et al., 2000; Heck and Seitz, 2007; Wild-Pfeiffer, 2008; Grombein et al., 2013).

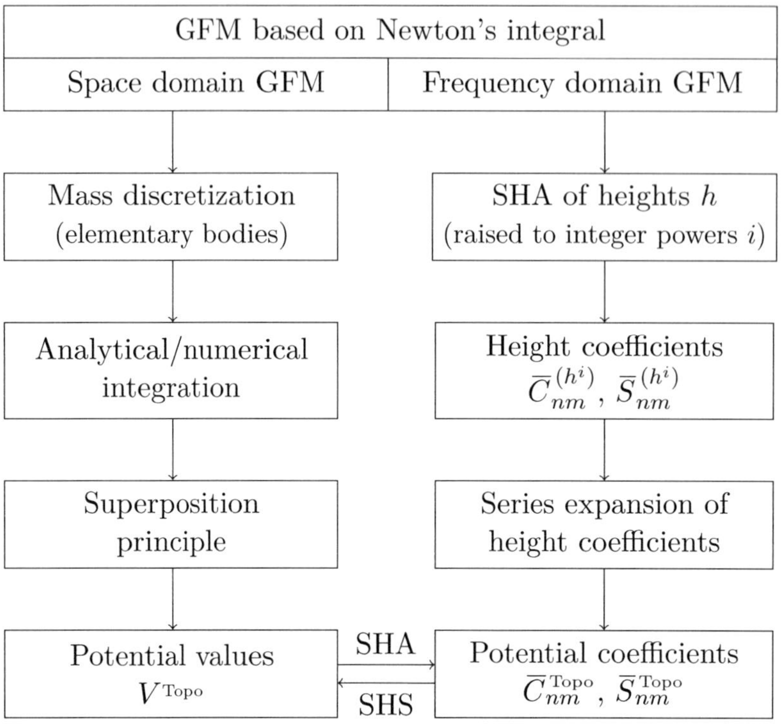

Fig. 1. General processing scheme for GFM in the space and frequency domain. The results of both GFM methods, i.e., potential values or coefficients, can be converted to each other by means of spherical harmonic analysis (SHA) and synthesis (SHS).

Usually, each DTM grid element is represented by one mass body. In the second step, the gravitational potential of each individual mass body is calculated analytically or by approximate solutions. In the third step, the total topographic potential V^{Topo} is calculated as the sum of the potential values of all individual mass bodies (superposition principle).

For frequency domain GFM, Newton's integral in Eq. (1) is evaluated through a transformation into the frequency domain (Rummel et al., 1988; Wieczorek, 2007; Hirt and Kuhn, 2012). In the first step, surface SH coefficients of global gridded DTM heights and its integer powers are calculated via spherical harmonic analysis (SHA) for each integer power. In the second step, the solid SH coefficients ($\overline{C}_{nm}^{\text{Topo}}$, $\overline{S}_{nm}^{\text{Topo}}$) of the topographic potential are calculated based on a series expansion of the inverse distance, involving the surface SH coefficients obtained from the DTM heights. In the third step, the derived set of SH coefficients ($\overline{C}_{nm}^{\text{Topo}}$, $\overline{S}_{nm}^{\text{Topo}}$) can be used to calculate the topographic potential via spherical harmonic synthesis (SHS) according to Eq. (2).

It is acknowledged that there are other GFM approaches that do not directly fit in this general classification, such as Gruber et al. (2014), Abd-Elmotaal and Kühtreiber (2014).

In order to generate topographic gravity field models, frequency domain GFM is commonly applied as these techniques directly provide the desired SH coefficients (Rummel et al., 1988; Balmino et al., 2012; Hirt and Kuhn, 2012). However, frequency domain methods suffer from some drawbacks, which are pointed out in the following. Due to the nature of spherical harmonics, frequency domain GFM is generally limited to a spherical approximation, i.e., topographic masses are arranged on a spherical reference surface. As this is insufficient for current high-resolution applications, Claessens and Hirt (2013) proposed a new spectral approach taking into account an ellipsoidal approximation.

Moreover, almost all frequency domain approaches are restricted to the use of one mass layer of constant density. In order to account for variable density values of different terrain types, the concept of rock-equivalent heights (REQ) is widely used, in which the DTM heights of water and ice masses are condensed to a constant rock density (e.g., Rummel et al., 1988; Kuhn and Seitz, 2005; Hirt et al., 2015). The shortcoming of such an approach is a changed geometry with considerable mass displacements, as noted by Tsoulis and Kuhn (2007). Due to the distance dependency, this has also an impact on any gravity field functional.

In contrast to the above-mentioned limitations, the consideration of an ellipsoidal mass arrangement and the use of different mass layers is straightforward when using GFM in the space domain. For a more realistic modeling, Grombein et al. (2010, 2011) developed the Rock-Water-Ice (RWI) approach based on GFM in the space domain. This method is characterized by a three-layer decomposition of the topography that enables a rigorous separate modeling of rock, water, and ice masses with layer-specific density values. In this context, SH coefficients of the topographic potential are obtained in a two-step sequence: In the first step, GFM is performed in the space domain by using tesseroid mass bodies (Grombein et al., 2013). In the second step, global gridded topographic potential values are transformed to the frequency domain by using SHA (e.g., Sneeuw, 1994; Abd-Elmotaal et al., 2014). By applying this RWI approach to the $5' \times 5'$ global topographic database DTM2006.0 (Pavlis et al., 2007), the topographic gravity field model RWI_TOPO_2012 up to degree and order (d/o) 1800 has been generated by Grombein et al. (2014a). Moreover, taking into account deeper lying compensation masses, a modified Airy-Heiskanen concept has been applied to calculate the consistent isostatic gravity field model, RWI_ISOS_2012, as well as a combined topographic-isostatic gravity field model, RWI_TOIS_2012.

According to Grombein et al. (2014b), the models of the RWI_2012 suite have originally been created for smoothing gravity gradients as measured by the satellite mission Gravity field and steady-state Ocean Circulation Explorer (GOCE, Rummel et al., 2011), but have also been used in a wide range of applications. For generating the combined gravity field model GOCO05c, Fecher et al. (2015, 2016) use topography-induced gravity anomalies of the RWI model to fill gaps in the global gravity coverage. To analyze the structure and state of stress of the Chilean subduction zone, Gutknecht et al. (2014) utilize the RWI model to derive topographically reduced gravity gradients of the satellite based GOCO03s model (Mayer-Gürr et al., 2012). Bouman et al. (2016) make use of the RWI model to derive topographic mass reductions for GOCE gravity gradient grids, as used in geophysical applications. Furthermore, in some studies the RWI model has been applied as a reference for an independent validation (e.g., Novák and Tenzer, 2013; Pitoňák et al., 2016).

In the present paper, the RWI approach is developed further and a new topographic gravity field model called RWI_TOPO_2015 is presented, which is based on updated topographic input data of the Earth2014 model (Hirt and Rexer, 2015). In contrast to RWI_TOPO_2012, the new model provides SH coefficients up to d/o 2190, additionally featuring separate SH coefficients for the rock, water, and ice proportions. Moreover, the ellipsoidal arrangement of the topography is refined by a geoid model (cf. Sect. 2.1). To see the additional benefit of the RWI method, a consistent rock-equivalent version of the model, called REQ_TOPO_2015, has been generated. This allows a detailed comparison of both approaches. In Table 1, the main features of the new RWI model are presented and compared to those of other topographic gravity field models.

The paper is organized as follows: in Sect. 2, the updated Rock-Water-Ice method is presented and adapted formulas for space domain GFM and SHA are derived. The used topographic input data is described in Sect. 3. While Sect. 4 provides detailed information on the processing, in Sect. 5 the main characteristics of the new RWI_TOPO_2015 model are presented and discussed, in both frequency and space domains. Furthermore, a comparison to the predecessor model RWI_TOPO_2012 and the rock-equivalent version REQ_TOPO_2015 is carried out. In Sect. 6, the performance of the new RWI topographic gravity field model is validated by gravity data of current GGMs, clearly showing the attained improvements. Finally, in Sect. 7, a summary and an outlook are provided.

Table 1. Characteristics of current topographic gravity field models based on GFM in the frequency domain (*upper three models*) and in the space domain (*lower three models*).

Model name	Topography	Mass layer	Ref. surface	$N_{\max}$	Reference
dV_SPH_RET2012	Earth2012	1 (REQ)	Sphere	2160	Hirt and Kuhn (2012)
dV_ELL_RET2012	Earth2012	1 (REQ)	Ellipsoid	2190	Claessens and Hirt (2013)
dV_ELL_RET2014	Earth2014	1 (REQ)	Ellipsoid	2190	Hirt et al. (2015)
RWI_TOPO_2012	DTM2006.0	3 (RWI)	Ellipsoid	1800	Grombein et al. (2014a)
RWI_TOPO_2015	Earth2014	3 (RWI)	Ellip. + Geoid	2190	This paper
REQ_TOPO_2015	Earth2014	3 (REQ)	Ellip. + Geoid	2190	This paper

2. Method

2.1. RWI-based topographic model

The RWI approach as proposed by Grombein et al. (2014a) is based on a three-layer decomposition of the Earth's topography with respect to its rock, water, and ice masses. The basic idea is that the masses of each terrain type, e.g., land topography, oceans, lakes, ice sheets and ice shelves can consistently be described by a vertically arranged (1) rock, (2) water, and (3) ice proportion. As schematically shown in Fig. 2, the sequence of these masses is the same for each terrain type; however, some mass types may be lacking. Such a RWI-based topographic model allows a rigorous separate GFM of the rock, water, and

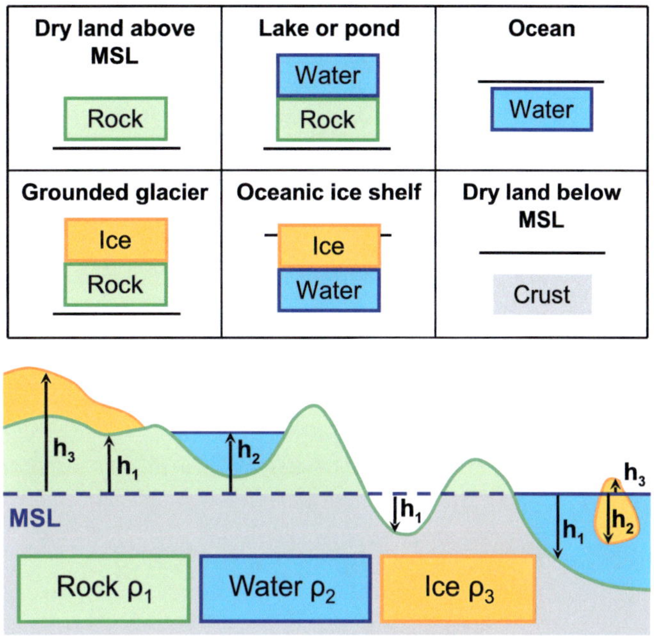

Fig. 2. Schematic representation of the RWI-based topographic model and the composition of different terrain types by rock, water, and ice masses. The reference surface is defined by the mean sea level (MSL).

ice masses with layer-specific density values. This prevents geometry changes and mass displacements, occurring in conventional rock-equivalent methods.

For the RWI approach, a global DTM is needed that provides heights ($h_1^{\mathrm{RWI}}, h_2^{\mathrm{RWI}}, h_3^{\mathrm{RWI}}$), corresponding to the upper boundary surfaces of the rock, water, and ice masses (see Fig. 2). While the previous RWI_TOPO_2012 model is based on the $5' \times 5'$ global topographic database DTM2006.0, the new model uses updated topographic information of the $1' \times 1'$ Earth2014 model, as described in Sect. 3. Furthermore, the applied layer-specific density values (ρ_1, ρ_2, ρ_3) are slightly modified, particularly accounting for the difference in salt- and freshwater density (see Table 2).

Table 2. Used density values ρ_s of the rock, water, and ice masses in the case of the RWI_TOPO_2012 and RWI_TOPO_2015 model. Note that for the new model there is a separation with respect to salt- and freshwater density. All values are specified in [$\mathrm{kg\,m^{-3}}$].

	RWI_TOPO_2012	RWI_TOPO_2015
ρ_1 (Rock)	2670	2670
ρ_2 (Water)	1000	1000 (Inland) / 1030 (Ocean)
ρ_3 (Ice)	920	917

To correctly locate the topographic masses in space, the ellipticity of the Earth's shape is taken into account. Therefore, the used reference surface is defined by

$$R_0 = r_E + N, \tag{3}$$

where, as in Heck (2003, p. 68),

$$r_E(\varphi) = \frac{a}{\sqrt{1 + e'^2 \sin^2 \varphi}} \quad (\varphi\text{: geocentric latitude}) \tag{4}$$

is the latitude-dependent radius of a reference ellipsoid (e.g., GRS80, Moritz, 1980), parameterized by the semi-major axis a and the square of the second numerical eccentricity e'^2. Moreover, N denotes an additionally considered geoid undulation as can be obtained from a GGM by using SHS.

By using Eq. (3), the geocentric radii of the upper boundary surfaces of the rock, water, and ice masses are approximated by

$$R_s^{\mathrm{RWI}} = R_0 + h_s^{\mathrm{RWI}}, \quad s \in \{1, 2, 3\}. \tag{5}$$

2.2. REQ-based topographic model

In addition to the RWI-based topographic model, a consistent rock-equivalent (REQ) variant is introduced. Here, DTM heights are condensed such that their associated mass elements correspond to an equivalent mass of rock density (e.g., Rummel et al., 1988; Kuhn and Seitz, 2005).

To allow a more detailed comparison, the rock-equivalent condensation is performed separately for the water and ice masses. Therefore, for each mass layer $s \in \{1, 2, 3\}$ the following mass conservation condition is applied:

$$m_s^{\mathrm{REQ}}(\rho_1) = m_s^{\mathrm{RWI}}(\rho_s), \tag{6}$$

where m_s^{RWI} are the masses of the rock, water, and ice proportions, and m_s^{REQ} are the corresponding condensed masses referring to the rock density value ρ_1.

Employing a spherical approximation, Eq. (6) can be formulated directly for geocentric radii as

$$\rho_1 \int_{R_{s-1}^{\mathrm{REQ}}}^{R_s^{\mathrm{REQ}}} r'^2 \,\mathrm{d}r' = \rho_s \int_{R_{s-1}^{\mathrm{RWI}}}^{R_s^{\mathrm{RWI}}} r'^2 \,\mathrm{d}r', \tag{7}$$

where $R_0^{\mathrm{REQ}} := R_0^{\mathrm{RWI}} := R_0$. Note that each rock-equivalent mass layer is situated on the surface of the layer underneath, thus preventing any gap in between. Solving Eq. (7), the rock-equivalent geocentric radii can be recursively calculated using

$$R_s^{\mathrm{REQ}} = \sqrt[3]{\left(R_{s-1}^{\mathrm{REQ}}\right)^3 + \frac{\rho_s}{\rho_1}\left[\left(R_s^{\mathrm{RWI}}\right)^3 - \left(R_{s-1}^{\mathrm{RWI}}\right)^3\right]}. \tag{8}$$

The corresponding rock-equivalent heights are then obtained by

$$h_s^{\mathrm{REQ}} = R_s^{\mathrm{REQ}} - R_0. \tag{9}$$

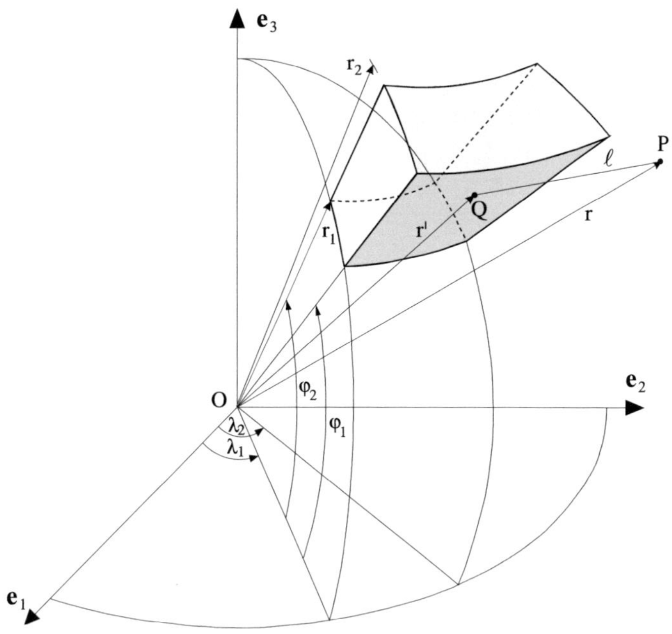

Fig. 3. Geometry of a tesseroid used for the space domain GFM (Heck and Seitz, 2007); the spherical coordinates (r, φ, λ) are referred to the geocentric Earth-fixed equatorial reference system defined by the base vectors $\boldsymbol{e_1}$, $\boldsymbol{e_2}$, $\boldsymbol{e_3}$.

2.3. Gravity forward modeling based on tesseroids

For space domain GFM, the information of the RWI- and REQ-based topographic models has to be represented by vertically arranged mass bodies for each grid element. For this purpose, tesseroid mass bodies as introduced by Anderson (1976) are used in the following. Considering geocentric spherical coordinates (r, φ, λ), tesseroids are bounded by a pair of concentric spheres ($r_1 =$ const., $r_2 =$ const.), a pair of meridional planes ($\lambda_1 =$ const., $\lambda_2 =$ const.), and a pair of coaxial circular cones, defined by the parallels $\varphi_1 =$ const., $\varphi_2 =$ const. (see Fig. 3).

Tesseroids directly take the curvature of the Earth into account, which is particularly beneficial for regional and global applications. Several publications that studied the use of tesseroids have shown its advantages in comparison to classical prism approaches (e.g., Heck and Seitz, 2007; Wild-Pfeiffer, 2008; Grombein et al., 2013).

Based on Newton's integral in Eq. (1), the gravitational potential $V^*$of a tesseroid with a constant mass density ρ can be represented by

$$V^*(P) = G\rho \int_{\lambda_1}^{\lambda_2} \int_{\varphi_1}^{\varphi_2} \int_{r_1}^{r_2} \frac{r'^2 \cos\varphi'}{\ell} \, \mathrm{d}r' \, \mathrm{d}\varphi' \, \mathrm{d}\lambda', \tag{10}$$

where $\ell = \ell(P, Q)$ denotes the Euclidean distance between the computation point $P(r, \varphi, \lambda)$ and the running integration point $Q(r', \varphi', \lambda')$.

As the integration with respect to λ' and φ' comprises elliptical integrals, Eq. (10) cannot be solved in closed analytical form. Alternatively, a numerical evaluation can be achieved by means of expanding the integral kernel of Eq. (10) in a Taylor series

$$K(P,Q) := \frac{r'^2 \cos\varphi'}{\ell} = \sum_{i,j,k\in\mathbb{N}_0} \frac{K_{ijk}(P,Q_0)}{i!\ j!\ k!} (r'-r_0)^i (\varphi'-\varphi_0)^j (\lambda'-\lambda_0)^k, \tag{11}$$

where

$$K_{ijk}(P,Q_0) := \left.\frac{\partial^{i+j+k} K}{\partial r'^i \partial\varphi'^j \partial\lambda'^k}\right|_{\substack{r'=r_0\\ \varphi'=\varphi_0\\ \lambda'=\lambda_0}} \tag{12}$$

are the partial derivatives of the integral kernel K evaluated at the Taylor point $Q_0(r_0, \varphi_0, \lambda_0)$ that will be fixed at the geometrical center of the tesseroid, i.e., $r_0 = (r_1 + r_2)/2$, $\varphi_0 = (\varphi_1 + \varphi_2)/2$, and $\lambda_0 = (\lambda_1 + \lambda_2)/2$.

Inserting Eq. (11) into Eq. (10) and using the substitutions $r^* = r' - r_0$, $\varphi^* = \varphi' - \varphi_0$, and $\lambda^* = \lambda' - \lambda_0$ yields

$$V^*(P) = G\rho \sum_{i,j,k\in\mathbb{N}_0} \frac{K_{ijk}(P,Q_0)}{i!\ j!\ k!} \int_{-\Delta\lambda/2}^{\Delta\lambda/2} \int_{-\Delta\varphi/2}^{\Delta\varphi/2} \int_{-\Delta r/2}^{\Delta r/2} (r^*)^i (\varphi^*)^j (\lambda^*)^k \, \mathrm{d}r^* \, \mathrm{d}\varphi^* \, \mathrm{d}\lambda^*, \tag{13}$$

where $\Delta r = r_2 - r_1$, $\Delta\varphi = \varphi_2 - \varphi_1$, and $\Delta\lambda = \lambda_2 - \lambda_1$ denote the dimensions of the tesseroid.

Due to the special choice of the Taylor point Q_0, any terms of odd order i, j, or k cancel out after performing the integration in Eq. (13), cf. Heck and Seitz (2007). For the integration with respect to r^*, and analogously φ^* and λ^*, it follows that

$$\int_{-\Delta r/2}^{\Delta r/2} (r^*)^i \, \mathrm{d}r^* = \frac{1-(-1)^{i+1}}{(i+1)2^{i+1}} (\Delta r)^{i+1} = \begin{cases} 0, & \text{if } i \text{ is odd}, \\ (\Delta r)^{i+1}/[(i+1)2^i], & \text{if } i \text{ is even}. \end{cases} \tag{14}$$

Therefore, after performing the integration in Eq. (13), only terms with even order remain in the summation. Finally, the tesseroid potential in a $(\tau+1)$-th order approximation can be calculated by

$$V^*_\tau(P) := G\rho \sum_{\substack{i,j,k\in 2\mathbb{N}_0\\ i+j+k\le\tau}} \Gamma_{ijk}(\Delta r, \Delta\varphi, \Delta\lambda) \cdot K_{ijk}(P,Q_0), \tag{15}$$

where

$$\Gamma_{ijk}(\Delta r, \Delta\varphi, \Delta\lambda) := \frac{(\Delta r)^{i+1} (\Delta\varphi)^{j+1} (\Delta\lambda)^{k+1}}{(i+1)!\ (j+1)!\ (k+1)!\ 2^{i+j+k}}, \tag{16}$$

and $2\mathbb{N}_0 = \{2m : m \in \mathbb{N}_0\}$ is the index set of all even natural numbers. In the following, Eq. (15) is evaluated with $\tau = 2$, achieving a fourth-order error in the spatial coordinates of the integration point. For this case, optimized evaluation rules and an analysis concerning the approximation error due to the omitted Taylor residual are presented in Grombein et al. (2013).

For the RWI approach, Eq. (15) is used in combination with the superposition principle under the assumption that the height information is parameterized in terms of geocentric spherical coordinates

$$\varphi_v = 90^\circ - (v - 1/2) \cdot \Delta\varphi, \qquad v = 1, \ldots, v_n \in \mathbb{N}, \tag{17}$$

$$\lambda_w = (w - 1/2) \cdot \Delta\lambda, \qquad w = 1, \ldots, w_m \in \mathbb{N}, \tag{18}$$

where $\Delta\varphi = 180^\circ / v_n$ and $\Delta\lambda = 360^\circ / w_m$.

The topographic potential generated by the RWI-based rock, water, and ice masses can then be calculated separately as the sum of the impact over all individual tesseroids of the specific layer $s \in \{1, 2, 3\}$:

$$V_s^{\mathrm{RWI}}(P) = G\rho_s \sum_{v=1}^{v_n} \sum_{w=1}^{w_m} \sum_{\substack{i,j,k \in 2\mathbb{N}_0 \\ i+j+k \leq 2}} \Gamma_{ijk}(\Delta r_s^{\mathrm{RWI}}, \Delta\varphi, \Delta\lambda) \cdot K_{ijk}(P, Q_0(r_s^{\mathrm{RWI}}, \varphi_v, \lambda_w)), \tag{19}$$

where for each grid element (v,w)

$$\Delta r_s^{\mathrm{RWI}} := R_s^{\mathrm{RWI}} - R_{s-1}^{\mathrm{RWI}} \tag{20}$$

defines the radial dimension of the masses and

$$r_s^{\mathrm{RWI}} := (R_s^{\mathrm{RWI}} + R_{s-1}^{\mathrm{RWI}})/2 \tag{21}$$

its geometrical center as required for the Taylor series approach. For readability, the indices v and w are omitted in Eqs. (20) and (21).

The total RWI-based topographic potential is calculated as the sum of their rock, water, and ice proportions:

$$V^{\mathrm{RWI}}(P) = \sum_{s=1}^{3} V_s^{\mathrm{RWI}}(P). \tag{22}$$

Analogously, the potential generated by the REQ topographic mass layers is obtained by

$$V_s^{\mathrm{REQ}}(P) = G\rho_1 \sum_{v=1}^{v_n} \sum_{w=1}^{w_m} \sum_{\substack{i,j,k \in 2\mathbb{N}_0 \\ i+j+k \leq 2}} \Gamma_{ijk}(\Delta r_s^{\mathrm{REQ}}, \Delta\varphi, \Delta\lambda) \cdot K_{ijk}(P, Q_0(r_s^{\mathrm{REQ}}, \varphi_v, \lambda_w)), \tag{23}$$

where for each grid element (v,w)

$$\Delta r_s^{\mathrm{REQ}} := R_s^{\mathrm{REQ}} - R_{s-1}^{\mathrm{REQ}}, \tag{24}$$

$$r_s^{\mathrm{REQ}} := (R_s^{\mathrm{REQ}} + R_{s-1}^{\mathrm{REQ}})/2. \tag{25}$$

The total REQ-based topographic potential is then determined by

$$V^{\mathrm{REQ}}(P) = \sum_{s=1}^{3} V_s^{\mathrm{REQ}}(P). \tag{26}$$

2.4. Spherical harmonic analysis of gridded data

According to Fig. 1, the GFM-derived topographic potential values can be converted to corresponding potential coefficients by means of SHA.

Due to the orthogonality of the surface spherical harmonic functions $\overline{Y}^C_{nm}$ and $\overline{Y}^S_{nm}$ on the unit sphere σ, the desired set of SH coefficients $(\overline{C}^{\,\text{Topo}}_{nm}, \overline{S}^{\,\text{Topo}}_{nm})$ according to Eq. (2) can be obtained analytically via the Legendre transformation (e.g., Jekeli, 1988):

$$\begin{Bmatrix} \overline{C}^{\,\text{Topo}}_{nm} \\ \overline{S}^{\,\text{Topo}}_{nm} \end{Bmatrix} = \frac{1}{4\pi} \iint\limits_{\sigma} \begin{Bmatrix} \overline{Y}^C_{nm} \\ \overline{Y}^S_{nm} \end{Bmatrix} V^{\,\text{Topo}}(r, \varphi, \lambda)\, \mathrm{d}\sigma, \tag{27}$$

where

$$\begin{Bmatrix} \overline{Y}^C_{nm} \\ \overline{Y}^S_{nm} \end{Bmatrix} = \begin{Bmatrix} \cos m\lambda \\ \sin m\lambda \end{Bmatrix} \overline{P}_{nm}(\sin\varphi), \tag{28}$$

and $\overline{P}_{nm}$ denote the fully normalized associated Legendre functions of the first kind (ALFs).

However, for a practical computation, Eq. (27) cannot be used directly, since $V^{\,\text{Topo}}(r, \varphi, \lambda)$ would be needed as a continuous function, defined over the whole globe. Therefore, many SHA approaches make use of a discretization of Eq. (27) and applying fast Fourier transform techniques, see Sneeuw (1994) for an overview of methods.

The SHA approach that is used in the following is a least-squares technique for gridded data as described and analyzed in Abd-Elmotaal et al. (2014). It requires data sampled on a regular global grid parameterized by I equally spaced parallels and J equally spaced meridians. In terms of geocentric spherical coordinates, the nodes of such a grid can be defined by the points $P_{ij}(r_i, \varphi_i, \lambda_j)$ with

$$\varphi_i = 90^\circ - (i - 1/2)\cdot\Delta\varphi, \qquad i = 1, \ldots, I \in \mathbb{N}, \tag{29}$$

$$\lambda_j = (j - 1/2)\cdot\Delta\lambda, \qquad j = 1, \ldots, J \in \mathbb{N}, \tag{30}$$

where $\Delta\varphi = 180^\circ/I$ and $\Delta\lambda = 360^\circ/J$. The surface on which the data is given can be a sphere, an ellipsoid, or theoretically any rotationally symmetric star-shaped surface. In the case of a sphere, the geocentric radii are set to a constant spherical radius $r_i = R$, while in the case of an ellipsoidal surface they are calculated by

$$r_i(\varphi_i) = \frac{a}{\sqrt{1 + e'^2 \sin^2\varphi_i}}. \tag{31}$$

In the first step of the SHA approach, a discrete Fourier transformation of the gridded data is performed for each parallel. The derived Fourier coefficients are then used as observations in a least-squares approach to estimate the desired SH coefficients. In this context, functional relations among the base functions of Fourier and spherical harmonic coefficients are utilized. For the explicit formulation of this approach, the reader is referred to Abd-Elmotaal et al. (2014).

The maximum degree $N_{\max}$ of the SH coefficients that can be derived by this method is limited by the number of meridians I and parallels J of the grid, i.e.,

$$N_{\max} \leq \min\left\{I-1, \frac{J-1}{2}\right\}. \tag{32}$$

Particularly, when performing SHA up to ultra-high degrees, the numerical stability of the used algorithm for the calculation of the ALFs may limit the maximum degree, as noted by Wittwer et al. (2008). For example, in the case of the previous RWI_TOPO_2012 model a maximum degree of $N_{\max} = 2190$ would have been possible; however, the derived SH coefficients were restricted to d/o 1800, mainly due to the numerical instability of the used ALF algorithm according to Thong (1989). Thus, for the new RWI model, an improved ALF algorithm based on Holmes and Featherstone (2002) is used that provides numerical stability for calculations up to d/o 2700.

3. Input data

The used global Earth2014 topography model (Hirt and Rexer, 2015) comprises information of the Shuttle Radar Topography Mission (SRTM) in terms of the $7.5'' \times 7.5''$ SRTM v4.1 model for continents and islands between $\pm 60°$ latitude (Jarvis et al., 2008) and ocean and lake bathymetry of the $30'' \times 30''$ SRTM30_PLUS v9 model (Becker et al., 2009). Moreover, bedrock and ice sheet data for Antarctica and Greenland are obtained from the 1-km resolution grids of the Bedmap2 (Fretwell et al., 2013) and GBT v3 model (Greenland Bedrock Topography, Bamber et al., 2013), respectively.

By combining this topographic information, Earth2014 provides $1' \times 1'$ global gridded data on surface elevations h^{SUR}, bedrock elevations h^{BED}, and ice thickness t^{ICE}. Each grid element of the model is classified into one of the following terrain types tt: (0) land topography above mean sea level (MSL), (1) land topography below MSL, (2) ocean bathymetry, (3) inland lake (bedrock above MSL), (4) inland lake (bedrock below MSL), (5) ice cover (bedrock above MSL), (6) ice cover (bedrock below MSL), (7) ice shelf, (8) ice covered lake.

In Fig. 4, the global bedrock elevations h^{BED} contained in the Earth2014 model are compared to those of the formerly used $5' \times 5'$ DTM2006.0 model (Pavlis et al., 2007). For this comparison, the heights and depths of DTM2006.0 are interpolated to a $1' \times 1'$ resolution. As can be seen from the differences, Earth2014 provides regionally refined bathymetric depths mainly due to new ship sounding measurements and significantly improved bedrock information under the ice masses of Greenland and Antarctica. For the remaining continental parts, the differences are considerably smaller as both models are based on the same SRTM data for surface elevations. However, the effect of the higher spatial resolution in the case of Earth2014 can clearly be seen in regions with highly variable topography such as the Himalaya and the Andes, but also in the deepest continental rift forming Lake Baikal.

As can be expected and seen from Table 3, Earth2014 provides significantly larger extreme values due to its higher spatial resolution, while the mean values and standard

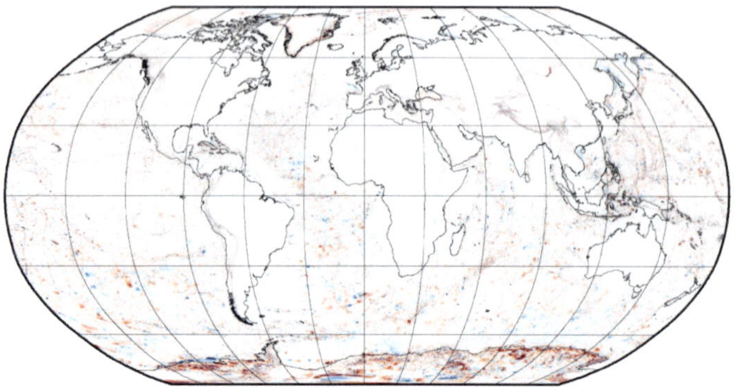

Fig. 4. Difference of the $1' \times 1'$ Earth2014 model and an interpolated version of the $5' \times 5'$ DTM2006.0 in terms of bedrock elevations. Robinson projection centered at 0° longitude.

Table 3. Statistics of bedrock elevations of the $1' \times 1'$ Earth2014 model, an interpolated version of the $5' \times 5'$ DTM2006.0 model, and their difference. All values are specified in [m].

	Min	Max	Mean	STD
Earth2014	−10 847	8212	−2444	2421
DTM2006.0	−9857	6800	−2440	2420
Difference	−4389	4144	−4	143

deviations (STDs) for both models are nearly equal. Regionally, the differences reach maximum amplitudes up to ±4000 m, while the global mean value is just −4 m with a standard deviation of 131 m. The Earth2014 height information is parameterized in terms of geodetic coordinates

$$B_v = 90° - (v - 1/2) \cdot \Delta B, \qquad v = 1, \ldots, v_n \in \mathbb{N}, \tag{33}$$

$$L_w = (w - 1/2) \cdot \Delta L, \qquad w = 1, \ldots, w_m \in \mathbb{N}, \tag{34}$$

with respect to the GRS80 reference ellipsoid (Moritz, 1980), where $\Delta B = \Delta L = 1'$, $v_n = 10\,800$, and $w_m = 21\,600$. In order to use the information of the Earth2014 model for the RWI approach, the geodetic coordinates are transformed to corresponding geocentric spherical coordinates (e.g., Heck, 2003, p. 69):

$$\varphi_v = \arctan\left(\frac{1}{1 + e'^2} \tan B_v\right), \qquad \lambda_w = L_w, \tag{35}$$

where e'^2 is set according to the parameter of the GRS80 ellipsoid.

Furthermore, to obtain the reference surface radius R_0 as defined by Eq. (3), the parameters of the GRS80 ellipsoid are used as well as a global grid of geoid undulations N

derived by SHS. For this purpose, the EGM96 geopotential model (Lemoine et al., 1998) is used as it also provides the reference for the SRTM elevation data (cf. Farr et al., 2007).

The height information of the Earth2014 grid layers can then be used to describe the geocentric radii of the upper boundary surfaces of the rock, water, and ice masses by setting

$$\begin{aligned} R_1^{\mathrm{RWI}} &= R_0 + h^{\mathrm{BED}}, \\ R_2^{\mathrm{RWI}} &= R_0 + h^{\mathrm{SUR}} - t^{\mathrm{ICE}}, \\ R_3^{\mathrm{RWI}} &= R_0 + h^{\mathrm{SUR}}. \end{aligned} \tag{36}$$

In combination with the specified density values in Table 2, the assignment in Eq. (36) is globally valid. Hence, no further distinction of terrain types is necessary for calculating the rock, water, and ice proportions of the RWI-based topographic potential via Eq. (19).

However, it proves helpful to have a closer look at grid elements over the ocean. Here, R_1^{RWI} defines the sea floor, while $R_2^{\mathrm{RWI}} = R_0$ represents the MSL, as $h^{\mathrm{SUR}} = 0$. According to Eqs. (20) and (21), the tesseroids associated with rock and water masses then have the same thickness and geometrical center, i.e., $|\Delta r_1^{\mathrm{RWI}}| = |\Delta r_2^{\mathrm{RWI}}|$ and $r_1^{\mathrm{RWI}} = r_2^{\mathrm{RWI}}$. Both tesseroids only differ in the sign of their radial dimension and their density values, i.e., ρ_1 and ρ_2. The effect on the potential can be interpreted as follows. In a first step, the water masses of the ocean are removed with ρ_2 (positive effect), while in the second step, the resulting mass deficit is filled with rock of density ρ_1 (negative effect). Therefore, by using the assignment in Eq. (36), the gravitational impact of the oceans depends on both, the rock and the water signal.

As an alternative, the ocean masses can be represented by a single tesseroid using the rock-water density contrast. The resulting effect is then assigned to the water proportion, achieving a more meaningful separation of the impact of the rock and water masses. Moreover, this is also computationally more efficient than evaluating two tesseroids. By using the Earth2014 terrain types tt, this strategy can be implemented by setting

$$\begin{aligned} R_1^{\mathrm{RWI}} &= R_0 + \begin{cases} h^{\mathrm{SUR}}, & \text{if } \; tt = 2, \\ h^{\mathrm{BED}}, & \text{otherwise}, \end{cases} \\ R_2^{\mathrm{RWI}} &= R_0 + \begin{cases} h^{\mathrm{BED}}, & \text{if } \; tt = 2, \\ h^{\mathrm{SUR}} - t^{\mathrm{ICE}}, & \text{otherwise}, \end{cases} \\ R_3^{\mathrm{RWI}} &= R_0 + \begin{cases} h^{\mathrm{BED}}, & \text{if } \; tt = 2, \\ h^{\mathrm{SUR}}, & \text{otherwise}, \end{cases} \end{aligned} \tag{37}$$

and using the following layer-specific density values:

$$\begin{aligned} \rho_1 &= \; 2670\,\mathrm{kg\,m^{-3}}, \\ \rho_2 &= \begin{cases} 1640\,\mathrm{kg\,m^{-3}}, & \text{if } \; tt = 2, \\ 1000\,\mathrm{kg\,m^{-3}}, & \text{if } \; tt = 3 \text{ or } tt = 4 \text{ or } tt = 8, \\ 1030\,\mathrm{kg\,m^{-3}}, & \text{otherwise}, \end{cases} \\ \rho_3 &= \; 917\,\mathrm{kg\,m^{-3}}. \end{aligned} \tag{38}$$

This assignment that is used in the following differs from Eq. (36) only in the separation of the rock, water, and ice masses. Thus, it is important to note that the total RWI-based topographic potential as the sum of the three components according to Eq. (22) is equal for both assignments.

Figure 5 illustrates the RWI-based radial dimensions $\Delta r_s^{\mathrm{RWI}}$ of the rock, water, and ice masses. Additionally, the dimensions of the rock-equivalent masses $\Delta r_s^{\mathrm{REQ}}$ as calculated by Eqs. (8) and (24) are plotted and the corresponding statistics are provided in Table 4.

In the case of the rock masses, as shown in Fig. 5a, the RWI and REQ methods provide the same radial dimensions, i.e., $\Delta r_1^{\mathrm{RWI}} = \Delta r_1^{\mathrm{REQ}}$. They range from -2727 to 8212 m for the continents and have a zero value for the oceans. Positive values can be attributed to rock masses above MSL, while negative values indicate regions where the bedrock surface lies below MSL, e.g., for larger parts of Antarctica and Greenland as well as the major inland lakes.

The radial dimensions of the RWI-based water masses $\Delta r_2^{\mathrm{RWI}}$ are plotted in Fig. 5b. For the oceans, they correspond to the bathymetric depths, having a minimum value of $-10\,847$ m. Moreover, positive values can be associated to ocean water masses under ice shelves or the water masses of the major inland lakes, where a maximum value of 1630 m is reached in the case of Lake Baikal. In contrast, the radial dimensions of the REQ-based water masses $\Delta r_2^{\mathrm{REQ}}$ provide considerably smaller extreme values of -6658 and 611 m (see Table 4). Due to the rock-equivalent condensation, the sea floor of the oceans rises by about 60 %, as can clearly be seen in Fig. 5c. On average, the RWI- and REQ-based water masses differ by about 1000 m in radial dimension.

A similar mass displacement can be observed when comparing the radial dimensions of the RWI- and REQ-based ice masses $\Delta r_3^{\mathrm{RWI}}$ (Fig. 5d) and $\Delta r_3^{\mathrm{REQ}}$ (Fig. 5e). Due to the density contrast between ice and rock, the thickness of the Antarctic ice sheet is reduced by about 30 %. This can also be detected when considering the change in the maximum values of $\Delta r_3^{\mathrm{RWI}}$ and $\Delta r_3^{\mathrm{REQ}}$ (see Table 4).

4. Processing

In this section, the derived formulas for space domain GFM and SHA are applied to the topographic input data as preprocessed in the previous section.

In the first step, the RWI- and REQ-based topographic potential V_s^{RWI} and V_s^{REQ} of each mass layer is calculated. In order to utilize the least-squares SHA approach as described in Sect. 2.4, topographic potential values are needed for each node $P_{ij}(r_i, \varphi_i, \lambda_j)$ of an equally spaced grid as specified in Eq. (30), i.e.,

$$\{V^{\mathrm{RWI}}\}_{ij} := V^{\mathrm{RWI}}(P_{ij}) \quad \text{and} \quad \{V^{\mathrm{REQ}}\}_{ij} := V^{\mathrm{REQ}}(P_{ij}). \tag{39}$$

For this purpose, a grid with $I = 2700$ parallels and $J = 5400$ meridians is used, equivalent to a $4' \times 4'$ spatial resolution. According to Eq. (32), SH coefficients with a maximum degree of 2699 can then be derived, which is just within the numerical stability range of the Holmes and Featherstone (2002) ALF algorithm.

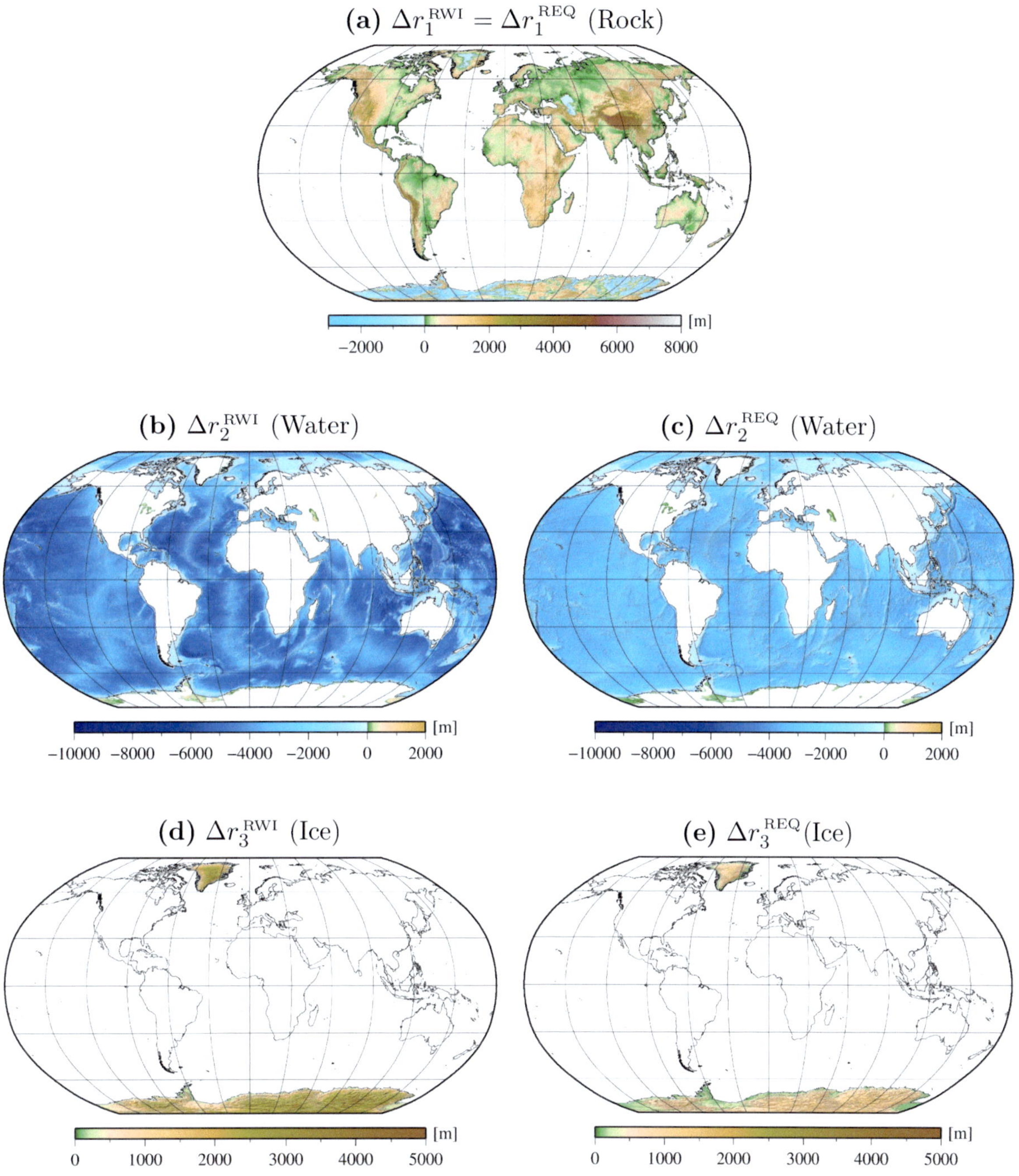

Fig. 5. RWI- and REQ-based radial dimensions of **(a)** the rock masses $\Delta r_1^{\mathrm{RWI}} = \Delta r_1^{\mathrm{REQ}}$, **(b)** and **(c)** the water masses $\Delta r_2^{\mathrm{RWI}}$ and $\Delta r_2^{\mathrm{REQ}}$, and **(d)** and **(e)** the ice masses $\Delta r_3^{\mathrm{RWI}}$ and $\Delta r_3^{\mathrm{REQ}}$. For all plots, the same color bar is used, where grid elements with a zero dimension are indicated by *white color*. Robinson projection centered at 0° longitude.

Table 4. Statistics of the RWI- and REQ-based radial dimensions $\Delta r_s^{\mathrm{RWI}}$ and $\Delta r_s^{\mathrm{REQ}}$ of the rock, water, and ice masses and their differences $\Delta r_s^{\mathrm{RWI}} - \Delta r_s^{\mathrm{REQ}}$. All values are specified in [m].

	Min	Max	Mean	STD
$\Delta r_1^{\mathrm{RWI}} = \Delta r_1^{\mathrm{REQ}}$	−2727	8212	172	532
$\Delta r_2^{\mathrm{RWI}}$	−10 847	1630	−2615	2164
$\Delta r_3^{\mathrm{RWI}}$	0	4613	58	375
$\Delta r_2^{\mathrm{REQ}}$	−6658	611	−1606	1328
$\Delta r_3^{\mathrm{REQ}}$	0	1585	20	129
$\Delta r_2^{\mathrm{RWI}} - \Delta r_2^{\mathrm{REQ}}$	−4189	1019	−1009	835
$\Delta r_3^{\mathrm{RWI}} - \Delta r_3^{\mathrm{REQ}}$	0	3028	38	246

In the following, two variants for the reference surface of the grid are analyzed, a spherical and an ellipsoidal grid. As already done for creating the RWI_TOPO_2012 model (Grombein et al., 2014a), a spherical grid with radius $R = a^{\mathrm{GRS80}} + 20\,\mathrm{km}$ is selected, which ensures that all grid points are safely outside the topographic masses. However, toward the poles, the radial distance between these spherical grid points and the ellipsoidal arranged topography rises up to 40 km.

As an alternative, an ellipsoidal grid is used for the new RWI_TOPO_2015 model. In this case, the semi-major and semi-minor axes of the used ellipsoid are set to $a = a^{\mathrm{GRS80}} + 20\,\mathrm{km}$ and $b = b^{\mathrm{GRS80}} + 20\,\mathrm{km}$, where the second numerical eccentricity as needed for Eq. (31) can be obtained by $e'^2 = (a^2 - b^2)/b^2$. In contrast to the spherical surface, all grid points are approximately 20 km above the GRS80 surface. This places the grid nearer to the topographic masses in polar regions. In order to maintain an equiangular spacing for the SHA, the ellipsoidal grid is also parameterized in terms of geocentric spherical coordinates, thus preventing numerical instabilities as noted by Holmes and Pavlis (2007).

For calculating the RWI- and REQ-based topographic potential values for each mass layer, Eqs. (19) and (23) have to be evaluated for $2700 \times 5400 = 14\,580\,000$ grid points P_{ij}. According to the used topographic input data, for each grid point the gravitational effect of about 265 million tesseroids has to be calculated. This number consists of about 80 million tesseroids associated with rock masses, 160 million with water masses, and 25 million with ice masses (tesseroids with zero radial dimension are not considered).

As the computations for each grid point are independent of each other, techniques of parallel computing are most efficient. Thus, in order to handle the enormous computational demand of the GFM, a software optimized for parallel computing on high-performance computer systems has been developed using the message passing interface (MPI). Using parallel computations on 480 processors (Intel Xeon E5540, 2.53 GHz), a run-time of about

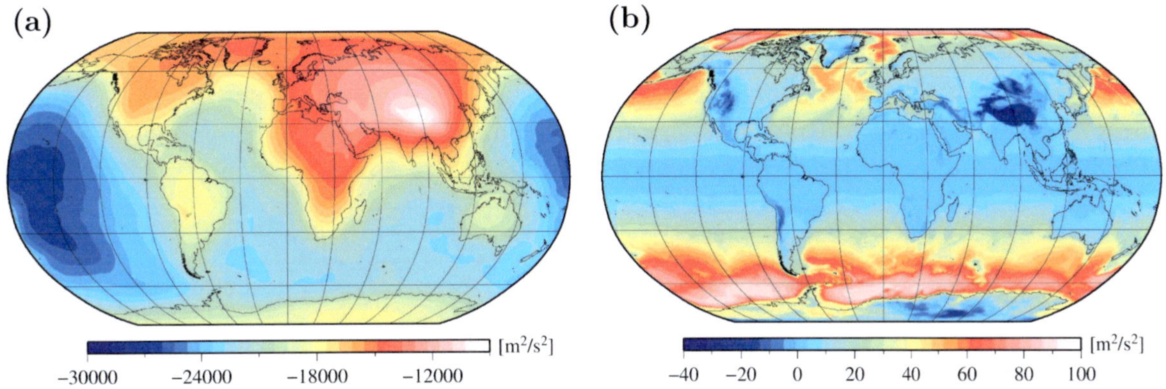

Fig. 6. **(a)** GFM-derived topographic potential values $\{V^{\text{RWI}}\}_{ij}$ calculated at a $4' \times 4'$ spherical grid ($R = a^{\text{GRS80}} + 20\,\text{km}$). **(b)** Differences to topographic potential values calculated at a $4' \times 4'$ ellipsoidal grid ($a = a^{\text{GRS80}} + 20\,\text{km}$, $b = b^{\text{GRS80}} + 20\,\text{km}$). Robinson projection centered at 0° longitude.

10 days is required for the calculation of each grid, spherical or ellipsoidal. Finally, the total RWI- and REQ-based topographic potential values $\{V^{\text{RWI}}\}_{ij}$ and $\{V^{\text{REQ}}\}_{ij}$ are obtained by Eqs. (22) and (26).

Taking the RWI-based topographic potential as an example, the difference in using the spherical and ellipsoidal grid for the SHA is analyzed in the following. In Fig. 6a, the GFM-derived topographic potential values $\{V^{\text{RWI}}\}_{ij}$ are plotted in the case of the spherical grid. The topographic signal is systematically negative and ranges from about $-27\,000$ to $-9000\,\text{m}^2\,\text{s}^{-2}$. It shows mostly long-wavelength structures and is strongly dominated by the influence of the high mountain ranges of the Himalayan region and deep ocean trenches.

In Fig. 6b, the differences with respect to the ellipsoidal grid are illustrated. As both grids have the same geocentric radius for the case of the equator, the differences are nearly zero in this region. Toward the poles, the ellipsoidal grid provides a stronger signal, as its distance to the ellipsoidally arranged topographic masses is smaller. These differences reach extreme values of -40 to $90\,\text{m}^2\,\text{s}^{-2}$ and are negative for the oceans and positive for the continents.

By applying the least-squares SHA approach, both grids are used to obtain solid SH coefficients of the topographic potential up to d/o 2699, denoted as SHC_N2699_Sph_grid and SHC_N2699_Ell_grid. They refer to $R = 6\,378\,136.3\,\text{m}$ and are scaled by GM/R with $GM = 3.986\,004\,415 \cdot 10^{14}\,\text{m}^3\,\text{s}^{-2}$. The software implementation for the SHA used is originally based on Seitz and Heck (1991) and has been further developed.

In order to investigate the differences of both sets of SH coefficients in the frequency domain, their (dimensionless) signal degree variances are evaluated as defined by Eq. (40), see below. As shown in Fig. 7, the coefficients belonging to the spherical and ellipsoidal grid (*blue* and *green curve*) show a strong agreement over most of the spectrum. Although the signal of the coefficient differences (*red curve*) continuously rises with increasing degree n, its order of magnitude is considerably smaller compared to the signals of the absolute values. As illustrated in the close-up, significant differences of both SH coefficients can be detected for the highest degrees ($n > 2670$). While initially the coefficients of the ellipsoidal grid provide slightly more energy, the degree variance abruptly drops by about half an order of magnitude. In contrast, the signal of the spherical grid remains of the same order of magnitude and only slightly decreases for the last few degrees.

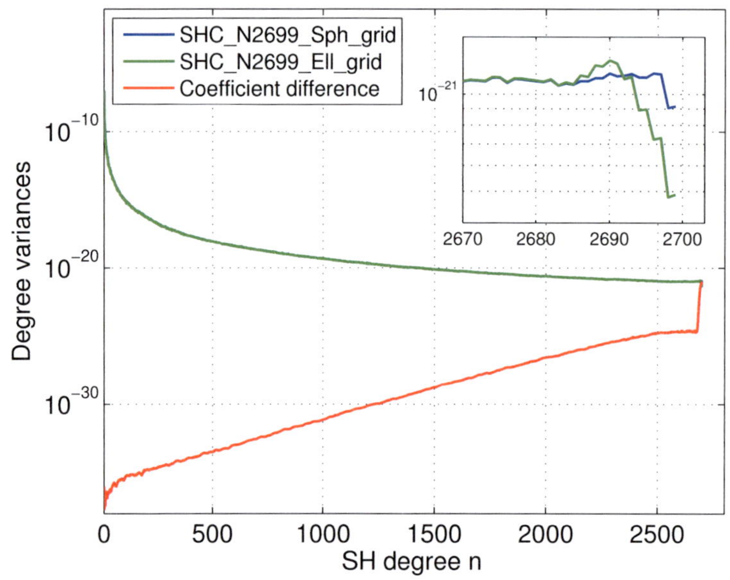

Fig. 7. Degree variances of SHC_N2699_Sph_grid (*blue curve*), SHC_N2699_Ell_grid (*green curve*), and their coefficient difference (*red curve*). Note that the vertical axis is plotted on a logarithmic scale.

To further analyze these differences in the space domain, the SH coefficients are used to derive gravity disturbances $\delta g = -\partial V/\partial r$ based on Eq. (42), see below. The required SHS is performed for the spherical harmonic bands (SH bands) of 721–2670, 2671–2699, and 721–2699. Moreover, the gravity disturbances are evaluated on the surface of the GRS80 ellipsoid as well as on the spherical reference surface of the coefficients with $R = 6\,378\,136.3\,\text{m}$.

For the regional area of Europe, Fig. 8 illustrates the results from the SHS. In the *first column*, gravity disturbances on the GRS80 ellipsoid from SHC_N2699_Ell_grid are plotted. Besides the actual topographic signal as for example induced by the Alps, the plot for the SH band 721–2670 clearly shows artifacts in higher latitudes. These high-frequency oscillations are strongly amplified toward the poles and even reach amplitudes up to ±1700 mGal. The subsequent SH band 2671–2699 generates nearly identical oscillations of opposite sign. Therefore, in the case of the combined SH band 721–2699, these effects are completely removed, which makes the topographic signal also visible for polar latitudes.

In the case of SHC_N2699_Sph_grid (*second column*), this compensation due to the SH band 2671–2699 fails, so that latitude-dependent patterns remain in the complete SH band 721–2699. This clearly illustrates the different effect of the SH band 2671–2699, already visible in terms of the degree variances.

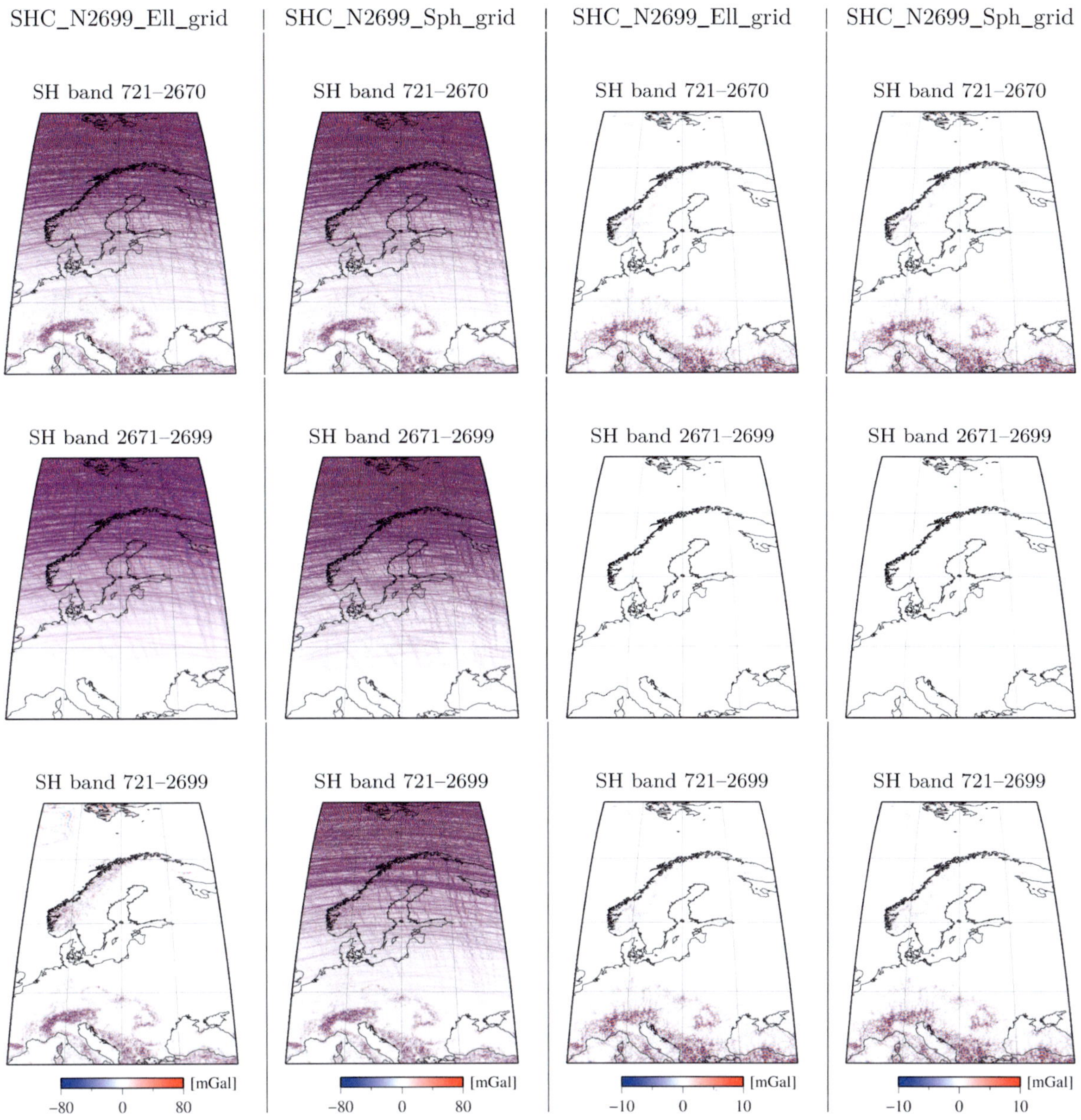

Fig. 8. Gravity disturbances δg from SHC_N2699_Sph_grid and SHC_N2699_Ell_grid evaluated on the surface of the GRS80 ellipsoid (*column* 1–2) and as well as on the the spherical reference surface of the coefficients with $R = 6\,378\,136.3\,\mathrm{m}$ (*column* 3–4). Gravity disturbances δg are presented for the SH bands 721–2670 (*row* 1), 2671–2699 (*row* 2), and 721–2699 (*row* 3), in the area of Europe ($40° \leq \varphi \leq 80°$, $0° \leq \lambda \leq 40°$). Robinson projection centered at 20° longitude; meridians and parallels are 10° apart.

In contrast to the GRS80 ellipsoid, both sets of SH coefficients produce nearly the same results, when they are evaluated on the spherical reference surface (*third* and *fourth column*). In this case, no oscillations occur and the SH band 2671–2699 does not add any significant signal.

In fact, the detected oscillations are not restricted to the European area as shown in Fig. 8. Generally, they crisscross over both polar regions. Similar effects have also been observed in several publications (e.g., Holmes and Pavlis, 2007; Claessens and Hirt, 2013) and can be explained as follows:

As the field generating topographic masses are arranged on an ellipsoidal reference surface as defined by Eq. (3), the GFM-derived topographic potential values essentially generate an ellipsoidal harmonic model. Due to the applied SHA approach, this ellipsoidal model is then mapped to the spherical spectrum of spherical harmonics, which are basically only valid outside the used reference sphere with $R = 6\,378\,136.3\,\mathrm{m}$.

When then evaluating these SH coefficients inside the reference sphere as in the case of the GRS80 ellipsoid where $r < R$, the high-frequency oscillations that occur can be associated with an erroneous downward continuation. This downward continuation is regulated by the term $(R/r)^n$ in Eq. (42), showing that the effect (i) increases toward the poles and (ii) is strongly amplified for higher degrees n.

While these downward continuation errors can be compensated by the coefficients of higher degrees when using the set obtained from the ellipsoidal grid (SHC_N2699_Ell_grid), this is not the case for the coefficients gained from the spherical grid (SHC_N2699_Sph_grid).

However, this means that similar to the Hotine–Jekeli transformation between ellipsoidal and spherical harmonic coefficients (Jekeli, 1988), the resulting SH coefficients (SHC_N2699_Ell_grid) of higher degrees are correlated with those of lower degrees. Therefore, to accurately represent the topographic potential on the GRS80 ellipsoid, the SH coefficients need to be evaluated up to its maximum degree $N_{\mathrm{max}} = 2699$ and should not be truncated. From a practical point of view, however, it seems not to be feasible for all applications (or users) to perform a SHS up to such ultra-high degrees.

The investigation is repeated for the same $4' \times 4'$ ellipsoidal grid of topographic potential values, but restricting the SHA up to d/o 2190. By using the resulting degree-2190 SH coefficients, Fig. 9 shows comparable results to Fig. 8, now using the SH bands 721–2160, 2160–2190, and 721–2190. Analogously, high-latitude oscillations of about $\pm 100\,\mathrm{mGal}$ occur in the SH band 721–2160 that are compensated by the SH band 2160–2190, so that the combined SH band 721–2190 is free from any error. Thus, the same conclusions hold true for the case of the SHA up to d/o 2190.

In fact, the last few degrees generally seem to provide crucial information for an accurate SHS on the GRS80 ellipsoid, regardless up to which maximum d/o the SH coefficients are derived. Therefore, the coefficients should be evaluated up to their full resolution. Theoretically, this means that a separate set of SH coefficients would be needed for an accurate SHS up to a specific d/o, which is not practicable. As a compromise, the SH coefficients up to d/o 2190 are selected, thus matching the resolution of EGM2008 (Pavlis et al., 2007) and dV_ELL_RET2014 (Hirt et al., 2015). Note that both of these models are affected by the same issue and hence also need to be evaluated up to their full resolution.

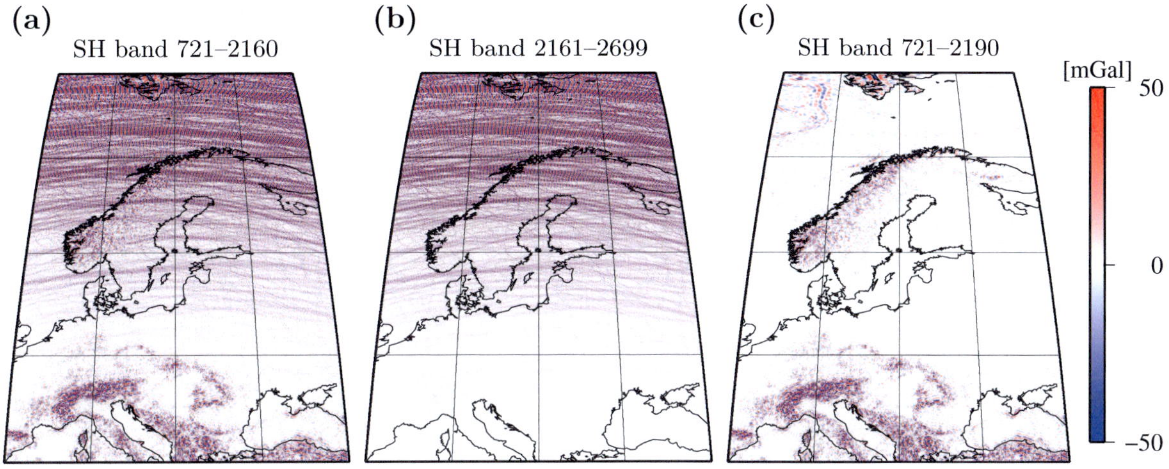

Fig. 9. Gravity disturbances δg from SHC_N2190_Ell_grid evaluated on the surface of the GRS80 ellipsoid. Gravity disturbances δg are presented for the SH bands **(a)** 721–2160, **(b)** 2161–2190, and **(c)** 721–2190, in the area of Europe ($40° \leq \varphi \leq 80°$, $0° \leq \lambda \leq 40°$). Robinson projection centered at 20° longitude; meridians and parallels are 10° apart.

Consequently, based on the $4' \times 4'$ ellipsoidal grids, the following sets of SH coefficients up to d/o 2190 are generated for the RWI model

- RWI_TOPO_2015 (RWI-based topographic potential)
- RWI_TOPO_2015_Rock (rock proportion)
- RWI_TOPO_2015_Water (water proportion)
- RWI_TOPO_2015_Ice (ice proportion)

and the REQ model

- REQ_TOPO_2015 (REQ-based topographic potential)
- REQ_TOPO_2015_Water (condensed water proportion)
- REQ_TOPO_2015_Ice (condensed ice proportion)

A separate set of SH coefficients for the REQ-based rock proportion is not defined as it is identical to RWI_TOPO_2015_Rock.

5. Results and discussion

In this section, the SH coefficients of the RWI_TOPO_2015 model are analyzed and compared to the coefficients of the previous version RWI_TOPO_2012 and the rock-equivalent version REQ_TOPO_2015. For this purpose, the (dimensionless) signal degree variances

$$\sigma_n^2\left(\bar{C}_{nm}, \bar{S}_{nm}\right) = \sum_{m=0}^{n} \left(\bar{C}_{nm}^2 + \bar{S}_{nm}^2\right) \tag{40}$$

are used in the frequency domain. Furthermore, to analyze effects in the space domain, the SH coefficients are used to derive global grids of different gravity field functionals by SHS. Being representative for various terrestrial and satellite-based applications, the following three commonly used functionals are evaluated:

1. The height anomaly that is derived by dividing the potential V according to Eq. (2) by the normal gravity γ

$$\zeta = \frac{GM}{r\gamma} \sum_{n=0}^{N} \left(\frac{R}{r}\right)^n \sum_{m=0}^{n} \left(\bar{C}_{nm}\, \bar{Y}_{nm}^C + \bar{S}_{nm}\, \bar{Y}_{nm}^S\right). \tag{41}$$

2. The first radial derivative of the potential $-\partial V/\partial r$, known as gravity disturbance

$$\delta g = \frac{GM}{r^2} \sum_{n=0}^{N} (n+1) \left(\frac{R}{r}\right)^n \sum_{m=0}^{n} \left(\bar{C}_{nm}\, \bar{Y}_{nm}^C + \bar{S}_{nm}\, \bar{Y}_{nm}^S\right). \tag{42}$$

3. The second radial derivative of the potential $\partial^2 V/\partial r^2$, called the radial-radial gravity gradient

$$M_{33} = \frac{GM}{r^3} \sum_{n=0}^{N} (n+1)(n+2) \left(\frac{R}{r}\right)^n \sum_{m=0}^{n} \left(\bar{C}_{nm}\, \bar{Y}_{nm}^C + \bar{S}_{nm}\, \bar{Y}_{nm}^S\right). \tag{43}$$

The height anomaly ζ and the gravity disturbance δg, mostly related to terrestrial applications, are evaluated on a $5' \times 5'$ grid on the surface of the GRS80 ellipsoid. The gravity gradient M_{33} as measured by the GOCE gravity field mission is evaluated on a $5' \times 5'$ spherical grid at a mean satellite altitude of 254.9 km. The SHS according to Eqs. (41) – (43) is carried out by using an adapted version of the freely available *harmonic_synth* software (Holmes and Pavlis, 2006) that also utilizes the ALF algorithm based on Holmes and Featherstone (2002).

While in the following section the analysis and discussion exemplarily concentrates on the gravity disturbance δg, the additional results for the height anomaly ζ and the gravity gradient M_{33} are presented in the Appendix.

5.1. RWI_TOPO_2015

In Fig. 10, the spectral energy of the RWI_TOPO_2015 SH coefficients is shown by means of degree variances. With increasing degree n, the energy of the topographic signal (*red curve*) decreases from about 10^{-7} to 10^{-21}.

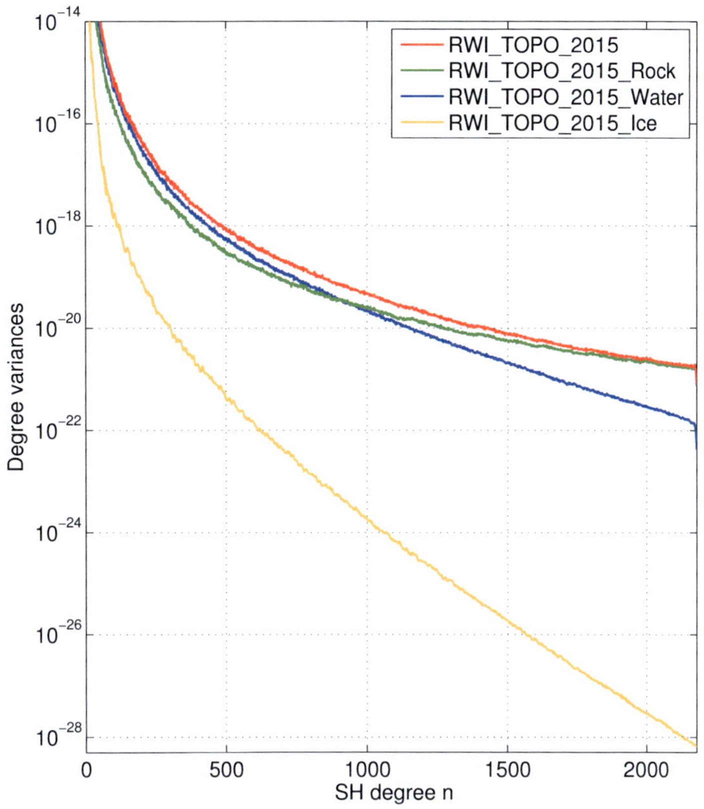

Fig. 10. Degree variances of RWI_TOPO_2015 (*red curve*), RWI_TOPO_2015_Rock (*green curve*), RWI_TOPO_2015_Water (*blue curve*), and RWI_TOPO_2015_Ice (*yellow curve*). Note that the vertical axis is plotted on a logarithmic scale.

As the new RWI_TOPO_2015 model offers separate SH coefficients for the rock, water, and ice proportion, there is the opportunity to have a more detailed look on the composition of the RWI-based topographic signal. Therefore, Fig. 10 additionally provides the corresponding degree variances for the SH coefficients of the rock, water, and ice components.

While for lower degrees ($n < 500$) the topographic signal is mainly dominated by the water proportion (*blue curve*), the influence of the rock signal (*green curve*) becomes dominant for degrees higher than about 900. For $n > 2000$, the topographic signal more or less coincides with the rock signal, while the water signal is about one order of magnitude smaller. This characteristic can be explained by the mass distribution of the Earth's rock and water masses as well as their spatial arrangement. On the one hand, there is much more water than rock masses, which results in a stronger water signal in the case of lower frequencies. On the other hand, the roughness of the oceanic water masses induced by the sea floor is much farer away from the used reference surface than the most highly variable rock masses. Therefore, a considerably smoother signal is induced by the water masses, resulting in a lower energy at higher spectral scales.

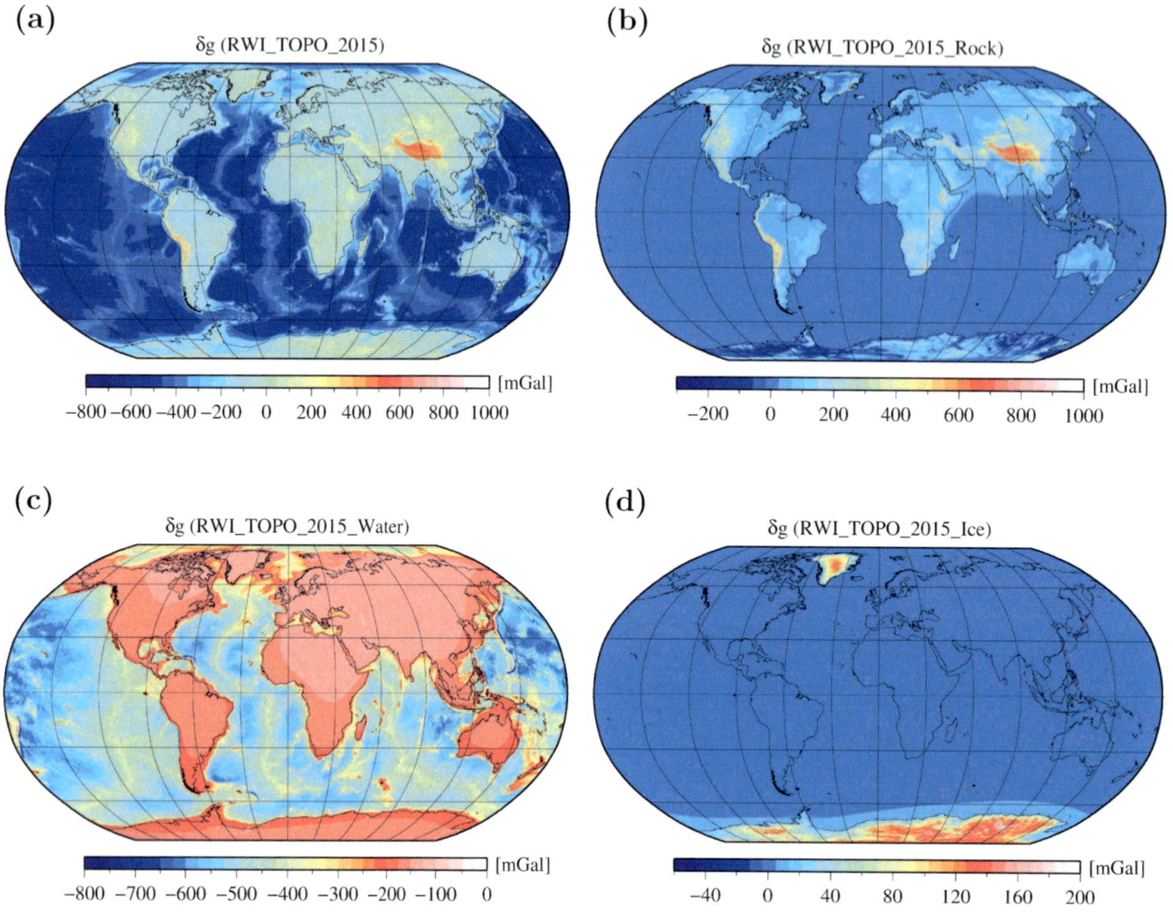

Fig. 11. Topographic signal of **(a)** RWI_TOPO_2015, **(b)** RWI_TOPO_2015_Rock, **(c)** RWI_TOPO_2015_Water, and **(d)** RWI_TOPO_2015_Ice in terms of gravity disturbances δg evaluated on the surface of the GRS80 ellipsoid. Robinson projection centered at 0° longitude.

The ice proportion (*yellow curve*) only significantly contributes to the very low degrees, since its energy strongly decreases from 10^{-7} to 10^{-26}. At degree $n = 500$, the ice signal is already four orders of magnitude smaller than the topographic signal; for $n > 2000$, this gap even reaches seven orders of magnitude. Besides the actual mass distribution of the Earth's rock, water, and ice proportions, the behavior of the degree variances may also reflect the influence of differing spatial resolutions of the input topography for various regions, e.g., a lower resolution for most parts of the oceans, Antarctica, and Greenland.

In Fig. 11a, the topographic signal of the RWI_TOPO_2015 model is shown in terms of the gravity disturbance δg. Moreover, the corresponding statistics are presented in Table 5. The topographic signal ranges from −804.425 to 1014.430 mGal with a mean value of −317.158 mGal and a standard deviation of 207.874 mGal ($1\,\text{mGal} = 10^{-5}\,\text{m}\,\text{s}^{-2}$). The signal clearly shows the influence of the Earth's main topographic features like continental mountain ranges (positive values) or oceanic ridges and trenches (negative values). Besides the signal peaks induced by the Himalaya and the Andes, also smaller mountain ranges like the European Alps or the mid-oceanic ridges can be clearly distinguished. The minimum values are caused by the Mariana Trench located in the western Pacific Ocean and the Tonga trench north of New Zealand.

Table 5. Statistics of the RWI_TOPO_2015 topographic signal and its rock, water, and ice components in terms of gravity disturbances δg evaluated on the surface of the GRS80 ellipsoid. All values are specified in [mGal].

	Min	Max	Mean	STD
RWI	−804.425	1014.430	−317.158	207.874
Rock	−258.245	1153.033	38.695	65.329
Water	−818.907	−67.594	−360.327	168.477
Ice	−11.587	180.322	4.474	15.756

Analogous to the frequency domain representation, Fig. 11b–d separately illustrates the rock, water, and ice proportions within δg, which are denoted by δg_1, δg_2, and δg_3, respectively. The rock proportion δg_1 exhibits extreme values of −258.245 and 1153.033 mGal with a mean value of 38.695 mGal and a standard deviation of 65.329 mGal (cf. Table 5). While positive values can be associated with rock masses above MSL, negative values indicate mass deficits below MSL. This can nicely be seen in Antarctica, where the bedrock surface, i.e., the interface between the rock and ice mass layer, is partially located above as well as below MSL, generating a highly variable signal ranging from about −200 to 500 mGal. As mainly affected by the mass deficits of the oceans, the water proportion δg_2 provides a systematic negative contribution to the topographic signal that ranges from −818.907 to −67.594 mGal with a mean value of −360.327 mGal and a standard deviation of 168.477 mGal. The maximum value is reached in the area of Lake Baikal.

Having a closer look on Fig. 11b and 11c, it can be seen that the effects of rock and water masses are not restricted to the continents and oceans, but considerably superpose each other. For example, over the Earth's continents, the water proportion δg_2 still causes long-wavelength structures with signal strengths of about −200 to −130 mGal. This also explains the discrepancy of about 140 mGal between the maximum values of the total topographic and rock signal in Table 5. Similarly, due to the large positive influence of the Himalayan region, the rock proportion δg_1 also contributes up to 40 mGal in the northern part of the Indian ocean.

In contrast to rock and water, the effect of the ice constituent δg_3 (Fig. 11d) is basically limited to Greenland and Antarctica and has no significant influence on other regions. It provides considerably smaller magnitudes with a maximum value of 180.322 mGal, a mean value of 4.474 mGal and a standard deviation of 15.756 mGal.

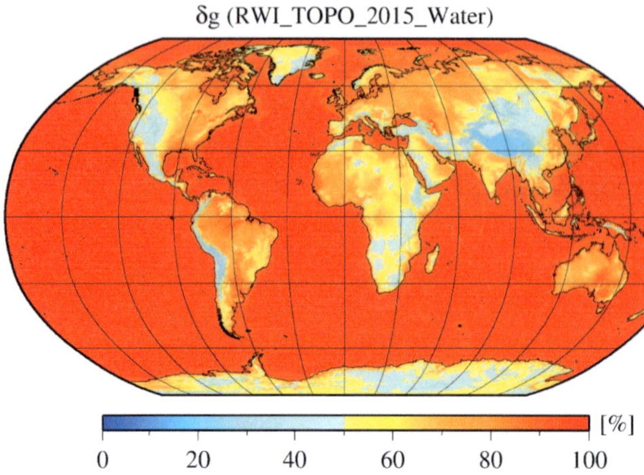

Fig. 12. Percentage contribution p_2 of the water proportion to the RWI-based topographic signal. Robinson projection centered at 0° longitude.

To quantify the impact of the rock, water, and ice proportions, their percentage contribution to the RWI-based topographic signal per grid element is determined by

$$p_s := \frac{|\delta g_s|}{\sum\limits_{t=1}^{3} |\delta g_t|}. \tag{44}$$

In Fig. 12, the contribution p_2 is plotted globally, showing that the RWI-based topographic signal is strongly dominated by the water signal. With the exception of regions with high continental mountain ranges or larger ice sheets, the contribution of the water can globally be estimated to be above 50 %. For further analysis, Table 6 provides mean values $\overline{p}_i$ averaged over different areas (global, land, oceans, Antarctica). On average, the water proportion contributes 86 % to the topographic signal, while the rock and ice masses have only a smaller impact of 12 % and 1 %, respectively. When analyzing these values for land and oceans separately, the dominating effect of the water proportion becomes more visible. While for the oceans, it reaches nearly 100 %, its influence over land is still about twice as large as the rock signal (2/3 water to 1/3 rock signal).

Table 6. Mean percentage contributions $\overline{p}_s$ of the rock, water, and ice proportion averaged over different areas. All values are specified in [%].

	Global	Land	Oceans	Antarctica
$\overline{p}_1$	12.4	32.5	4.2	18.1
$\overline{p}_2$	86.2	64.2	95.3	56.3
$\overline{p}_3$	1.3	3.3	0.5	25.6

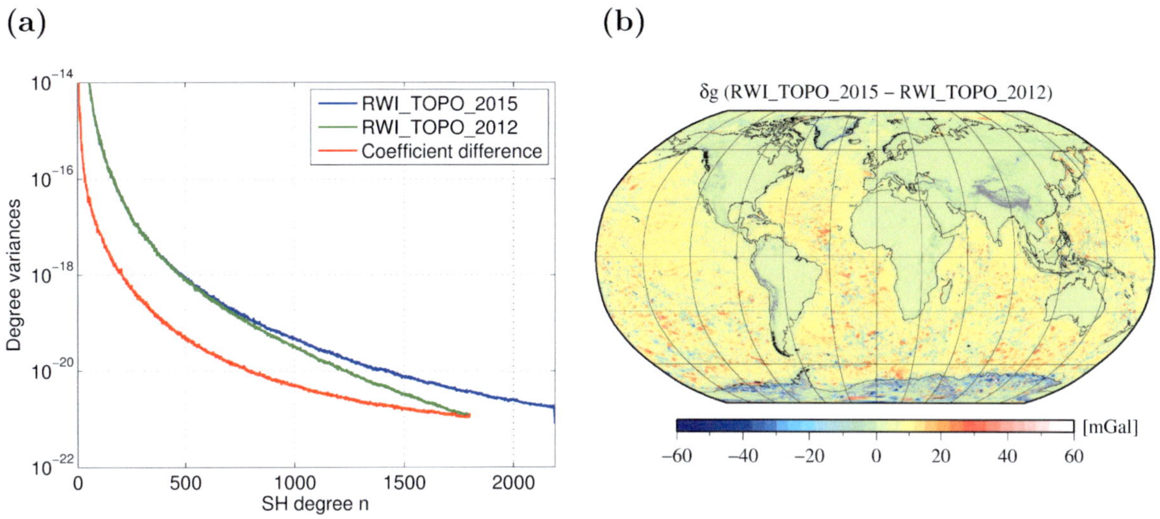

Fig. 13. **(a)** Degree variances of RWI_TOPO_2015 (*blue curve*), RWI_TOPO_2012 (*green curve*), and their coefficient difference (*red curve*). Note that the vertical axis is plotted on a logarithmic scale. **(b)** Difference of the RWI_TOPO_2015 and RWI_TOPO_2012 topographic signal in terms of gravity disturbances δg evaluated on the surface of the GRS80 ellipsoid. Robinson projection centered at 0° longitude.

5.2. Comparison to RWI_TOPO_2012

In order to see the improvement of the new topographic gravity field model, it is compared to its predecessor model RWI_TOPO_2012 (Grombein et al., 2014a), in both frequency and space domains.

In Fig. 13a, the degree variances of the RWI_TOPO_2015 model (*blue curve*), its predecessor model (*green curve*), and the coefficient difference of both versions (*red curve*) is shown. With increasing degree n, the model of 2012 loses more spectral energy than the new model. At the maximum degree ($N = 1800$), its energy is almost half an order of magnitude smaller and in the same range as the coefficient difference. The main reason for this behavior can be seen in the different spatial resolution of the used topographic input data ($5' \times 5'$ DTM2006.0 vs. $1' \times 1'$ Earth2014).

In the space domain, the comparison between both models allows a more differentiated analysis (see Fig. 13b; Table 7). Note that both models are evaluated up to their maximum degree, with $N = 1800$ for the 2012 model and $N = 2190$ for the 2015 model. The differences reach amplitudes of about ±400 mGal with a mean value of 5.981 mGal and a standard deviation of 9.914 mGal. The maximum values can be detected in regions with highly variable topography, such as the Himalaya, the Andes, and the Rocky Mountains. This is a consequence of both, the refined spatial resolution of the used topographic data and the increased maximum degree of the new RWI model. In contrast, for most of the remaining continental parts (except Greenland and Antarctica), the differences are nearly zero as the topographic model of both RWI versions is based on the same SRTM data.

Table 7. Statistics of the difference of RWI_TOPO_2015 and RWI_TOPO_2012 in terms of gravity disturbances δg evaluated on the surface of the GRS80 ellipsoid. All values are specified in [mGal].

Min	Max	Mean	STD
−399.640	383.188	5.981	9.914

For the oceans, the differences are in a range of ±200 mGal and basically reflect the regionally refined bathymetric depths of the Earth2014 topographic model (cf. Fig. 4 in Sect. 4). Moreover, the global mean value of the difference is strongly affected by a systematic positive offset of about 8 to 10 mGal that can be observed in the oceans. This is a consequence of the modified ocean water density from $1000\,\mathrm{kg\,m^{-3}}$ to $1030\,\mathrm{kg\,m^{-3}}$ (cf. Table 2). The newly determined bedrock surface in Antarctica and Greenland incorporated in the Earth2014 causes considerable differences with maximum values up to 350 mGal.

5.3. Comparison to REQ_TOPO_2015

In order to realize the significance of the more sophisticated RWI approach, the new RWI_TOPO_2015 model is compared to the consistent REQ_TOPO_2015 version based on a conventional rock-equivalent approach.

As can be seen in Fig. 14a, the degree variances of both models are in good agreement and show nearly the same behavior (*blue* and *green curve*). The signal of the coefficient difference (*red curve*) is about two to three orders of magnitude smaller. For the mid- and high-frequencies between degree 500 and 1700, the SH coefficients of the rock-equivalent version provide slightly more energy. This effect is due to the geometry changes induced by the rock-equivalent condensation and can be explained more accurately when looking at the degree variances for the water and ice constituents separately, see Fig. 14b–c.

While the water REQ-signal generates more energy than the corresponding water RWI-signal for the mid- to high-frequencies, this is contrasted in the case of the ice signal. Due to the rock-equivalent condensation, the sea floor is effectively moved upward (cf. Fig. 5 in Sect. 2), which shifts the masses closer to the used reference surface. Due to the distance dependency of gravity, a stronger signal is induced, so that the water proportion provides more energy. In contrast, in the case of the ice masses, the condensation causes a larger distance of these masses to the reference surface. As the water signal has more influence than the ice signal, the REQ-based topographic signal provides more energy than the RWI-based one.

The space domain difference of the RWI and REQ model, as plotted in Fig. 14d, shows the effect of the geometry changes and mass displacements. The magnitude of these differences strongly depends on the ratio between the actual mass density (water or ice) and the constant rock density value used for the condensation. The differences are nearly symmetrically distributed around the zero mean value and have a standard deviation

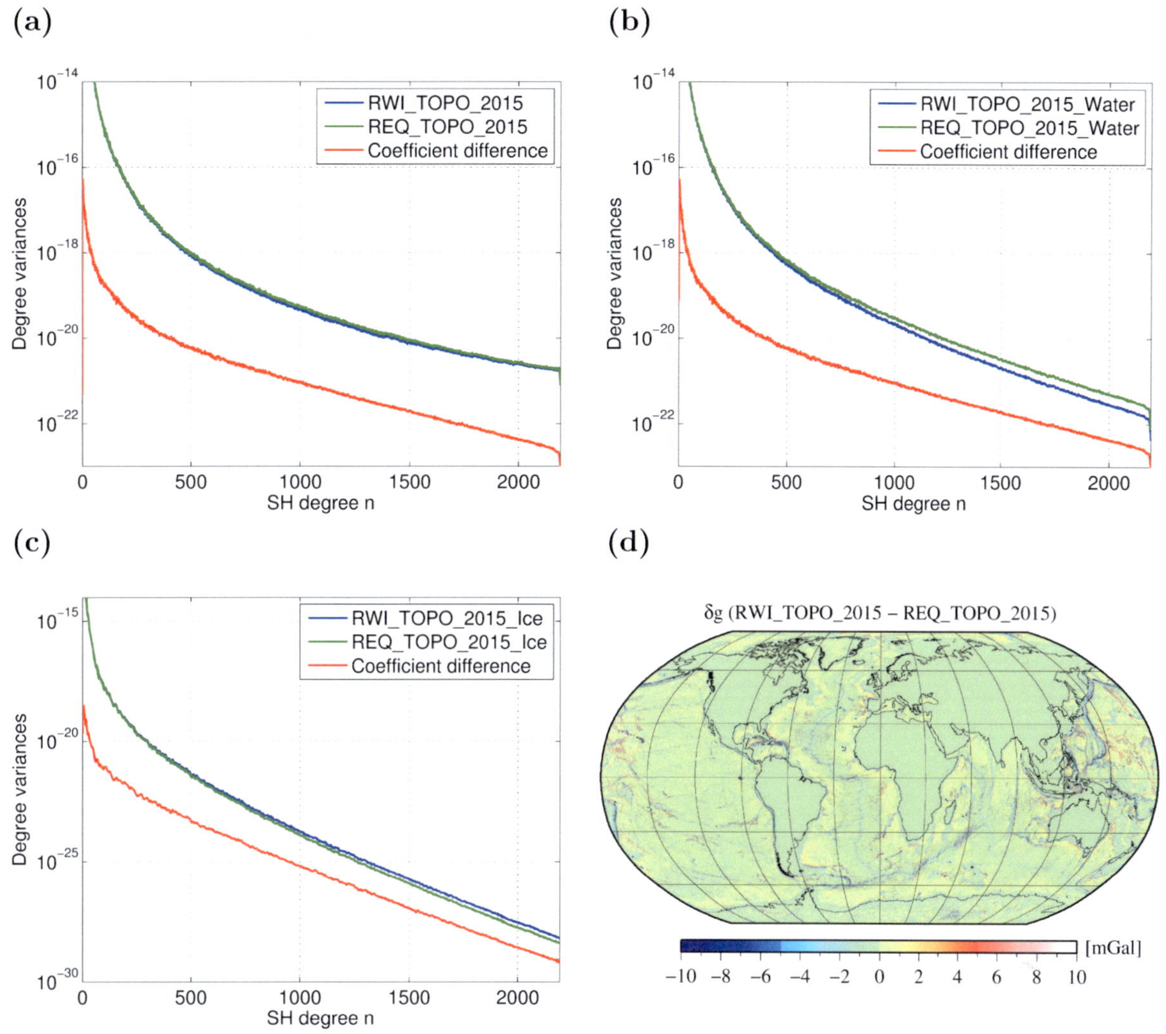

Fig. 14. Degree variances of RWI_TOPO_2015 (*blue curve*), REQ_TOPO_2015 (*green curve*), and their coefficient difference (*red curve*) shown for **(a)** the topographic signal, **(b)** the water proportion, and **(c)** the ice proportion. Note that the vertical axis is plotted on a logarithmic scale. **(d)** Difference of the RWI_TOPO_2015 and REQ_TOPO_2015 topographic signal in terms of gravity disturbances δg evaluated on the surface of the GRS80 ellipsoid. Robinson projection centered at 0° longitude.

of 1.916 mGal (cf. Table 8). While the extreme values of −46.723 and 36.410 mGal are attained in the oceans, the ice condensation in Greenland and Antarctica causes maximum amplitudes of about ±15 mGal. The condensation effect of the ocean water masses also causes differences of a few milligal in coastal areas, up to −10 mGal in case of Big Island (Hawaii) that is surrounded by a deep ocean.

In order to estimate the effect of the RWI–REQ difference for different kind of terrestrial and satellite-based applications, it is calculated as a function of the ellipsoidal height h

Table 8. Statistics of the difference of the RWI_TOPO_2015 and REQ_TOPO_2015 topographic signal and their water and ice proportions in terms of gravity disturbances δg evaluated on the surface of the GRS80 ellipsoid. All values are specified in [mGal].

	Min	Max	Mean	STD
RWI – REQ	−46.723	36.410	0.000	1.916
RWI – REQ (Water)	−46.728	36.392	0.000	1.899
RWI – REQ (Ice)	−14.252	16.649	0.000	0.250

above the GRS80 ellipsoid, ranging from 0 to 600 km in steps of 5 km. For this interval, Fig. 15 illustrates the behavior of the range and the standard deviation of the RWI–REQ difference for the height anomaly ζ, the gravity disturbance δg, and the gravity gradient M_{33}. As the differences are nearly symmetrically centered on zero in the case of all functionals, the range value also provides an impression of the maximum amplitudes.

In Fig. 15a, the range is plotted for the height anomaly ζ. While a maximum range of 1.2 m is reached on the GRS80 surface ($h = 0$ km), it strongly decreases to about 0.4 m at $h = 100$ km. From there on, the range of the RWI–REQ difference stays at nearly the same order of magnitude of a few decimeters. The standard deviation of the height anomaly ζ as shown in Fig. 15b decreases relatively uniformly from 13 to 8.5 cm. Generally, it can be seen that the RWI–REQ difference has a large influence on the height anomaly ζ compared to the goal of cm to mm accuracy in different applications.

In the case of the gravity disturbance δg (Fig. 15c) and the gravity gradient M_{33} (Fig. 15d), the range and standard deviation are both shown in the same panel. Furthermore, they are plotted on a logarithmic scale. For δg, the range (*blue curve*) and standard deviation (*green curve*) show nearly the same decreasing behavior for an increasing ellipsoidal height. For the GRS80 surface and lower altitudes up to a few kilometers, the range and standard deviation of the RWI–REQ difference stay above the level of 1 mGal. Thus, the difference provides significantly larger magnitudes compared to the accuracy of a few microgal in the case of terrestrial and airborne gravimetry applications. Even in the case of a GRACE (Gravity Recovery And Climate Experiment) satellite orbit of about 450 – 500 km, the differences still reach magnitudes of $3 \cdot 10^{-1}$ mGal (range) and $7 \cdot 10^{-2}$ mGal (standard deviation).

For the gravity gradient M_{33} as illustrated in Fig. 15d, range and standard deviation again show a comparable behavior. Considering a GOCE satellite altitude of about 250 km, the RWI–REQ difference provides a range of 45 mE (*blue curve*) and a standard deviation of 4 mE (*green curve*, $1\,\mathrm{mE} = 10^{-12}\,\mathrm{s}^{-2}$). Not before $h = 600$ km, the standard deviation reaches the level of the GOCE measurement accuracy of 1 mE. Therefore, as already mentioned in Grombein et al. (2010, 2014a), the difference between the RWI and REQ method can reach significant amplitudes even at the GOCE satellite orbit.

To summarize the findings obtained from Fig. 15, it should be noted that the RWI–REQ difference reaches significant magnitudes in the case of various terrestrial, airborne and satellite-based applications. Therefore, a more sophisticated modeling as provided by the RWI approach should be considered when calculating topographic effects.

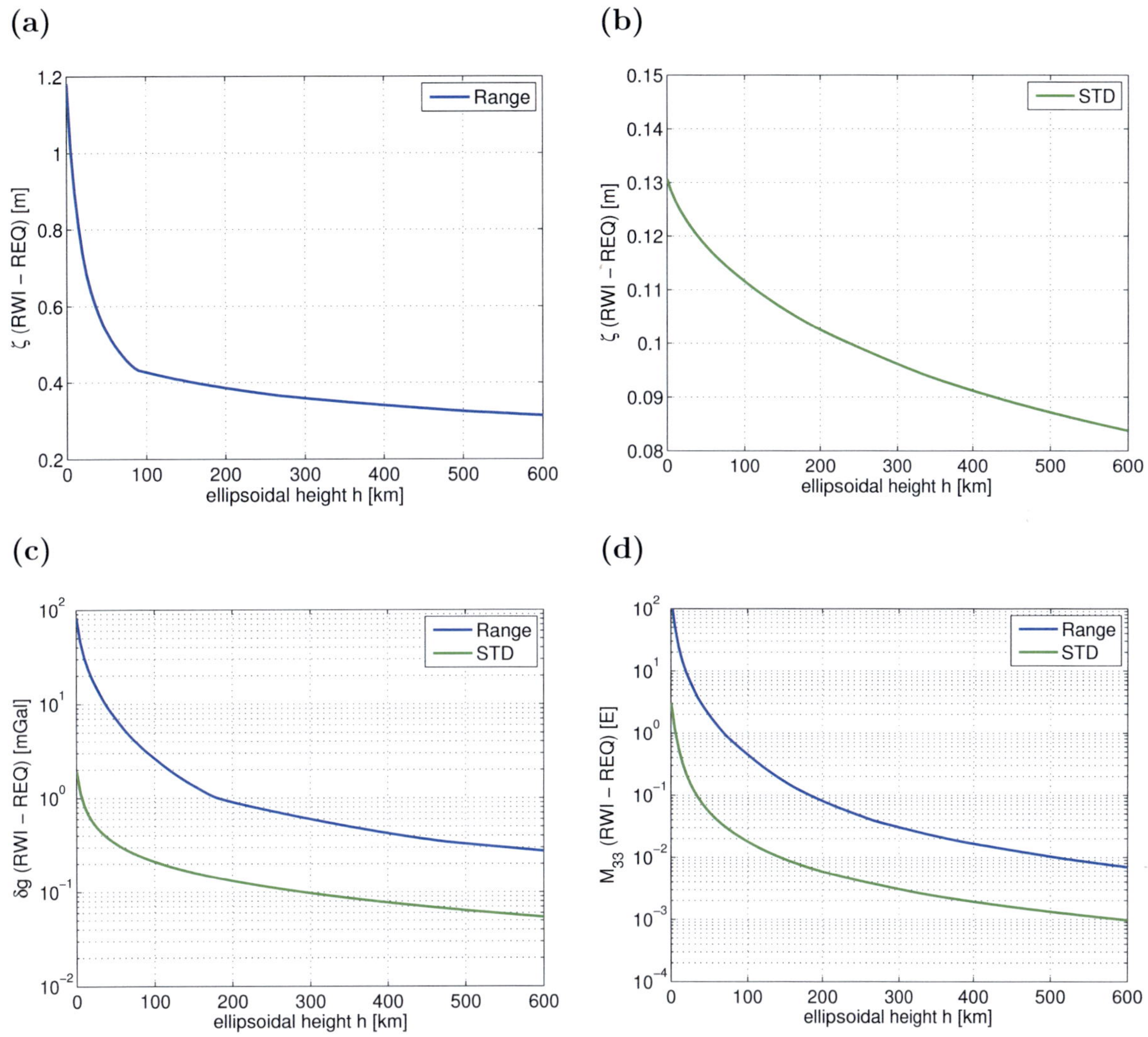

Fig. 15. Difference of the RWI_TOPO_2015 and REQ_TOPO_2015 model in dependence of the ellipsoidal height h above the GRS80 ellipsoid. Shown are the range and standard deviation (STD) of the difference for **(a)**–**(b)** the height anomaly ζ, **(c)** the gravity disturbance δg, and **(d)** the gravity gradient M_{33}.

6. Validation

In order to assess their performance, the new RWI_TOPO_2015 model, its predecessor model RWI_TOPO_2012, and the rock-equivalent version REQ_TOPO_2015 are validated against the gravity information of current GGMs. For this validation, the high-resolution

EGM2008 model (Pavlis et al., 2012) and the latest GOCE GGM based on the time-wise approach GOCE TIM R5 model (Brockmann et al., 2014) are used.

EGM2008 combines satellite-based gravity information of the GRACE satellite mission with terrestrial, airborne and altimetry-derived gravity data that are partially supplemented with topography-implied gravity information. In contrast, GOCE TIM R5 is a satellite-only GGM providing homogeneous and independent gravity information. This model is based on the observations of the complete GOCE mission lifetime and supplies SH coefficients up to d/o 280.

To compare the performance of different global models, a wide range of assessment tools have been developed (e.g., Tscherning, 1985; Rummel et al., 1988; Hirt et al., 2012; Tsoulis and Patlakis, 2013). A commonly used performance indicator is the degree correlation (cf. Rummel et al., 1988, p. 21) that allows a direct comparison between two models in the frequency domain. However, as noted by Tscherning (1985) and Tsoulis and Patlakis (2013), the correlation is not suitable to detect different scale factors in the models.

To overcome this disadvantage, the percentage agreement is introduced as

$$PA(n) = \frac{\sigma_n\left(\overline{C}_{nm}^{\,\mathrm{GGM}}, \overline{S}_{nm}^{\,\mathrm{GGM}}\right) - \sigma_n\left(\overline{C}_{nm}^{\,\mathrm{GGM}} - \overline{C}_{nm}^{\,\mathrm{Topo}}, \overline{S}_{nm}^{\,\mathrm{GGM}} - \overline{S}_{nm}^{\,\mathrm{Topo}}\right)}{\sigma_n\left(\overline{C}_{nm}^{\,\mathrm{GGM}}, \overline{S}_{nm}^{\,\mathrm{GGM}}\right)}, \tag{45}$$

where $(\overline{C}_{nm}^{\,\mathrm{Topo}}, \overline{S}_{nm}^{\,\mathrm{Topo}})$ are the SH coefficients of the topographic gravity field model to be evaluated, $(\overline{C}_{nm}^{\,\mathrm{GGM}}, \overline{S}_{nm}^{\,\mathrm{GGM}})$ those of the reference GGM, and σ_n denotes the square root of the coefficients' degree variances according to Eq. (40). This indicator is the complement of the percentage difference as used by Rummel et al. (1988, p. 20) and is similar to the smoothing per degree as defined by Tscherning (1985).

In Fig. 16, the percentage agreement of RWI_TOPO_2012, RWI_TOPO_2015, and REQ_TOPO_2015 with respect to EGM2008 is illustrated. As the topographic potential does not contain the gravitational impact of the whole Earth's mass, no agreement between the topographic gravity field models and EGM2008 can be detected for lower frequencies ($n < 100$). In the subsequent SH band, the percentage agreement of all three models continuously rises and reaches a level of about 55 % at degree $n = 500$; only some minor differences between the models can be seen.

At degree $n = 900$, a considerable jump to higher agreements above 60 % can bee seen. This can be explained by the compilation of the gravity data set used for generating EGM2008. For larger parts of the continental areas, the available gravity data was of proprietary nature. For this reason, they were only used up to d/o 900 and supplemented by topography-derived gravity values beyond this degree (cf. Pavlis et al., 2013). Therefore, this jump, which is also visible in the degree variances of EGM2008, nicely marks the beginning of topographic information incorporated into EGM2008.

Beyond $n = 900$, the percentage agreements start to diverge and significant differences are visible for the three models. In contrast to both RWI models, the REQ version provides a considerable lower agreement to EGM2008. This is remarkable, since EGM2008 only uses topography-derived gravity values over some land areas, where both approaches (RWI and REQ) provide the same signal. Therefore, the higher agreement suggests a better fit

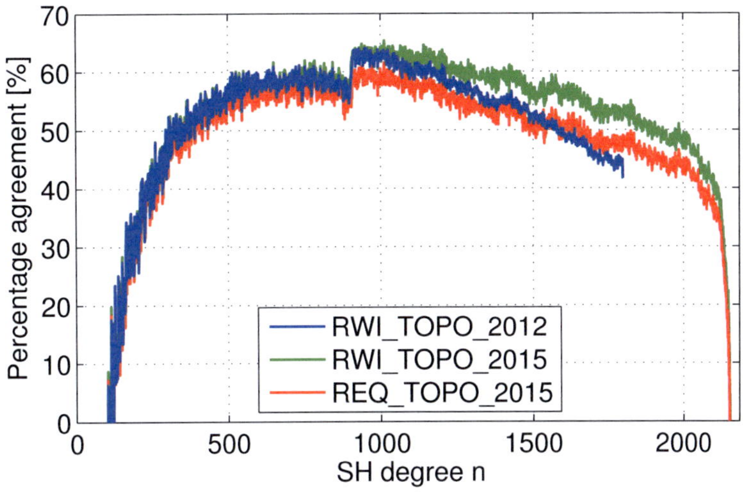

Fig. 16. Percentage agreement $PA(n)$ of RWI_TOPO_2012 (*blue curve*), RWI_TOPO_2015 (*green curve*), and REQ_TOPO_2015 (*red curve*) with respect to EGM2008.

Table 9. Mean percentage agreement of topographic gravity field models with respect to EGM2008, averaged over different SH bands. All values are specified in [%].

SH band	100–500	500–900	900–1800	1800–2160
RWI_TOPO_2012	40.8	58.1	55.0	–
RWI_TOPO_2015	40.9	58.3	59.0	45.6
REQ_TOPO_2015	39.5	56.3	53.8	41.2

of the RWI-based topographic signal to the measured gravity data included in EGM2008. This clearly highlights the benefit of the RWI approach compared to the REQ method.

For $n > 1000$, the percentage agreements gradually decrease for all models. While a stronger decrease can be detected for RWI_TOPO_2012 toward its maximum degree ($N_{\max} = 1800$), the offset between the new RWI and REQ models is nearly constant (about 5 %). For the last SH band 2160–2190, the percentage agreements drop abruptly, which can be explained as follows: Although the new RWI/REQ models and EGM2008 have the same maximum degree of $N_{\max} = 2190$, the latter is only complete to d/o 2159, containing some additional coefficients up to degree 2190 and order 2159 (cf. Pavlis et al., 2012).

In order to further quantify the improvement of the new RWI_TOPO_2015 model, Table 9 presents mean values of the percentage agreements for the SH bands 100–500, 500–900, 900–1800, and 1800–2160. For each of these SH bands, the RWI_TOPO_2015 model consistently provides the largest values. Its advantage is particularly emphasized in the SH band 900–1800. In this band, the improvement of RWI_TOPO_2015 can be quantified by about 7 % and 10 % compared to RWI_TOPO_2012 and REQ_TOPO_2015, respectively.

For the purpose of an additional validation in the space domain, the SH coefficients of the topographic gravity field models and the GGM are used to synthesize gravity disturbances δg. Instead of performing the SHS for each degree separately, narrow SH bands of five degrees $[5k+1,\ 5(k+1)]$ with $k \in \mathbb{N}$ are utilized, i.e., 6–10, 11–15, 16–20, etc. The use of such SH bands is common practice in order to prevent that the signal of the applied assessment indicator is affected by noisy oscillations (cf. Hirt et al., 2015).

In analogy to Eq. (45), the consistency rate of both models for the k-th SH band is introduced as

$$CR(k) = \frac{\mathrm{STD}(\delta g_k^{\mathrm{GGM}}) - \mathrm{STD}(\delta g_k^{\mathrm{GGM}} - \delta g_k^{\mathrm{Topo}})}{\mathrm{STD}(\delta g_k^{\mathrm{GGM}})}, \tag{46}$$

where the gravity disturbances of the k-th SH band of the topographic gravity field model and the GGM are denoted by $\delta g_k^{\mathrm{Topo}}$ and $\delta g_k^{\mathrm{GGM}}$, respectively. Such a kind of indicator was also used by Grombein et al. 2014a to detect the degree of smoothing of topographically-isostatically reduced GOCE gravity gradients.

An advantage of a space domain assessment is that the analysis can be restricted to different areas. Such a subdivision is carried out for the validation with respect to the GOCE TIM R5 model. Figure 17 shows the results for the consistency rate separated by different areas (global, land, oceans, Antarctica). In all cases, the calculation is limited to grid values with a latitude $|\varphi| < 83.3°$. This excludes data points not captured by the GOCE satellite due to its orbital inclination of 96.7°.

Generally, as can be seen from the global results in Fig. 17a, maximum consistency rates of up to 35 % are reached. Analogous to the validation with respect to EGM2008, no or only smaller consistency can be observed for lower frequencies. Due to an increasing signal-to-noise ratio, the SH coefficients of higher degrees cannot properly estimated by the GOCE measurements (cf. Brockmann et al., 2014). Therefore, lower consistency rates are also visible for higher frequencies.

In the case of land areas shown in Fig. 17b, all three topographic gravity field models provide quite similar consistency rates and only some minor differences can be detected. This can be expected, as (i) all three models are based on the same SRTM elevation data and (ii) the RWI and REQ approaches produce the same signal in the case of rock masses.

Concerning the oceans, displayed in Fig. 17c, a nearly systematic offset between the consistency rates of the new RWI and REQ model can be seen. In comparison to the REQ model that is based on the improved bathymetric depths contained in Earth2014, the RWI_TOPO_2015 model shows a quite good performance. That suggests that in the case of the oceans the choice of the approach (RWI vs. REQ) even has a larger influence than the used topographic input data (DTM2006.0 vs Earth2014).

In Fig. 17d, the results for Antarctica are illustrated. In contrast to the former RWI model, both new RWI and REQ models provide significantly larger consistency rates over most of the spectrum. This impressively demonstrates the improvement of the new Bedmap2 data incorporated in the Earth2014 topography model in comparison to Beadmap1 data used for DTM2006.0.

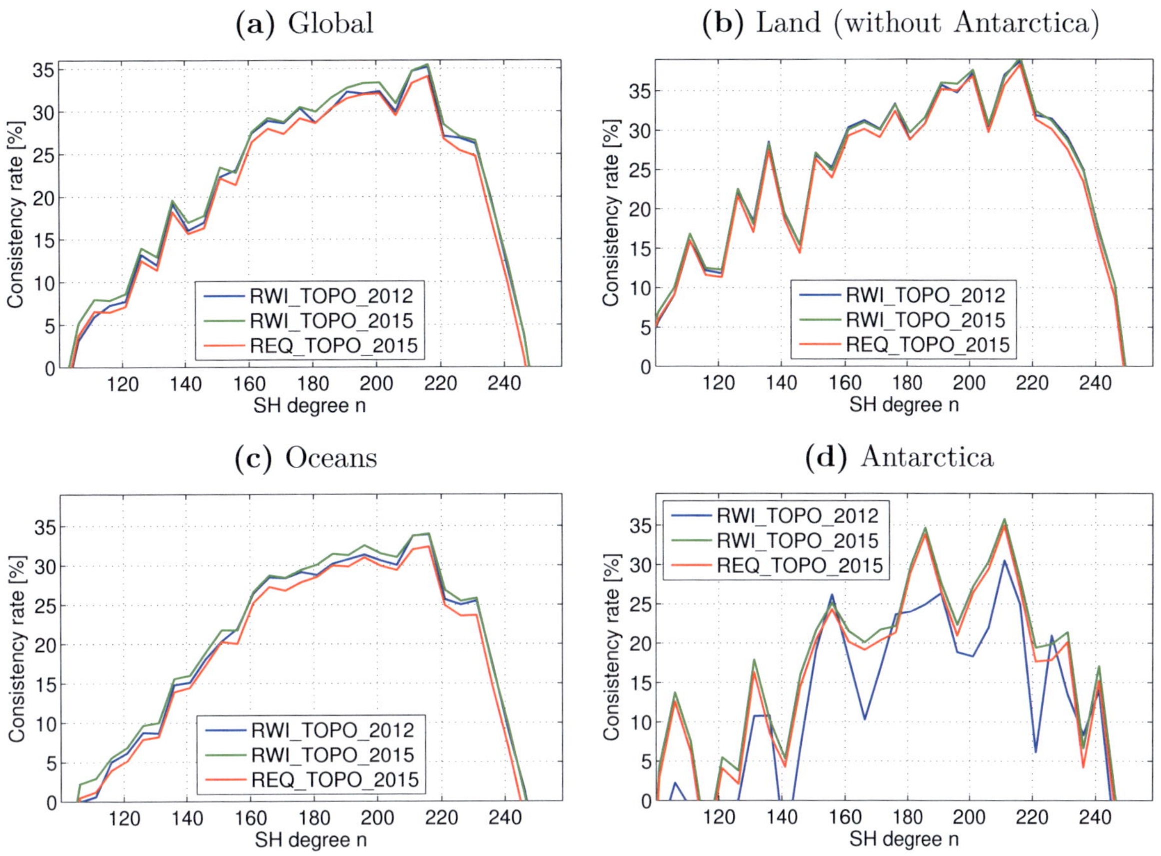

Fig. 17. Consistency rate $CR(k = (n-1)/5)$ of RWI_TOPO_2012 (*blue curve*), RWI_TOPO_2015 (*green curve*), and REQ_TOPO_2015 (*red curve*) with respect to GOCE TIM R5: **(a)** global, **(b)** land (without Antarctica), **(c)** oceans, and **(d)** Antarctica.

7. Conclusions and outlook

In this paper, a new spherical harmonic representation of the Earth's topographic gravitational potential up to degree and order 2190 has been presented. The topographic gravity field model RWI_TOPO_2015 is based on a refined Rock-Water-Ice approach that accounts for a rigorous separate modeling of the Earth's rock, water, and ice masses. While the predecessor model RWI_TOPO_2012 was based on the $5' \times 5'$ global DTM2006.0, the new RWI model uses height information of the $1' \times 1'$ Earth2014 topography suite.

For the processing of RWI_TOPO_2015, GFM is performed in the space domain by using tesseroids. To correctly locate these mass bodies in space, a GRS80 ellipsoidal reference surface is utilized that is extended by an additional geoid model. The gravitational effect of each mass layer (rock, water, and ice) is calculated separately, applying layer-specific density values. In order to obtain SH coefficients, a global grid of GFM-derived topographic potential values is transformed to the frequency domain by using SHA based on a least-squares technique. As has been shown, to correctly represent these topographic potential values by spherical harmonics, it is essential to consider an ellipsoidal calculation grid for the SHA, particularly when deriving coefficients up to higher degree and order. Furthermore, to ensure an accurate SHS in the polar regions, the coefficients of the new RWI model need to be evaluated up to their full resolution and should not be truncated.

Besides SH coefficients for the total RWI-based topographic potential, separate coefficients for the rock, water, and ice components have been determined. By analyzing these SH coefficients in the frequency and space domain, the strongly dominating effect of the water proportion becomes visible. For comparison, a consistent rock-equivalent version called REQ_TOPO_2015 has been generated, in which the heights of water and ice masses are condensed to a constant rock density. By comparing the topographic signal of the RWI and REQ model in terms of various gravity field functionals, differences with significant magnitudes can be detected in the case of terrestrial, airborne and satellite-based applications. To assess the performance of the RWI_TOPO_2015 model, a validation by gravity information of the high-resolution EGM2008 and the satellite-only model GOCE TIM R5 has been carried out. This evaluation impressively demonstrates (i) significant improvements of the new RWI model compared to its predecessor model RWI_TOPO_2012 and (ii) the benefit of the RWI approach compared to the conventional rock-equivalent condensation as used for REQ_TOPO_2015.

As described in the introduction, the RWI model can be used for various applications in geodesy and geophysics. For this purpose, the SH coefficients of different model variants are freely available from the website https://www.gik.kit.edu/rwi_model.php and via the database of the International Centre for Global Earth Models (http://icgem.gfz-potsdam.de/ICGEM).

As future work, a detailed comparison of the new RWI and REQ models to the topographic potential model dV_ELL_RET2014, which also relies on the Earth2014 topography, but uses frequency domain GFM, is in progress and will be subject of a further publication.

Acknowledgements. The authors acknowledge the financial support provided by the German Research Foundation (DFG) under grant number HE1433/20-2. Furthermore, we would like to thank Christian Hirt and Sten Claessens for valuable discussions. The Steinbuch Centre for Computing at the Karlsruhe Institute of Technology is acknowledged for the allocation of computing time on the high-performance parallel computer system HC3. Finally, Dimitrios Tsoulis and one anonymous reviewer as well as the Editor-in-Chief are acknowledged for their valuable comments.

Appendix

In addition to the investigations for the gravity disturbance δg as presented in Sect. 5, this appendix provides further results in the case of the height anomaly ζ (Eq. (41)) and the gravity gradient M_{33} (Eq. (43)). In Figs. 18 and 19, the topographic signal of RWI_TOPO_2015, RWI_TOPO_2015_Rock, RWI_TOPO_2015_Water, and RWI_TOPO_2015_Ice is plotted in terms of ζ and M_{33}, respectively, while corresponding statistics are presented in Tables 10 and 11. For the comparison of RWI_TOPO_2015 to RWI_TOPO_2012 and REQ_TOPO_2015, Figs. 20 and 21 show differences in terms of ζ and M_{33}, respectively. Corresponding statistics for these cases can be found in Tables 12 – 15.

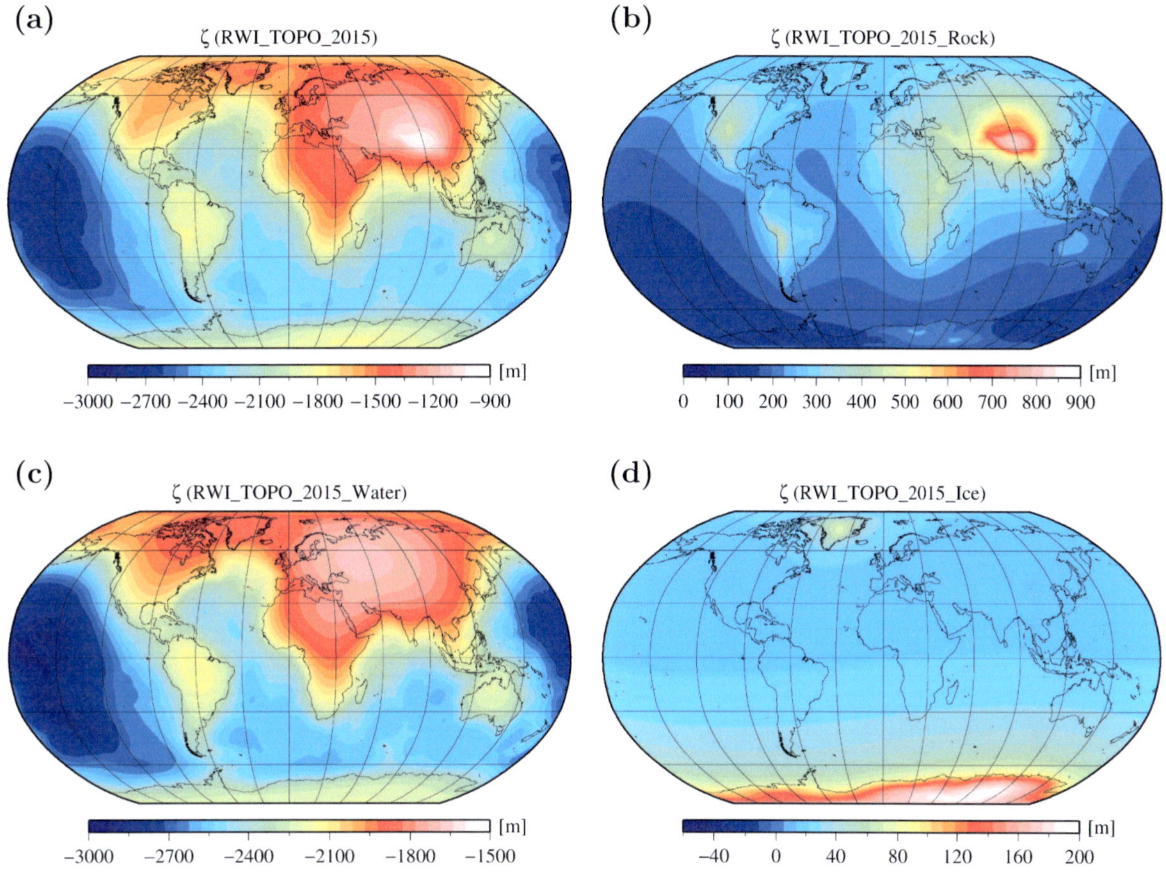

Fig. 18. Topographic signal of **(a)** RWI_TOPO_2015, **(b)** RWI_TOPO_2015_Rock, **(c)** RWI_TOPO_2015_Water, and **(d)** RWI_TOPO_2015_Ice in terms of height anomalies ζ evaluated on the surface of the GRS80 ellipsoid. Robinson projection centered at 0° longitude.

Table 10. Statistics of the RWI_TOPO_2015 topographic signal and its rock, water, and ice proportions in terms of height anomalies ζ evaluated on the surface of the GRS80 ellipsoid. All values are specified in [m].

	Min	Max	Mean	STD
RWI	−2732.564	−947.169	−2062.235	401.159
Rock	103.586	812.072	251.613	89.897
Water	−2924.319	−1669.936	−2342.957	328.921
Ice	17.288	189.362	29.109	22.458

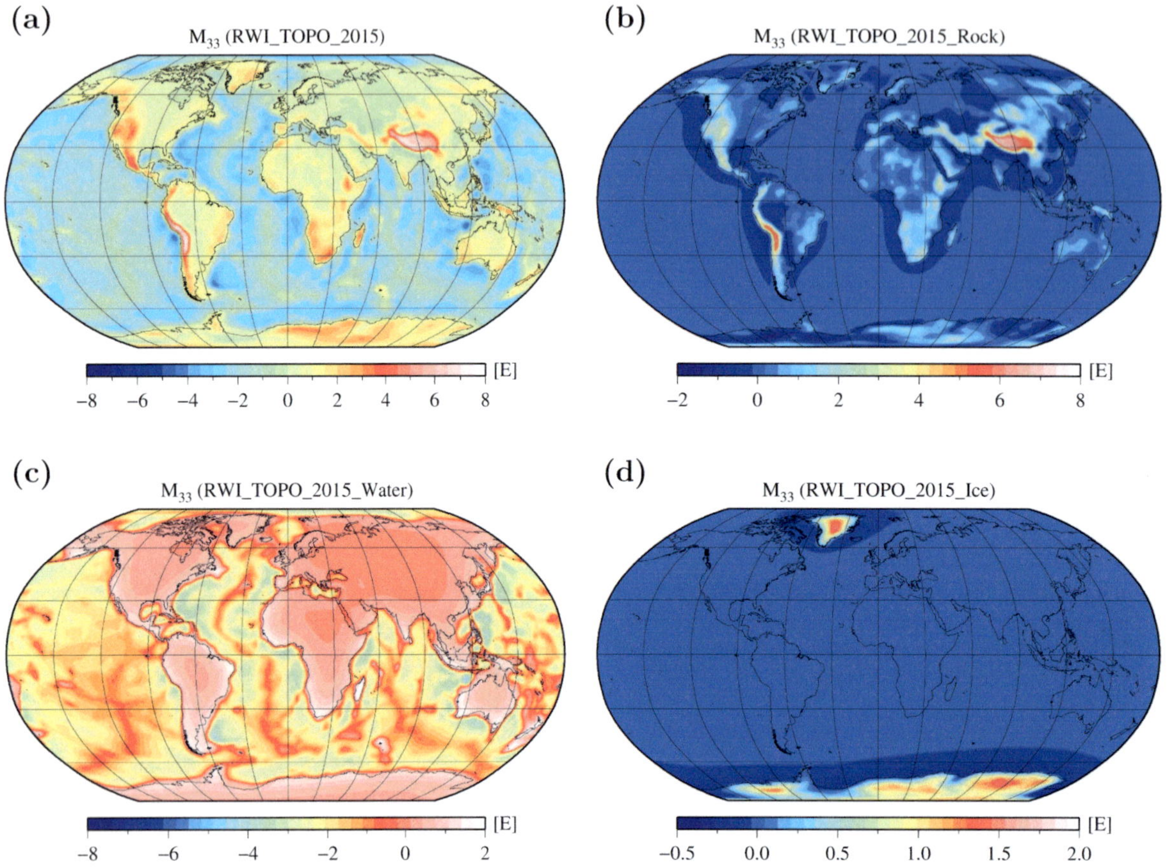

Fig. 19. Topographic signal of **(a)** RWI_TOPO_2015, **(b)** RWI_TOPO_2015_Rock, **(c)** RWI_TOPO_2015_Water, and **(d)** RWI_TOPO_2015_Ice in terms of the gravity gradient M_{33} evaluated on a spherical grid at a mean GOCE satellite altitude (254.9 km). Robinson projection centered at 0° longitude.

Table 11. Statistics of the RWI_TOPO_2015 topographic signal and its rock, water, and ice proportions in terms of the gravity gradient M_{33} evaluated on a spherical grid at a mean GOCE satellite altitude (254.9 km). All values are specified in [E].

	Min	Max	Mean	STD
RWI	−4.562	6.832	−0.882	1.612
Rock	−1.669	6.722	0.108	0.587
Water	−4.484	1.807	−1.002	1.303
Ice	−0.301	1.461	0.012	0.137

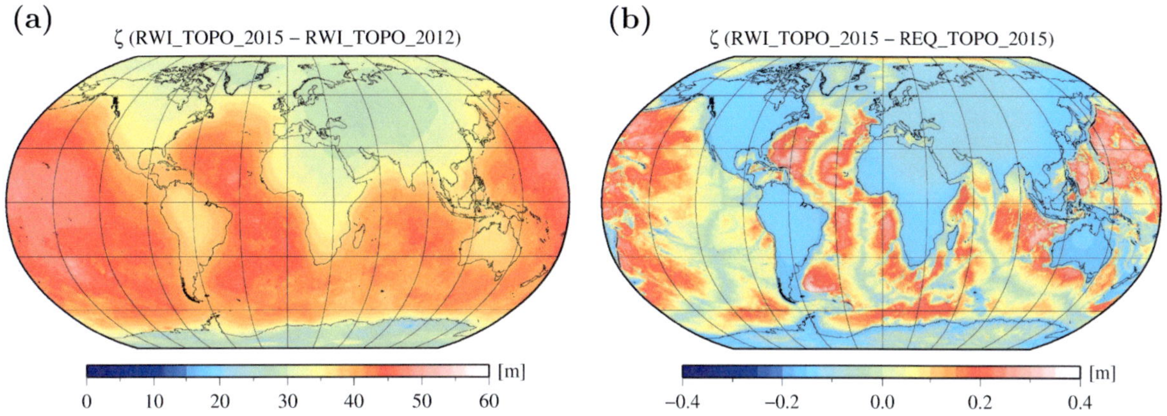

Fig. 20. Difference of the new RWI_TOPO_2015 model to **(a)** RWI_TOPO_2012 and **(b)** REQ_TOPO_2015 in terms of height anomalies ζ evaluated on the surface of the GRS80 ellipsoid. Robinson projection centered at 0° longitude.

Table 12. Statistics of the difference of RWI_TOPO_2015 and RWI_TOPO_2012 in terms of height anomalies ζ evaluated on the surface of the GRS80 ellipsoid. All values are specified in [m].

Min	Max	Mean	STD
17.445	51.776	38.891	5.539

Table 13. Statistics of the difference of the RWI_TOPO_2015 and REQ_TOPO_2015 topographic signal and their water and ice proportions in terms of height anomalies ζ evaluated on the surface of the GRS80 ellipsoid. All values are specified in [m].

	Min	Max	Mean	STD
RWI – REQ	−0.195	0.987	0.000	0.131
RWI – REQ (Water)	−0.192	0.990	0.002	0.132
RWI – REQ (Ice)	−0.030	0.268	−0.002	0.012

Table 14. Statistics of the difference of RWI_TOPO_2015 and RWI_TOPO_2012 in terms of the gravity gradient M_{33} evaluated on a spherical grid at a mean GOCE satellite altitude (254.9 km). All values are specified in [mE].

Min	Max	Mean	STD
−470.724	354.954	16.631	42.635

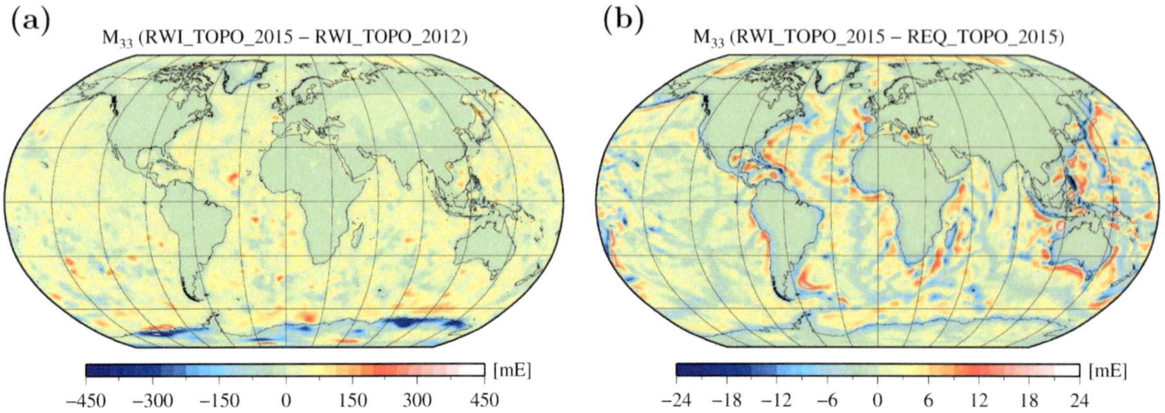

Fig. 21. Difference of the new RWI_TOPO_2015 model to **(a)** RWI_TOPO_2012 and **(b)** REQ_TOPO_2015 in terms of the gravity gradient M_{33} evaluated on a spherical grid at a mean GOCE satellite altitude (254.9 km). Robinson projection centered at 0° longitude.

Table 15. Statistics of the difference of the RWI_TOPO_2015 and REQ_TOPO_2015 topographic signal and their water and ice proportions in terms of the gravity gradient M_{33} evaluated on a spherical grid at a mean GOCE satellite altitude (254.9 km). All values are specified in [mE].

	Min	Max	Mean	STD
RWI – REQ	−19.336	22.589	0.000	3.833
RWI – REQ (Water)	−19.335	22.591	0.001	3.814
RWI – REQ (Ice)	−2.760	5.930	−0.001	0.336

References

Abd-Elmotaal, H. and Kühtreiber, N. (2014): The effect of DHM resolution in computing the topographic-isostatic harmonic coefficients within the window technique. *Studia Geophysica et Geodætica* 58(1):41–55. DOI: 10.1007/s11200-012-0231-6.

Abd-Elmotaal, H., Seitz, K., Abd-Elbaky, M., and Heck, B. (2014): Comparison among three harmonic analysis techniques on the sphere and the ellipsoid. *Journal of Applied Geodesy* 8(1):1–19. DOI: 10.1515/jag-2013-0008.

Abd-Elmotaal, H., Seitz, K., Abd-Elbaky, M., and Heck, B. (2016): Tailored reference geopotential model for Africa. In: Rizos, C., and Willis, P. (eds.) IAG 150 years. Proceedings of the IAG Scientific Assembly, Postdam, Germany, Sept. 1–6, 2013. *International Association of Geodesy Symposia*, vol. 143. Springer Berlin Heidelberg, pp. 383–390. DOI: 10.1007/1345_2015_84.

Anderson, E. G. (1976): The effect of topography on solutions of Stokes' problem. Unisurv S-14, Report. School of Surveying, University of New South Wales, Australia.

Balmino, G., Vales, N., Bonvalot, S., and Briais, A. (2012): Spherical harmonic modelling to ultra-high degree of Bouguer and isostatic anomalies. *Journal of Geodesy* 86(7):499–520. DOI: 10.1007/s00190-011-0533-4.

Bamber, J. L., Griggs, J. A., Hurkmans, R. T. W. L., Dowdeswell, J. A., Gogineni, S. P., Howat, I., Mouginot, J., Paden, J., Palmer, S., Rignot, E., and Steinhage, D. (2013): A new bed elevation dataset for Greenland. *The Cryosphere* 7(2):499–510. DOI: 10.5194/tc-7-499-2013.

Baran, I., Kuhn, M., Claessens, S. J., Featherstone, W. E., Holmes, S. A., and Vaníček, P. (2006): A synthetic Earth gravity model designed specifically for testing regional gravimetric geoid determination algorithms. *Journal of Geodesy* 80(1):1–16. DOI: 10.1007/s00190-005-0002-z.

Barthelmes, F. (2013): Definition of functionals of the geopotential and their calculation from spherical harmonic models. Scientific Technical Report STR09/02. German Research Centre for Geosciences (GFZ), Potsdam, Germany. URL: http://icgem.gfz-potsdam.de/ICGEM/theory/theory.html.

Becker, J. J., Sandwell, D. T., Smith, W. H. F., Braud, J., Binder, B., Depner, J., Fabre, D., Factor, J., Ingalls, S., Kim, S.-H., Ladner, R., Marks, K., Nelson, S., Pharaoh, A., Trimmer, R., von Rosenberg, J., Wallace, G., and Weatherall, P. (2009): Global bathymetry and elevation data at 30 arc seconds resolution: SRTM30_PLUS. *Marine Geodesy* 32(4):355–371. DOI: 10.1080/01490410903297766.

Bouman, J., Ebbing, J., Fuchs, M., Sebera, J., Lieb, V., Szwillus, W., Haagmans, R., and Novák, P. (2016): Satellite gravity gradient grids for geophysics. *Nature Scientific Reports* 6:21050. DOI: 10.1038/srep21050.

Bouman, J., Ebbing, J., Meekes, S., Fattah, R. A., Fuchs, M., Gradmann, S., Haagmans, R., Lieb, V., Schmidt, M., Dettmering, D., and Bosch, W. (2015): GOCE gravity gradient data for lithospheric modeling. *International Journal of Applied Earth Observation and Geoinformation*, Special issue on GOCE Earth science applications and models 35(Part A):16–30. DOI: 10.1016/j.jag.2013.11.001.

Brockmann, J. M., Zehentner, N., Höck, E., Pail, R., Loth, I., Mayer-Gürr, T., and Schuh, W.-D. (2014): EGM_TIM_RL05: an independent geoid with centimeter accuracy purely based on the GOCE mission. *Geophysical Research Letters* 41(22):8089–8099. DOI: 10.1002/2014GL061904.

Claessens, S. J. and Hirt, C. (2013): Ellipsoidal topographic potential: new solutions for spectral forward gravity modeling of topography with respect to a reference ellipsoid. *Journal of Geophysical Research* 118(11):5991–6002. DOI: 10.1002/2013JB010457.

Ebbing, J., Braitenberg, C., and Götze, H.-J. (2001): Forward and inverse modelling of gravity revealing insight into crustal structures of the Eastern Alps. *Tectonophysics* 337(3–4):191–208. DOI: 10.1016/S0040-1951(01)00119-6.

Farr, T. G., Rosen, P. A., Caro, E., Crippen, R., Duren, R., Hensley, S., Kobrick, M., Paller, M., Rodriguez, E., Roth, L., Seal, D., Shaffer, S., Shimada, J., Umland, J., Werner, M., Oskin, M., Burbank, E., and Alsdorf, D. (2007): The shuttle radar topography mission. *Reviews of Geophysics* 45(2):RG2004. DOI: 10.1029/2005RG000183.

Fecher, T., Pail, R., and Gruber, T. (2015): Global gravity field modeling based on GOCE and complementary gravity data. *International Journal of Applied Earth Observation and Geoinformation*, Special issue on GOCE Earth science applications and models 35(Part A):120–127. DOI: 10.1016/j.jag.2013.10.005.

Fecher, T., Pail, R., Gruber, T., and the GOCO project team (2016): The combined gravity field model GOCO05c. In: General Assembly of the European Geosciences Union 2016. Vienna, Austria, Apr. 17–22, 2016. *Geophysical Research Abstracts*, vol. 18. EGU2016-7696.

Forsberg, R. (1984): A study of terrain reductions, density anomalies and geophysical inversion methods in gravity field modelling. Report 355. Department of Geodetic Science and Surveying, The Ohio State University, Columbus, USA.

Forsberg, R. and Tscherning, C. C. (1997): Topographic effects in gravity field modelling for BVP. In: Sansò, F., and Rummel, R. (eds.) Geodetic boundary value problems in view of the one centimeter geoid, *Lecture Notes in Earth Sciences*, vol. 65. Springer Berlin Heidelberg, pp. 239–272. DOI: 10.1007/BFb0011707.

Fretwell, P., Pritchard, H. D., Vaughan, D. G., Bamber, J. L., Barrand, N. E., Bell, R., Bianchi, C., Bingham, R. G., Blankenship, D. D., Casassa, G., Catania, G., Callens, D., Conway, H., Cook, A. J., Corr, H. F. J., Damaske, D., Damm, V., Ferraccioli, F., Forsberg, R., Fujita, S., Gim, Y., Gogineni, P., Griggs, J. A., Hindmarsh, R. C. A., Holmlund, P., Holt, J. W., Jacobel, R. W., Jenkins, A., Jokat, W., Jordan, T., King, E. C., Kohler, J., Krabill, W., Riger-Kusk, M., Langley, K. A., Leitchenkov, G., Leuschen, C., Luyendyk, B. P., Matsuoka, K., Mouginot, J., Nitsche, F. O., Nogi, Y., Nost, O. A., Popov, S. V., Rignot, E., Rippin, D. M., Rivera, A., Roberts, J., Ross, N., Siegert, M. J., Smith, A. M., Steinhage, D., Studinger, M., Sun, B., Tinto, B. K., Welch, B. C., Wilson, D., Young, D. A., Xiangbin, C., and Zirizzotti, A. (2013): Bedmap2: improved ice bed, surface and thickness datasets for Antarctica. *The Cryosphere* 7(1):375–393. DOI: 10.5194/tc-7-375-2013.

Grombein, T., Luo, X., Seitz, K., and Heck, B. (2014a): A wavelet-based assessment of topographic-isostatic reductions for GOCE gravity gradients. *Surveys in Geophysics* 35(4):959–982. DOI: 10.1007/s10712-014-9283-1.

Grombein, T., Seitz, K., and Heck, B. (2010): Modelling topographic effects in GOCE gravity gradients. In: Münch, U., and Dransch, W. (eds.) Observation of the system Earth from space, *GEOTECHNOLOGIEN Science Report*, vol. 17, pp. 84–93. DOI: 10.2312/GFZ.gt.17.13.

Grombein, T., Seitz, K., and Heck, B. (2011): Smoothing GOCE gravity gradients by means of topographic-isostatic reductions. In: Ouwehand, L. (ed.) Proceedings of the 4th International GOCE User Workshop, Munich, Germany, Mar. 31–Apr. 1, 2011. *ESA Special Publication*, vol. 696. ESA Publications Division, ESTEC, Noordwijk, The Netherlands.

Grombein, T., Seitz, K., and Heck, B. (2013): Optimized formulas for the gravitational field of a tesseroid. *Journal of Geodesy* 87(7):645–660. DOI: 10.1007/s00190-013-0636-1.

Grombein, T., Seitz, K., and Heck, B. (2014b): Topographic-isostatic reduction of GOCE gravity gradients. In: Rizos, C., and Willis, P. (eds.) Earth on the edge: science for a sustainable planet. Proceedings of the IAG General Assembly, Melbourne, Australia, June 28–July 2, 2011. *International Association of Geodesy Symposia*, vol. 139. Springer Berlin Heidelberg, pp. 349–356. DOI: 10.1007/978-3-642-37222-3_46.

Gruber, C., Novák, P., Flechtner, F., and Barthelmes, F. (2014): Derivation of the topographic potential from global DEM models. In: Rizos, C., and Willis, P. (eds.) Earth on the edge: science for a sustainable planet. Proceedings of the IAG General Assembly, Melbourne, Australia, June 28–July 2, 2011. *International Association of Geodesy Symposia*, vol. 139. Springer Berlin Heidelberg, pp. 535–542. DOI: 10.1007/978-3-642-37222-3_71.

Gutknecht, B. D., Götze, H.-J., Jahr, T., Jentzsch, G., Mahatsente, R., and Zeumann, S. (2014): Structure and state of stress of the Chilean subduction zone from terrestrial and satellite-derived gravity and gravity gradient data. *Surveys in Geophysics* 35(6):1417–1440. DOI: 10.1007/s10712-014-9296-9.

Heck, B. (2003): Rechenverfahren und Auswertemodelle der Landesvermessung. Klassische und moderne Methoden. 3rd ed. Wichmann-Verlag, Heidelberg, Germany.

Heck, B. and Seitz, K. (2007): A comparison of the tesseroid, prism and point-mass approaches for mass reductions in gravity field modelling. *Journal of Geodesy* 81(2):121–136. DOI: 10.1007/s00190-006-0094-0.

Heiskanen, W. A. and Moritz, H. (1967): Physical geodesy. W. H. Freeman & Co., San Francisco, USA.

Hirt, C. and Kuhn, M. (2012): Evaluation of high-degree series expansions of the topographic potential to higher-order powers. *Journal of Geophysical Research* 117:B12407. DOI: 10.1029/2012JB009492.

Hirt, C., Kuhn, M., Featherstone, W. E., and Göttl, F. (2012): Topographic/isostatic evaluation of new-generation GOCE gravity field models. *Journal of Geophysical Research* 117:B05407. DOI: 10.1029/2011JB008878.

Hirt, C. and Rexer, M. (2015): Earth2014: 1 arc-min shape, topography, bedrock and ice-sheet models – available as gridded data and degree-10,800 spherical harmonics. *International Journal of Applied Earth Observation and Geoinformation* 39:103–112. DOI: 10.1016/j.jag.2015.03.001. URL: http://ddfe.curtin.edu.au/models/Earth2014.

Hirt, C., Rexer, M., and Claessens, S. (2015): Topographic evaluation of fifth-generation GOCE gravity field models – globally and regionally. In: Huang, J., Reguzzoni, M., and Gruber, T. (eds.) Assessment of GOCE geopotential models, *Newton's Bulletin*, no. 5. International Association of Geodesy and International Gravity Field Service, pp. 163–186. URL: http://www.isgeoid.polimi.it/Newton/Newton_5/12_Hirt_163_186.html.

Holmes, S. A. and Featherstone, W. E. (2002): A unified approach to the Clenshaw summation and the recursive computation of very high degree and order normalised associated Legendre functions. *Journal of Geodesy* 76(5):279–299. DOI: 10.1007/s00190-002-0216-2.

Holmes, S. A. and Pavlis, N. K. (2006): Spherical harmonic synthesis software harmonic_synth_v02.f. URL: http://earth-info.nga.mil/GandG/wgs84/gravitymod/new_egm/new_egm.html.

Holmes, S. A. and Pavlis, N. K. (2007): Some aspects of harmonic analysis of data gridded on the ellipsoid. In: Kiliçoglu, A., and Forsberg, R. (eds.) Proceedings of the 1st International Symposium of the International Gravity Field Service (IGFS), Istanbul, Turkey, Aug. 28–Sept. 1, 2006. Harita Dergisi, Special Issue 18, pp. 151–156.

Janák, J., Pitoňák, M., and Minarechová, Z. (2014): Regional quasigeoid from GOCE and terrestrial measurements. *Studia Geophysica et Geodætica* 58(4):626–649. DOI: 10.1007/s11200-013-0543-1.

Jarvis, A., Reuter, H. I., Nelson, A., and Guevara, E. (2008): Hole-filled seamless SRTM data v4. International Centre for Tropical Agriculture (CIAT). URL: http://srtm.csi.cgiar.org.

Jekeli, C. (1988): The exact transformation between ellipsoidal and spherical harmonic expansions. *Manuscripta Geodaetica* 13(2):106–113.

Kuhn, M. and Featherstone, W. E. (2005): Construction of a synthetic Earth gravity model by forward gravity modelling. In: Sansò, F. (ed.) A window on the future of geodesy. Proceedings of the IAG General Assembly, Sapporo, Japan, June 30–July 11, 2003. *International Association of Geodesy Symposia*, vol. 128. Springer Berlin Heidelberg, pp. 350–355. DOI: 10.1007/3-540-27432-4_60.

Kuhn, M. and Seitz, K. (2005): Comparison of Newton's integral in the space and frequency domains. In: Sansò, F. (ed.) A window on the future of geodesy. Proceedings of the IAG General Assembly, Sapporo, Japan, June 30–July 11, 2003. *International Association of Geodesy Symposia*, vol. 128. Springer Berlin Heidelberg, pp. 386–391. DOI: 10.1007/3-540-27432-4_66.

Lemoine, F. G., Kenyon, S. C., Factor, J. K., Trimmer, R. G., Pavlis, N. K., Chinn, D. S., Cox, C. M., Klosko, S. M., Luthcke, S. B., Torrence, M. H., Wang, Y. M., Williamson, R. G., Pavlis, E. C., Rapp, R. H., and Olson, T. R. (1998): The development of the joint NASA GSFC and the National Imagery and Mapping Agency (NIMA) geopotential model EGM96. Technical report 1998-206861. NASA Goddard Space Flight Center, Greenbelt, Maryland, USA. URL: http://ntrs.nasa.gov/search.jsp?R=19980218814.

Mayer-Gürr, T., Rieser, D., Höck, E., Brockmann, J. M., Schuh, W.-D., Krasbutter, I., Kusche, J., Maier, A., Krauss, S., Hausleitner, W., Baur, O., Jäggi, A., Meyer, U., Prange, L., Pail, R., Fecher, T., and Gruber, T. (2012): The new combined satellite only model GOCO03s. In: International Symposium on Gravity, Geoid and Height Systems 2012. Venice, Italy, Oct. 9–12, 2012.

Moritz, H. (1980): Geodetic Reference System 1980. *Bulletin Géodésique* 54(3):395–405. DOI: 10.1007/BF02521480.

Nagy, D., Papp, G., and Benedek, J. (2000): The gravitational potential and its derivatives for the prism. *Journal of Geodesy* 74(7–8):552–560. DOI: 10.1007/s001900000116.

Novák, P., Kern, M., Schwarz, K.-P., and Heck, B. (2003): Evaluation of band-limited topographical effects in airborne gravimetry. *Journal of Geodesy* 76(11–12):597–604. DOI: 10.1007/s00190-002-0282-5.

Novák, P. and Tenzer, R. (2013): Gravitational gradients at satellite altitudes in global geophysical studies. *Surveys in Geophysics* 34(5):653–673. DOI: 10.1007/s10712-013-9243-1.

Omang, D. O. C. and Forsberg, R. (2000): How to handle topography in practical geoid determination: three examples. *Journal of Geodesy* 74(6):458–466. DOI: 10.1007/s001900000107.

Pavlis, N. K., Factor, J. K., and Holmes, S. A. (2007): Terrain-related gravimetric quantities computed for the next EGM. In: Kiliçoglu, A., and Forsberg, R. (eds.) Proceedings of the 1st International Symposium of the International Gravity Field Service (IGFS), Istanbul, Turkey, Aug. 28–Sept. 1, 2006. Harita Dergisi, Special Issue 18, pp. 318–323.

Pavlis, N. K., Holmes, S. A., Kenyon, S. C., and Factor, J. K. (2012): The development and evaluation of the Earth Gravitational Model 2008. *Journal of Geophysical Research* 117:B04406. DOI: 10.1029/2011JB008916.

Pavlis, N. K., Holmes, S. A., Kenyon, S. C., and Factor, J. K. (2013): Correction to "The development and evaluation of the Earth Gravitational Model 2008 (EGM2008)". *Journal of Geophysical Research* 118(5):2633. DOI: 10.1002/jgrb.50167.

Pitonák, M., Šprlák, M., Hamácková, E., and Novák, P. (2016): Regional recovery of the disturbing gravitational potential by inverting satellite gravitational gradients. *Geophysical Journal International* 205(1):89–98. DOI: 10.1093/gji/ggw008.

Rummel, R., Rapp, R. H., Sünkel, H., and Tscherning, C. C. (1988): Comparisons of global topographic/isostatic models to the Earth's observed gravity field. Report 388. Department of Geodetic Science and Surveying, The Ohio State University, Columbus, USA.

Rummel, R., Yi, W., and Stummer, C. (2011): GOCE gravitational gradiometry. *Journal of Geodesy* 85(11):777–790. DOI: 10.1007/s00190-011-0500-0.

Seitz, K. and Heck, B. (1991): Harmonic analysis on the sphere. Internal report. Geodetic Institute, University of Karlsruhe, Germany.

Sneeuw, N. (1994): Global spherical harmonic analysis by least-squares and numerical quadrature methods in historical perspective. *Geophysical Journal International* 118(3):707–716. DOI: 10.1111/j.1365-246X.1994.tb03995.x.

Tenzer, R., Gladkikh, V., Novák, P., and Vajda, P. (2012): Spatial and spectral analysis of refined gravity data for modelling the crust-mantle interface and mantle-lithosphere structure. *Surveys in Geophysics* 33(5):817–839. DOI: 10.1007/s10712-012-9173-3.

Thong, N. C. (1989): Simulation of gradiometry using the spheroidal harmonic model of the gravitational field. *Manuscripta Geodaetica* 14(6):404–417.

Tscherning, C. C. (1985): On the long-wavelength correlation between gravity and topography. In: 5th International Symposium Geodesy and Physics of the Earth. Magdeburg, German Democratic Republic, Sept. 23–29, 1984. *Veröffentlichungen des Zentralinstituts Physik der Erde*, vol. 81, part II, pp. 134–143.

Tsoulis, D. and Kuhn, M. (2007): Recent developments in synthetic Earth gravity models in view of the availability of digital terrain and crustal databases of global coverage and increased resolution. In: Kiliçoglu, A., and Forsberg, R. (eds.) Proceedings of the 1st International Symposium of the International Gravity Field Service (IGFS), Istanbul, Turkey, Aug. 28–Sept. 1, 2006. Harita Dergisi, Special Issue 18, pp. 354–359.

Tsoulis, D. and Patlakis, K. (2013): A spectral assessment review of current satellite-only and combined Earth gravity models. *Reviews of Geophysics* 51(2):186–243. DOI: 10.1002/rog.20012.

Wieczorek, M. A. (2007): Gravity and topography of the terrestrial planets. *Treatise on Geophysics*, Planets and Moons 10(5):165–206. DOI: 10.1016/B978-044452748-6.00156-5.

Wild-Pfeiffer, F. (2008): A comparison of different mass elements for use in gravity gradiometry. *Journal of Geodesy* 82(10):637–653. DOI: 10.1007/s00190-008-0219-8.

Wittwer, T., Klees, R., Seitz, K., and Heck, B. (2008): Ultra-high degree spherical harmonic analysis and synthesis using extended-range arithmetic. *Journal of Geodesy* 82(4):223–229. DOI: 10.1007/s00190-007-0172-y.

Chapter V.

On high-frequency topography-implied gravity signals for height system unification using GOCE-based global geopotential models

Thomas Grombein, Kurt Seitz, Bernhard Heck
Surveys in Geophysics 38(2):443–477, 2017. DOI: 10.1007/s10712-016-9400-4.

Submitted: 28 June 2016 / Accepted: 01 November 2016 / Published online: 17 December 2016

Abstract. National height reference systems have conventionally been linked to the local mean sea level, observed at individual tide gauges. Due to variations in the sea surface topography, the reference levels of these systems are inconsistent, causing height datum offsets of up to ± 1–2 m. For the unification of height systems, a satellite-based method is presented that utilizes global geopotential models (GGMs) derived from ESA's satellite mission Gravity field and steady-state Ocean Circulation Explorer (GOCE). In this context, height datum offsets are estimated within a least squares adjustment by comparing the GGM information with measured GNSS/leveling data. While the GNSS/leveling data comprises the full spectral information, GOCE GGMs are restricted to long wavelengths according to the maximum degree of their spherical harmonic representation. To provide accurate height datum offsets, it is indispensable to account for the remaining signal above this maximum degree, known as the omission error of the GGM. Therefore, a combination of the GOCE information with the high-resolution Earth Gravitational Model 2008 (EGM2008) is performed. The main contribution of this paper is to analyze the benefit, when high-frequency topography-implied gravity signals are additionally used to reduce the remaining omission error of EGM2008. In terms of a spectral extension, a new method is proposed that does not rely on an assumed spectral consistency of topographic heights and implied gravity as is the case for the residual terrain modeling (RTM) technique. In the first step of this new approach, gravity forward modeling based on tesseroid mass bodies is performed according to the Rock-Water-Ice (RWI) approach. In a second step, the resulting full spectral RWI-based topographic potential values are reduced by the effect of the topographic gravity field model RWI_TOPO_2015, thus, removing the long to medium wavelengths. By using the latest GOCE GGMs, the impact of topography-implied gravity signals on the estimation of height datum offsets is analyzed for GNSS/leveling data sets in Germany, Austria, and Brazil. Besides considerable changes in the estimated offset of up to 3 cm, the conducted analyses show significant improvements of 30 – 40 % in terms of a reduced standard deviation and range of the least squares adjusted residuals.

Keywords Height system unification · GOCE · Gravity forward modeling · Rock-Water-Ice (RWI) approach · Tesseroids

1. Introduction

National height reference systems are traditionally based on leveling networks, where height differences are observed by a combination of spirit leveling and gravimetry. For each leveling network, a fixed datum point has been selected that defines the height reference level. In order to realize a physical meaning, the zero level of height systems is conventionally linked to the mean sea level (MSL). For this purpose, tide gauges usually serve as datum points, where sea level observations over a specific time period are used to fix the zero level. Due to location- and time-dependent variations in the sea surface topography, i.e., the height of the sea surface above a global equipotential surface like the geoid, different tide gauges and observation periods generally do not realize the same height reference level. As a consequence, national height systems that are mostly connected to individual tide gauges are based on their own local vertical datum (LVD). The resulting discrepancies between different LVD zones cause height datum offsets of about ±1–2 m (Heck, 1990), making it impossible to directly compare the physical heights of different, disconnected countries.

From a regional perspective, this can cause various problems for international engineering projects or flooding control, when height information from more than one LVD zone needs to be integrated (Gerlach and Fecher, 2012). Furthermore, a consistent physical height reference level seems to be indispensable for the accurate assessment of global geodynamic and climatological processes in the Earth's system (Blewitt et al., 2010). This includes environmental studies related to the monitoring of the global sea level rise or researches on coastal hazards. Moreover, a consistent connection between geometric and physical reference frames is relevant for many geodetic applications, such as the unification of national gravity databases or the realization of the Global Geodetic Observing System (GGOS, Ihde and Sánchez, 2005). Therefore, the current activities of GGOS Theme 1 are concentrated on the unification of height systems (Kutterer and Neilan, 2016).

For the purpose of height system unification, different strategies and methods have been proposed and discussed in various publications, see, for example, Colombo (1980), Rapp (1983), Heck and Rummel (1990), Sansò and Venuti (2002), and Sánchez (2009, 2015).

According to Rummel (2002), they can generally be classified into three categories: (i) the spirit leveling approach, (ii) the oceanographic approach, and (iii) the gravity field approach. A discussion of the advantages and drawbacks of these approaches, as well as a comparison of their practical realization in Europe, can be found in Rülke et al. (2016).

The spirit leveling approach is based on a joint adjustment of different national leveling networks by using cross-border connections. As an example, this has been done for the realization of the European Vertical Reference Frame (EVRF 2007, Sacher et al., 2009). However, this method is susceptible to systematic errors, has a low redundancy, and is restricted to only a continent-wide connection of local height systems.

In the case of the oceanographic approach, a connection of different tide gauges is realized by determining the sea surface topography between these points (e.g., Woodworth et al., 2012). For this purpose, oceanographic models are needed, which can be based on dynamic leveling or satellite altimetry (cf. Heck and Rummel, 1990; Luz et al., 2009a).

The gravity field approach uses information of GNSS/leveling benchmarks, i.e., observation points that combine ellipsoidal heights derived from global navigation satellite system (GNSS) observations and physical heights in the LVD. Height datum offsets are then estimated by comparing biased undulations resulting from these GNSS/leveling data with corresponding unbiased undulations derived from independent gravity field information. Depending on the accuracy requirements, this information can be obtained from (i) the solution of a Geodetic Boundary Value Problem (GBVP, Heiskanen and Moritz, 1967, p. 36 f.), (ii) a global geopotential model (GGM), or (iii) a combination of both. In the case of the GBVP approach (Rummel and Teunissen, 1988), respective boundary values in terms of terrestrial gravity data are required. In this context, investigations have been mainly concentrated on the use of the scalar-free GBVP (e.g., Xu, 1992; Gerlach and Rummel, 2013; Amjadiparvar et al., 2016), but have also been conducted for the case of the fixed GBVP (e.g., Grombein et al., 2016b).

In the following, a satellite-based method will be considered that utilizes gravity field information derived from ESA's satellite mission Gravity field and steady-state Ocean Circulation Explorer (GOCE, ESA, 1999). A primary geodetic mission objective of GOCE is to contribute to global height system unification by providing a precise global geoid with an accuracy of $1-2\,\mathrm{cm}$. As such a satellite-based geoid is not affected by any height datum offset, it can be used as a global homogeneous reference surface for the connection of different height datum zones (Rummel, 2002; Gatti et al., 2013). However, due to the measurement at satellite altitude and the limited bandwidth of the GOCE gradiometer, the derived geoid is restricted to a spatial resolution of about 100 km (Rummel et al., 2011). This corresponds to a spherical harmonic (SH) expansion of degree and order (d/o) 200 in terms of a GGM. The remaining signal above this degree is known as the omission error of the GGM and has to be taken into account in order to match the full spectral information of the GNSS/leveling data. Therefore, in order to provide accurate height datum offsets, the main issue of such an approach is to reduce the omission error of the satellite-based GGM by additionally considering the mid- and high-frequency terms of the gravity field, which are not captured by the model. A common method to reduce the influence of the omission error is to combine the long to medium wavelength information of a GOCE GGM with the high-resolution Earth Gravitational Model 2008 (EGM2008, Pavlis et al., 2012). Such a procedure has been used in the context of the validation of GOCE GGMs (e.g., Gruber et al., 2011; Šprlák et al., 2015; Voigt and Denker, 2015), as well as for the unification of height systems (e.g., Gruber et al., 2012; Barzaghi et al., 2016; Ferreira et al., 2016).

In many studies, a pure concatenation (or complementation) of the SH coefficients at a specific degree (e.g., 180 or 200) is performed. However, such a procedure might cause a spectral gap between both models. To ensure a smooth transition, a Hanning window (Blackman and Tukey, 1958) is used in this paper. In this context, a systematic analysis will be carried out to find optimal values for the combination degree and the window size.

In contrast to other publications, the main focus of this paper is the additional consideration of high-frequency gravity signals induced by the Earth's topographic masses. In terms of a spectral extension, residual terrain modeling (RTM) is a widely used tool, where gravity forward modeling is applied to a residual topography between a high-resolution digital terrain model (DTM) and a smoothed reference topography, whose spectral information is assumed to be already contained in the GGM. However, this procedure principally depends on the spectral consistency of topographic heights and implied gravity, which is not the case in general (cf. Hirt and Kuhn, 2014). To overcome this issue, a new approach is proposed, where the forward-modeled topographic gravity effect of a DTM is reduced by the information of the topographic gravity field model RWI_TOPO_2015 (Grombein et al., 2016a). Therefore, the required high-pass filtering is performed directly in the gravity domain, which allows a more consistent spectral combination with the GGM information.

The paper is organized as follows: in Sect. 2, the proposed satellite-based method for height system unification is introduced. Furthermore, formulas are derived for the applied combination of the GOCE and EGM2008 information, as well as the new (residual) gravity forward modeling approach. While Sect. 3 describes the study areas and used data sets, Sect. 4 provides numerical investigations concerning their uncertainties. In Sect. 5, the results of the height datum offset estimation are presented and discussed with a special focus on (i) an optimal combination of GOCE and EGM2008 and (ii) the benefit of high-frequency topographic signals. Finally, Sect. 6 concludes with a summary and an outlook.

2. Satellite-based height system unification

2.1. General considerations

Following the theory of Molodensky (Heiskanen and Moritz, 1967, p. 291ff.), the presented approach is based on the fundamental relation

$$h = H + \zeta, \tag{1}$$

where h is the ellipsoidal height, H is the normal height, and ζ is the corresponding height anomaly. By considering a biased normal height H' in some LVD zone, the height datum offset δH of this zone due to the discrepancies of the three quantities can be derived by

$$\delta H = h - H' - \zeta. \tag{2}$$

Note that an equivalent relation holds true in the case of the orthometric height and the geoid undulation with respect to the theory of Stokes. However, a formulation in the sense of Molodensky will be used throughout this paper.

For the practical evaluation of Eq. (2), a set of GNSS/leveling benchmarks $P_j, j = 1, \ldots, J$, is needed that combines unbiased GNSS-based ellipsoidal heights $h(P_j)$ and biased normal heights $H'(P_j)$ derived from spirit leveling and gravity measurements in the LVD zone. Furthermore, the unbiased height anomalies $\zeta(P_j)$ are obtained from gravity information of a GGM. In this way, the GGM provides a reference surface to which different LVD zones can be connected, hence, allowing a unification of the corresponding height systems.

As outlined in Sjöberg (2011), methods based on GNSS/leveling benchmarks are generally not able to provide (absolute) height datum offsets with respect to the unknown global vertical datum. However, from a practical point of view, this is not absolutely necessary for a global unification of height systems. Instead, the derived height datum offsets are defined relatively with respect to the GGM-based reference surface. Thus, an "absolute" vertical datum is defined by convention (cf. Heck, 2004).

According to Hirt et al. (2010), the SH series of a GGM is evaluated at the Earth's surface. In terms of geocentric spherical coordinates (r, φ, λ), the required spherical harmonic synthesis in the SH band $[n_1, n_2] \subset \mathbb{N}_0$ can be performed by

$$\zeta^{\mathrm{GGM}}_{n_1,n_2}(P) = \frac{GM}{r\gamma} \sum_{n=n_1}^{n_2} \left(\frac{R}{r}\right)^n \sum_{m=0}^{n} \left(\Delta\overline{C}_{nm} \cos m\lambda + \Delta\overline{S}_{nm} \sin m\lambda\right) \overline{P}_{nm}(\sin\varphi), \tag{3}$$

where GM is the geocentric gravitational constant (product of Newton's gravitational constant G and the Earth's mass M) and R the reference radius of the GGM, $\gamma = \gamma(P)$ is the normal gravity value at the Earth's surface point P, and $\overline{P}_{nm}$ denotes the fully normalized associated Legendre functions of degree n and order m. The fully normalized SH coefficients $\overline{C}^{\mathrm{GGM}}_{nm}$ and $\overline{S}^{\mathrm{GGM}}_{nm}$ of the GGM are reduced by the coefficients $\overline{C}^{\mathrm{REF}}_{nm}$ and $\overline{S}^{\mathrm{REF}}_{nm}$ of a normal field by setting

$$\begin{Bmatrix} \Delta\overline{C}_{nm} \\ \Delta\overline{S}_{nm} \end{Bmatrix} = \begin{Bmatrix} \overline{C}^{\mathrm{GGM}}_{nm} \\ \overline{S}^{\mathrm{GGM}}_{nm} \end{Bmatrix} - \frac{GM_0}{GM} \cdot \left(\frac{a}{R}\right)^n \cdot \begin{Bmatrix} \overline{C}^{\mathrm{REF}}_{nm} \\ \overline{S}^{\mathrm{REF}}_{nm} \end{Bmatrix}, \tag{4}$$

where GM_0 and a are the geocentric gravitational constant and the semi-major axis of the used reference ellipsoid, e.g., GRS80 (Geodetic Reference System 1980, Moritz, 1980).

As mentioned above, satellite-based GGMs derived from gravity field missions like GOCE are of special interest, as they provide independent and homogeneous gravity field information that is not affected by the discrepancies of LVD zones. For the accuracy of the height datum offset estimation, the omission and the commission error of the GGM plays an important role. While the omission error is induced due to the truncation of the series in Eq. (3) at a certain degree $n = n_2$, the commission error results from the uncertainties of the SH coefficients up to this degree.

In the present context, the GNSS/leveling data comprises the full spectral information of the gravity field. For spectral consistency, this would require an evaluation of the series in

Eq. (3) up to infinity. However, the latest satellite-based GOCE GGMs are restricted to a maximum degree $n_{\max}^{\text{GOCE}} \leq 300$. The resulting omission error may reach several decimeters in locations with rugged terrain. Hence, its influence has to be reduced by estimating the remaining signal above the spectral resolution of the GOCE GGM, using additional data. To this end, a spectral decomposition of the height anomaly ζ is performed by setting

$$\zeta = \zeta^{\text{LF}} + \zeta^{\text{MF}} + \zeta^{\text{HF}}, \tag{5}$$

where ζ^{LF} are the low-frequencies that are represented by a satellite-based GOCE GGM, and ζ^{MF} are the mid-frequencies that are derived from EGM2008 up to d/o 2190. The remaining high-frequency signals ζ^{HF} above the spectral resolution of EGM2008 are assumed to be mainly influenced by the Earth's topographic masses and are obtained by gravity forward modeling in this paper.

To achieve a suitable spectral extension, the three signal components have to be combined carefully in order to match their respective spectral signal content. Details on this aspect are provided in the following two subsections.

2.2. Combination of GOCE GGMs and EGM2008

The information of a satellite-based GOCE GGM and EGM2008 can be merged in the frequency domain by using a combination of their respective SH coefficients $(\overline{C}_{nm}^{\text{GOCE}}, \overline{S}_{nm}^{\text{GOCE}})$ and $(\overline{C}_{nm}^{\text{EGM}}, \overline{S}_{nm}^{\text{EGM}})$. To ensure a smooth transition between the two models, a transition window is used in the following, that is parameterized by its central degree N, called the SH degree of combination, and its half-bandwidth dN, denoted as transition bandwidth. Note that it is required that $N - dN \geq 0$ and $N + dN \leq n_{\max}^{\text{GOCE}}$.

The combined SH coefficients $(\overline{C}_{nm}^{\text{G/E}\,(N,\,dN)}, \overline{S}_{nm}^{\text{G/E}\,(N,\,dN)})$ can then be derived by a (convex) linear combination:

$$\begin{Bmatrix} \overline{C}_{nm}^{\text{G/E}\,(N,\,dN)} \\ \overline{S}_{nm}^{\text{G/E}\,(N,\,dN)} \end{Bmatrix} = \Big[1 - w_{(N,dN)}(n)\Big] \begin{Bmatrix} \overline{C}_{nm}^{\text{GOCE}} \\ \overline{S}_{nm}^{\text{GOCE}} \end{Bmatrix} + w_{(N,dN)}(n) \begin{Bmatrix} \overline{C}_{nm}^{\text{EGM}} \\ \overline{S}_{nm}^{\text{EGM}} \end{Bmatrix}, \tag{6}$$

where

$$w_{(N,dN)}(n) = \begin{cases} 0, & 0 \leq n \leq N - dN, \\ f_{(N,dN)}(n), & N - dN < n \leq N + dN, \\ 1, & N + dN < n \leq n_{\max}^{\text{EGM}}, \end{cases} \tag{7}$$

are degree-wise weight factors and $f_{(N,dN)}(n) \in (0,1)$ is the applied transition function.

In this paper, an adapted Hanning function (Blackman and Tukey, 1958) will be used, that is restricted to the interval $(0,1)$, i.e.,

$$f_{(N,dN)}(n) = \frac{1}{2}\Big[1 - \cos\Big(\frac{\pi(N - dN - n)}{2dN}\Big)\Big]. \tag{8}$$

Note that for the special case of $dN = 0$, Eq. (7) degenerates to a widely used concatenation of the SH coefficients, i.e.,

$$w_{(N,0)}(n) = \begin{cases} 0, & 0 \leq n \leq N, \\ 1, & N < n \leq n_{\max}^{\text{EGM}}. \end{cases} \tag{9}$$

It is acknowledged that there are other approaches like a stochastic combination based on the error variances of the SH coefficients, e.g., Huang and Véronneau (2013) and Ferreira et al. (2016). However, such a procedure assumes that the specified errors of both models are uncorrelated and realistic, which is not the case in general as will be shown in Sect. 4.3.

By using the combined SH coefficients ($\overline{C}_{nm}^{\text{G/E}\,(N,dN)}$, $\overline{S}_{nm}^{\text{G/E}\,(N,dN)}$) for the spherical harmonic synthesis in Eq. (3), the height anomaly $\zeta_{0,2190}^{\text{G/E}\,(N,dN)}$ is obtained that combines the low- and mid-frequency signal components as needed for Eq. (5), i.e.,

$$\zeta^{\text{LF}} + \zeta^{\text{MF}} := \zeta_{0,2190}^{\text{G/E}\,(N,dN)}. \tag{10}$$

2.3. Topography-implied gravity signals

Based on the information of a global high-resolution DTM, the impact of the Earth's topographic masses can generally be obtained by gravity forward modeling via the evaluation of Newton's integral (Heiskanen and Moritz, 1967, p. 3).

In the following, space domain gravity forward modeling based on the Rock–Water–Ice (RWI) approach is used that has been proposed by Grombein et al. (2014) and was further developed in Grombein et al. (2016a). This method is characterized by a three-layer decomposition of the Earth's topography with respect to its (1) rock, (2) water, and (3) ice masses. In contrast to condensation methods, e.g., the use of rock-equivalent heights, geometry changes and mass displacements are avoided due to a rigorous separate modeling of the rock, water, and ice masses with layer-specific density values (ρ_1, ρ_2, ρ_3).

The required geometrical information, i.e., the MSL heights of the upper boundary surfaces of the rock, water, and ice masses (h_1, h_2, h_3), is obtained from the $1' \times 1'$ global Earth2014 topography model (Hirt and Rexer, 2015) and parameterized in terms of geocentric spherical coordinates

$$\varphi_v = 90^\circ - (v - 1/2) \cdot \Delta\varphi, \qquad v = 1, \ldots, v_n \in \mathbb{N}, \tag{11}$$

$$\lambda_w = (w - 1/2) \cdot \Delta\lambda, \qquad w = 1, \ldots, w_m \in \mathbb{N}, \tag{12}$$

with $\Delta\varphi = \Delta\lambda = 1'$.

To correctly locate the topographic masses in space, the MSL is represented by a geocentric reference radius $R_0(\varphi, \lambda)$ that combines the latitude-dependent radius of the GRS80 ellipsoid and an additional geoid undulation, see Grombein et al. (2016a) for details. For each grid element, the geocentric radii of the upper boundary surfaces of rock, water, and ice masses are then approximated by $R_1 = R_0 + h_1$, $R_2 = R_0 + h_2$, and $R_3 = R_0 + h_3$, respectively. Although the sequence of these masses is the same for different terrain types, some mass types may be lacking (cf. Fig. 2 in Grombein et al., 2016a).

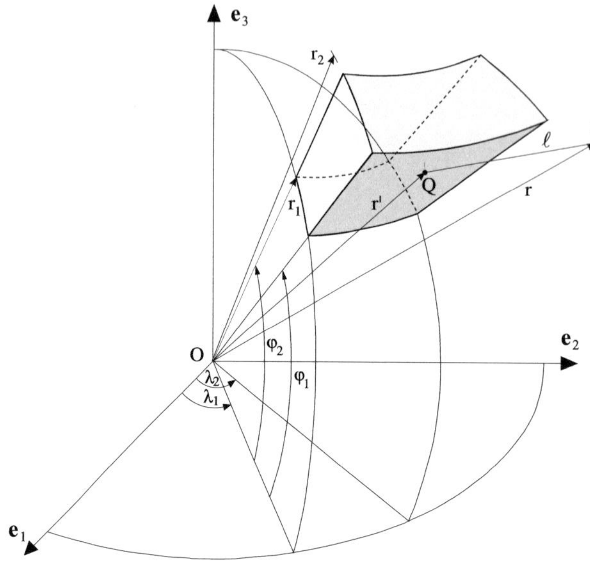

Fig. 1. Geometry of a tesseroid mass body used for mass discretization; the spherical coordinates (r, φ, λ) are referred to the geocentric Earth-fixed equatorial reference system defined by the base vectors $\boldsymbol{e_1}$, $\boldsymbol{e_2}$, $\boldsymbol{e_3}$.

For the mass discretization of Newton's integral, tesseroid mass bodies are used, which are shown in Fig. 1 and described in Heck and Seitz (2007) and Grombein et al. (2013). In terms of the height anomaly ζ, the gravitational impact of a single tesseroid with a constant mass density ρ can be calculated by

$$\zeta^*(P) = \frac{G\rho}{\gamma} \int_{r_1}^{r_2} \int_{\varphi_1}^{\varphi_2} \int_{\lambda_1}^{\lambda_2} \frac{r'^2 \cos\varphi'}{\ell} \, \mathrm{d}r' \, \mathrm{d}\varphi' \, \mathrm{d}\lambda', \tag{13}$$

where G denotes Newton's gravitational constant, and $\ell = \ell(P, Q)$ is the Euclidean distance between the attracted computation point $P(r, \varphi, \lambda)$ and the running integration mass point $Q(r', \varphi', \lambda')$, see Fig. 1.

The total RWI-based topographic effect ζ^{Topo} is then calculated as the sum of the impact over all individual tesseroids (superposition principle):

$$\zeta^{\text{Topo}}(P) = \frac{G}{\gamma} \sum_{v=1}^{v_n} \sum_{w=1}^{w_m} \sum_{s=1}^{3} \rho_s \int_{R_{s-1}}^{R_s} \iint_{\sigma^*_{vw}} \frac{r'^2 \cos\varphi'}{\ell} \, \mathrm{d}r' \, \mathrm{d}\varphi' \, \mathrm{d}\lambda', \tag{14}$$

where

$$\sigma^*_{vw} = \left[\varphi_v - \frac{\Delta\varphi}{2}, \varphi_v + \frac{\Delta\varphi}{2}\right] \times \left[\lambda_w - \frac{\Delta\lambda}{2}, \lambda_w + \frac{\Delta\lambda}{2}\right] \tag{15}$$

denotes the spherical tesseroid base surface associated with the grid element (v, w).

As the integration with respect to σ^*_{vw} comprises elliptical integrals, a numerical evaluation of Eq. (14) is achieved by means of expanding the integral kernel in a Taylor series and performing a subsequent term-wise integration (cf. Heck and Seitz, 2007). The applied evaluation rules that provide a fourth-order error in the spatial coordinates of the integration point are explicitly given in Grombein et al. (2013) and are adapted to the RWI approach in Grombein et al. (2016a).

As the height anomaly ζ^{Topo} in Eq. (14) contributes to all spectral scales of the gravity field, it cannot directly be used for the spectral extension of the GGM information. Thus, in addition, some kind of high-pass filtering has to be performed.

For this purpose, residual terrain modeling (RTM, Forsberg and Tscherning, 1997) is a widely used tool that has been utilized in various studies and for different kinds of applications (e.g., Hirt et al., 2010; Hirt, 2013; Šprlák et al., 2015). In an RTM-based approach, gravity forward modeling is applied to a residual topography between a high-resolution DTM and a smoothed reference topography, whose spectral information is assumed to be already contained in the GGM. In this way, the RTM method principally presupposes a spectral consistency of topographic heights and implied gravity. However, as shown by Hirt and Kuhn (2014), a band-limited topography can also generate a full spectrum gravity field information, which emphasizes the nonlinearity of the underlying problem. Therefore, the basic assumption of the RTM approach cannot be considered valid in general. Thus, it is recommended to perform the high-pass filtering not on the basis of topographic heights, but directly in the gravity domain.

For this purpose, the topographic gravity field model RWI_TOPO_2015 (Grombein et al., 2016a) is used in the following. The SH coefficients of this model have been obtained from the spherical harmonic analysis of a $4' \times 4'$ global grid of RWI-based topographic potential values, calculated according to the formalism specified above and the used topographic input data. Thus, it is expected that the height anomalies derived from the RWI_TOPO_2015 model are consistent with those of Eq. (14), and only differ in their spectral content. As a consequence, the long to medium wavelengths of ζ^{Topo} in Eq. (14) can be removed consistently by subtracting the effect of the RWI_TOPO_2015 model.

To augment the spectral resolution of $\zeta^{\text{G/E}\,(N,\,dN)}_{0,\,2190}$ in Eq. (5), the topography-implied height anomaly is then calculated by

$$\delta\zeta^{\text{Topo}} = \zeta^{\text{Topo}} - \zeta^{\text{RWI}}_{0,\,2159} =: \zeta^{\text{HF}}, \tag{16}$$

where $\zeta^{\text{RWI}}_{0,\,2159}$ is the synthesized height anomaly of the RWI model up to d/o 2159, which can be computed by spherical harmonic synthesis according to Eq. (3). Note that, in this case, a normal field in Eq. (4) must not be subtracted from the SH coefficients of RWI_TOPO_2015.

At this point, some additional remarks on the choice of d/o 2159 are appropriate. While EGM2008 provides a maximum degree of $n^{\text{EGM}}_{\text{max}} = 2190$, it is only complete to d/o 2159, containing some additional coefficients up to degree 2190 and order 2159 (Pavlis et al., 2012). These coefficients result from the transformation between ellipsoidal and spherical harmonics and provide crucial information, needed for an accurate spherical harmonic synthesis, particularly in polar regions. Therefore, it is recommended to consider them, although they do not add further spectral information beyond d/o 2159. Thus, in order to augment the spectral resolution of EGM2008, topographic potential values need to be high-pass filtered with respect to d/o 2160.

2.4. Estimation of height datum offsets by least squares adjustment

By inserting Eq. (5) into Eq. (2) and using

$$\zeta^{\,\text{GNSS/lev}} = h - H', \tag{17}$$

the height datum offset δH can be evaluated pointwise for each GNSS/leveling benchmark P_j by

$$\delta H(P_j) = \zeta^{\,\text{GNSS/lev}} - \zeta^{\,\text{G/E}\,(N,\,dN)}_{\,0,\,2190} - \delta\zeta^{\,\text{Topo}}\Big|_{P_j}. \tag{18}$$

However, due to approximation and measurement errors, offsets derived in this way will vary regionally, so that the accuracy and reliability can be increased by averaging over a larger area or the entire LVD zone.

While the errors in ellipsoidal heights and GGM-derived height anomalies are expected to be mainly random, physical heights can additionally be affected by systematic leveling errors and distortions in the leveling network. In order to compensate for these errors, the height datum offset δH will be estimated in a least squares adjustment (LSA) by applying an additional parametric model. In this context, different models have been analyzed and tested by Kotsakis et al. (2012) and Ferreira et al. (2016), in the case of the Hellenic islands and Brazil, respectively. In the following, a 3-parameter (plane) model will be used, which is in accordance with other studies, e.g., Rülke et al. (2012) and Amjadiparvar et al. (2016).

For each GNSS/leveling point P_j, parameterized by its geodetic coordinates (B_j, L_j), an observation equation is then formulated as

$$l_j = \zeta^{\,\text{GNSS/lev}} - \zeta^{\,\text{G/E}\,(N,\,dN)}_{\,0,\,2190} - \delta\zeta^{\,\text{Topo}}\Big|_{P_j} = \delta H + a_1(B_j - B_0) + a_2(L_j - L_0)\cos B_j, \tag{19}$$

where δH is the unknown height datum offset in the centroid (B_0, L_0), corresponding to the mean over all GNSS/leveling benchmarks P_j, and a_1 and a_2 are the associated unknown tilts in north-south and east-west direction. The functional model according to Eq. (19) is specified by

$$\underbrace{\begin{pmatrix} l_1 + v_1 \\ l_2 + v_2 \\ \vdots \\ l_J + v_J \end{pmatrix}}_{\boldsymbol{l}+\boldsymbol{v}} = \underbrace{\begin{pmatrix} 1 & (B_1 - B_0) & (L_1 - L_0)\cos B_1 \\ 1 & (B_2 - B_0) & (L_2 - L_0)\cos B_2 \\ \vdots & \vdots & \vdots \\ 1 & (B_J - B_0) & (L_J - L_0)\cos B_J \end{pmatrix}}_{\mathbf{A}} \cdot \underbrace{\begin{pmatrix} \delta H \\ a_1 \\ a_2 \end{pmatrix}}_{\boldsymbol{x}}, \tag{20}$$

where $\boldsymbol{l}$ is the observation vector, $\boldsymbol{v}$ the residual (or inconsistency) vector, and $\boldsymbol{x}$ the vector of unknowns. The design matrix $\mathbf{A}$ contains the partial derivatives of the observations with respect to the unknowns. Using a standard LSA, the unknown height datum offsets are estimated by

$$\hat{\boldsymbol{x}} = \mathbf{N}^{-1}\mathbf{A}^{\mathrm{T}}\mathbf{P}\,\boldsymbol{l}, \tag{21}$$

where $\mathbf{N} = \mathbf{A}^{\mathrm{T}}\mathbf{P}\mathbf{A}$ is the normal matrix and $\mathbf{P}$ is the weight matrix that can be specified, when reliable error information of the observations is available.

As noted by Gerlach and Fecher (2012), an optimal estimation of height datum offsets would theoretically require the full error variance-covariance matrix (VCM) of the observations $\mathbf{C}_{ll} = \mathbf{P}^{-1}$. Assuming that the errors of the involved terms in Eq. (19) are pairwise uncorrelated, this matrix can be specified by

$$\mathbf{C}_{ll} = \mathbf{C}_{hh} + \mathbf{C}_{H'H'} + \mathbf{C}_{\zeta\zeta}, \tag{22}$$

where $\mathbf{C}_{hh}$ and $\mathbf{C}_{H'H'}$ are the error VCMs of the measured ellipsoidal and normal heights, respectively, and $\mathbf{C}_{\zeta\zeta}$ is the error VCM of the GGM-derived height anomalies ζ.

On the one hand, $\mathbf{C}_{\zeta\zeta}$ might be derived from error propagation of the GGM's SH coefficients, cf. Haagmans and van Gelderen (1991). However, this (i) generally requires the full error VCM of the SH coefficients, which might not be available, and (ii) is computationally very demanding (e.g., Gerlach and Fecher, 2012). One the other hand, realistic uncertainties for the observed ellipsoidal and normal heights are generally not available, particularly for historical measurements (Amjadiparvar et al., 2016). Therefore, in most practical cases, it is not possible to compile $\mathbf{C}_{hh}$ and $\mathbf{C}_{H'H'}$, and thus $\mathbf{C}_{ll}$.

In this case, all observations are usually equally weighted by setting $\mathbf{P} = \mathbf{I}$, where $\mathbf{I}$ is the identity matrix. Although this is also done in this study, in Sect. 4, the uncertainties of all involved terms will be estimated in a separate LSA, assuming that they are uncorrelated.

3. Study areas and data sets

In Sect. 3.1, the three study areas and their respective national height systems are introduced, while Sect. 3.2 describes the used GNSS/leveling data sets in these countries. Furthermore, in Sect. 3.3, the applied global geopotential models are presented, and in Sect. 3.4, a homogenization of all data sets is performed.

3.1. Study areas

For the analysis in this paper, three representative study areas have been selected: Germany, Austria, and Brazil. They differ strongly in size, geographical latitude, topographic properties, as well as the accuracy of their geodetic measurements. In the following, the national height system in each of these countries is briefly described.

The German height system is currently materialized by the German Primary Leveling Network 1992 (Deutsches Haupthöhennetz 1992, DHHN92), which was established after the German reunification by combining the former networks of East and West Germany. With the implementation of DHHN92, gravity-related normal heights were introduced for the whole country. The LVD of DHHN92 is realized by the reference level of the Normaal

Amsterdams Peil (NAP) as observed in the time period 1683–1684. For further information on DHHN92, the reader is referred to Weber (1994) and Ihde (1995). Note that DHHN92 will soon be replaced by a new realization called DHHN2016, which is based on the latest precise levelings, measured between 2006 and 2012.

The national height system in Austria dates back to the spirit leveling measurements carried out by the Military Geographical Institute (MGI) in the end of the 19th century. Since 1949, the leveling network has been continuously supplemented by several densifications and updated due to re-measurements (Ruess and Mitterschiffthaler, 2015). Although a new orthometric height system with respect to NAP is planned to be implemented in several years (Höggerl and Ruess, 2004), the official physical heights in Austria are currently still normal-orthometric heights related to the mean sea surface of the Adriatic Sea as observed in 1875 at the tide gauge in Trieste. Additional information about the Austrian height system and the adjustment of the leveling network is provided in Höggerl (1986).

The Brazilian height system is based on the Brazilian Fundamental Vertical Network (BFVN) that was established in 1945 and consists of two independent parts, both connected to their own tide gauges. The major part, spreading over almost the whole country, is linked to the tide gauge in Imbituba in the south of Brazil. As separated by the estuary of the Amazon River, a second, smaller leveling network was established in the north of Brazil that is linked to the tide gauge in Santana. Hence, in the case of Brazil, there exist two independent LVD zones. While in the following the larger datum zone with the origin in Imbituba is considered, a study of the discrepancies between both Brazilian datum zones can be found in Montecino and de Freitas (2014). At the Imbituba tide gauge, sea level observations from 1949 to 1957 were used to define the local MSL (Gomez et al., 2016). As the spirit leveling in Brazil was not generally supplemented by gravity measurements, normal-orthometric heights were derived (Luz et al., 2009b). More information and a review of the Brazilian first-order leveling network are provided by Luz et al. (2002).

3.2. GNSS/leveling data sets

For Germany, a GNSS/leveling data set consisting of 675 benchmarks was made available by the German Federal Agency for Cartography and Geodesy (BKG). This data comprises ellipsoidal heights that originate from GNSS campaigns conducted between 1994 and 2001, and normal heights with respect to DHHN92. The Austrian Federal Office for Metrology and Surveying (BEV) provided a set of 198 GNSS/leveling points, consisting of GNSS-based ellipsoidal heights and MGI-based normal-orthometric heights. In the case of Brazil, the Brazilian Institute of Geography and Statistics (IBGE) delivered GNSS-derived ellipsoidal heights and normal-orthometric heights, corresponding to the latest adjustment of the BFVN in 2011. While the data set consists of 683 benchmarks, a subset of 666 points is used in this study that can be associated with the LVD defined by the Imbituba tide gauge.

In Fig. 2, the spatial distribution of the GNSS/leveling benchmarks and their physical heights are visualized for the three study areas. The coordinates (B_0, L_0) of the centroid of these benchmarks are (50.9182° N, 10.2230° E) for Germany, (47.5022° N, 13.9498° E) for Austria, and (15.3106° S, 46.4830° W) for Brazil. In the case of Germany and Austria, the GNSS/leveling points are more or less homogeneously scattered over the country. In contrast, the benchmarks in Brazil are predominantly concentrated in coastal regions and around major cities. The point distances in the back country are considerably larger, particularly in the poorly accessible area of the Amazon Rainforest in Northern Brazil.

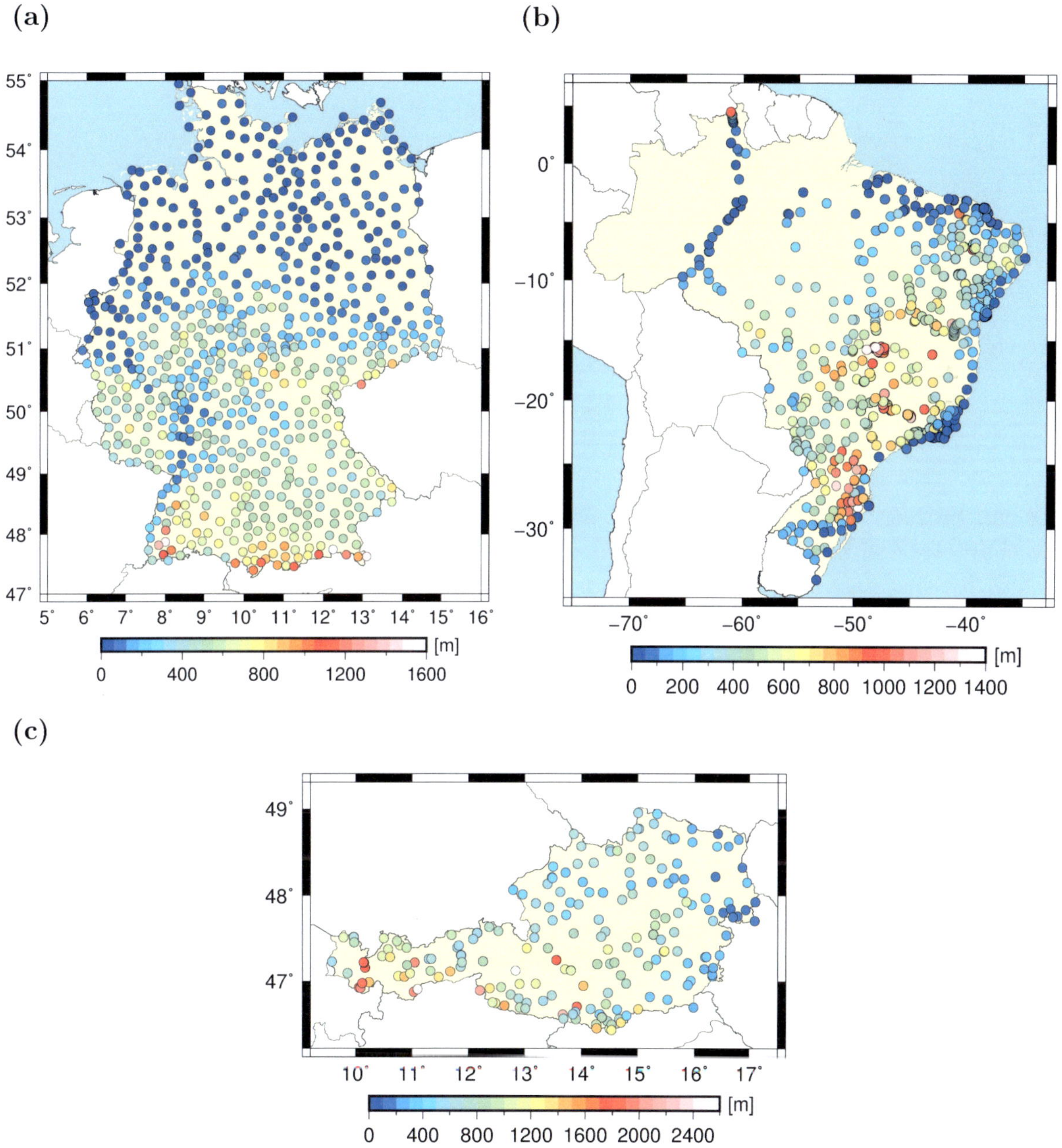

Fig. 2. Spatial distribution of the used GNSS/leveling data sets in **(a)** Germany, **(b)** Brazil, and **(c)** Austria. The color coding represents the physical heights in the corresponding LVD. Note the different scaling of the color bars.

Table 1. Statistics of the physical heights H', the ellipsoidal heights h, and the resulting height anomalies $\zeta^{\text{GNSS/lev}} = h - H'$ of the used GNSS/leveling benchmarks in Germany, Austria, and Brazil. All values are specified in [m].

		Min	Max	Mean	STD
	H'	0.361	1567.272	313.081	274.452
Germany	h	36.117	1613.603	357.903	276.822
	$\zeta^{\text{GNSS/lev}}$	35.676	50.265	44.822	3.466
	H'	113.464	2577.840	762.904	469.799
Austria	h	157.300	2627.882	810.323	471.420
	$\zeta^{\text{GNSS/lev}}$	42.742	52.243	47.419	2.100
	H'	1.726	1356.648	398.530	319.907
Brazil	h	−18.360	1361.470	390.953	321.247
	$\zeta^{\text{GNSS/lev}}$	−28.134	22.007	−7.577	8.594

As can be seen from Table 1 and the color coding in Fig. 2, the three study areas can be characterized by different magnitudes in physical height. In the case of the German GNSS/leveling data set, the heights show an almost continuously increasing behavior from the Northern Lowland via the Central Uplands toward South Germany, where the mountain range of the Black Forest and the northern edge of the Alps provide the highest elevations of up to 1600 m. Due to the high mountain ranges of the Alps, the physical heights of the Austrian GNSS/leveling data set provide significantly larger elevations and show a higher variability, particularly in West and South Austria. Thus, the mean and standard deviation (STD) values of the physical heights are about twice as large as in the case of the German data set. The Brazilian GNSS/leveling data set comprises lower heights in coastal regions and for the Amazon basin, and moderate elevations for the Brazilian Plateau in between. Particularly in the southeast, it is notable that the physical heights are rapidly increasing from the zero level at the coast up to 1000 m at the plateau. The highest elevations in the data set, about 1400 m, are reached in central Brazil, around the capital Brasília.

3.3. Global geopotential models

Three different GOCE GGMs of the latest (fifth) generation will be considered in this study: TIM R5 (Brockmann et al., 2014), DIR R5 (Bruinsma et al., 2014), and GOCO05s (Mayer-Gürr et al., 2015). These models are derived by using observations of the complete GOCE mission of about 42 months (Nov 2009 – Oct 2013), including the lower orbit phase at the end of the mission's lifetime.

TIM R5 expanded up to d/o 280 is the latest model of the time-wise approach and is the only GGM included that is based purely on GOCE data. In contrast, the DIR R5 model up to d/o 300 is based on the direct approach and combines GOCE data with observations of the satellite mission Gravity Recovery and Climate Experiment (GRACE, Tapley et al., 2004) and satellite laser ranging (SLR). Finally, the GOCO05s model up to d/o 280 basically consists of a combination of TIM R5 and the latest GRACE-only model ITSG-Grace2014s (Mayer-Gürr et al., 2014) and is also supplemented by SLR observations. Detailed information about the applied processing strategies of the different approaches is presented in Pail et al. (2010, 2011).

Additionally, the high-resolution EGM2008 model (Pavlis et al., 2012) up to d/o 2190 is used for the reduction of the omission error of the GOCE GGMs. EGM2008 combines satellite-based gravity information of GRACE with terrestrial, airborne and altimetry-derived gravity data that are partially supplemented with topography-implied gravity information.

For the spherical harmonic synthesis of GGMs according to Eq. (3), an adapted version of the *harmonic_synth* software (Holmes and Pavlis, 2006) is used. The contribution of the zero degree term $\zeta_{0,0}^{\mathrm{GGM}}$ is calculated separately, as shown in Lemoine et al. (1998, Sect. 11.2):

$$\zeta_{0,0}^{\mathrm{GGM}} = \frac{GM - GM_0}{r\gamma} - \frac{W_0 - U_0}{\gamma}, \tag{23}$$

where W_0 denotes the gravity potential of the global geoid and U_0 is the normal potential of the used reference ellipsoid. By applying the conventional value $W_0 = 62\,636\,856.0\,\mathrm{m^2\,s^{-2}}$ (IERS Conventions, 2010, p. 18) and using the normal potential of the GRS80 ellipsoid, $U_0 = 62\,636\,860.850\,\mathrm{m^2\,s^{-2}}$ (Moritz, 1980), Eq. (23) results in $\zeta_{0,0}^{\mathrm{GGM}} \approx -44.2\,\mathrm{cm}$.

It is acknowledged that due to the IAG Resolutions (2015), a potential value of $W_0 = 62\,636\,853.4\,\mathrm{m^2\,s^{-2}}$ has been defined for the realization of an international height reference system. However, as the present study started before this resolution has been adapted, the above specified W_0 value of the IERS Conventions (2010) is used throughout this paper.

3.4. Homogenization of the data

Before the different data sets can be used for the estimation of height datum offsets, they have to be homogenized and transformed to the same tidal system and reference frame. Moreover, in the following, the normal-orthometric heights of the Austrian and Brazilian data sets are used as approximations of normal heights. This is in accordance with Wolf (1974), where it has been demonstrated that normal-orthometric heights are closer to normal heights than to orthometric heights.

Regarding the treatment of permanent tides, various geodetic quantities are usually referred to different tidal systems (e.g., Mäkinen and Ihde, 2009). Since in all three study areas no tidal corrections have been applied to the leveling observations, the derived physical heights H' of the GNSS/leveling data can be referred to the mean tide system. In contrast, the GNSS-based ellipsoidal heights h are conventionally specified in the tide free system. This is also the case for the SH coefficients of the used GGMs with the exception of GOCO05s, where the coefficients are provided in the zero tide system. For more details on the relations between the different tidal systems, see IERS Conventions (2010, p. 15ff.).

In this study, the (conventional) tide-free system will consistently be used for all quantities. Note that this specific choice has no influence on the results, as the impact of the tidal system is eliminated due to the subtraction in Eq. (18). According to Ekman (1989), the physical heights H' are then transformed from the mean tide (MT) to the tide free (TF) system by using

$$H'^{(\mathrm{TF})} = H'^{(\mathrm{MT})} + (1 + \bar{k} - \bar{h}) \cdot (0.099 - 0.296 \sin^2 \varphi), \tag{24}$$

where $\bar{k} = 0.30$ and $\bar{h} = 0.62$ are the (semidiurnal) Love numbers that are assumed constant. In the case of the SH coefficients, only $\overline{C}_{20}$ is affected by the used tidal system. Hence, the $\overline{C}_{20}$ coefficient of GOCO05s is transformed from the zero tide (ZT) to the tide free system according to Rapp et al. (1991):

$$\overline{C}_{20}^{(\mathrm{TF})} = \overline{C}_{20}^{(\mathrm{ZT})} + 3.1108 \cdot 10^{-8} \cdot \frac{\bar{k}}{\sqrt{5}}. \tag{25}$$

Although the geodetic coordinates of all three GNSS/leveling data sets are consistently defined with respect to the GRS80 ellipsoid, the original GNSS-derived Cartesian coordinates, the GOCE GGMs, and EGM2008 are principally related to different reference frames and epochs (Barzaghi et al., 2016). Generally, the used reference frame of the GNSS coordinates is indicated, but mostly without an explicit specification of a corresponding epoch. The GOCE measurements basically refer to the Earth's center of mass, which might be best represented by the most recent solution of the International Terrestrial Reference Frame (ITRF). However, in the case of EGM2008, no information is available about the time reference of the used terrestrial gravity data.

Due to this lack of metadata, reference frame transformations can only be properly addressed by using suitable assumptions. Since the required analysis in this context is beyond the scope of this article, no additional reference frame transformation is applied in this study.

4. Error analysis

In Sect. 4.1 the omission error of the GOCE GGMs is analyzed in the three study areas, while Sect. 4.2 focuses on the commission error. In addition, Sect. 4.3 provides an estimation of the uncertainties of all terms involved in the calculation of height datum offsets.

4.1. Omission error

The omission error of GOCE GGMs is mainly affected by the roughness of the remaining (residual) gravity field signal. However, its actual impact on the offset estimation depends on various factors, such as the size of the study area, and the density as well as spatial distribution of GNSS/leveling benchmarks within the area (Gruber et al., 2012). When averaged over all GNSS/leveling benchmarks, the effect of the omission error decreases or may cancel out to a certain degree, if the study area is sufficiently large (e.g., Amjadiparvar et al., 2013, 2016; Rülke et al., 2016).

To quantify the magnitude of the GOCE omission error above d/o 200 in the three study areas, the effect of $\zeta^{\mathrm{EGM}}_{201,2190}$ and $\delta\zeta^{\mathrm{Topo}}$ is evaluated at the respective GNSS/leveling points P_j. While Fig. 3 illustrates the spatial distribution of the omission error, corresponding statistics are presented in Table 2. In all study areas, the omission error due to $\zeta^{\mathrm{EGM}}_{201,2190}$ mainly ranges between ±100 cm, where in the case of Brazil the largest amplitude with −142.9 cm is reached. While in Germany the mean omission error practically cancels out over the GNSS/leveling benchmarks (0.1 cm), in Austria and Brazil this is quite different with values of 3.3 and −7.1 cm, respectively. The remaining part of the omission error, modeled by the topography-implied gravity signals $\delta\zeta^{\mathrm{Topo}}$, is about one order of magnitude smaller and reaches values on a cm–dm level in the case of the three study areas. Due to the influence of the Alps, the largest amplitude of 22.5 cm as well as the highest variation in terms of a STD of 5.2 cm can be detected for the Austrian data set. Moreover, Austria is also the only study area, where the mean omission error due to $\delta\zeta^{\mathrm{Topo}}$ has a significant value of −3 cm. In contrast, in the two other study areas, the omission error averaged over the GNSS/leveling benchmarks only amounts to 0.1 cm. However, as will be shown below, although the consideration of the omission error might only slightly affect the mean value, it can help to significantly improve the STD of the residuals.

Table 2. Statistics of the omission error as obtained from $\zeta^{\mathrm{EGM}}_{201,2190}$ and $\delta\zeta^{\mathrm{Topo}}$, both evaluated at the GNSS/leveling benchmarks P_j in Germany, Austria, and Brazil. All values are specified in [cm].

		SH band	Min	Max	Mean	STD
Germany	EGM2008	201–2190	−105.4	126.7	0.1	30.3
	Topo	≥ 2160	−13.2	6.9	0.1	1.1
Austria	EGM2008	201–2190	−87.4	133.6	3.3	48.4
	Topo	≥ 2160	−22.5	9.0	−3.0	5.2
Brazil	EGM2008	201–2190	−142.9	110.4	−7.1	32.3
	Topo	≥ 2160	−8.4	3.5	−0.1	1.0

4.2. Commission error

The commission error of a GGM can be obtained from its error degree variances. In this context, the commission error corresponds to the square root of the cumulative error degree

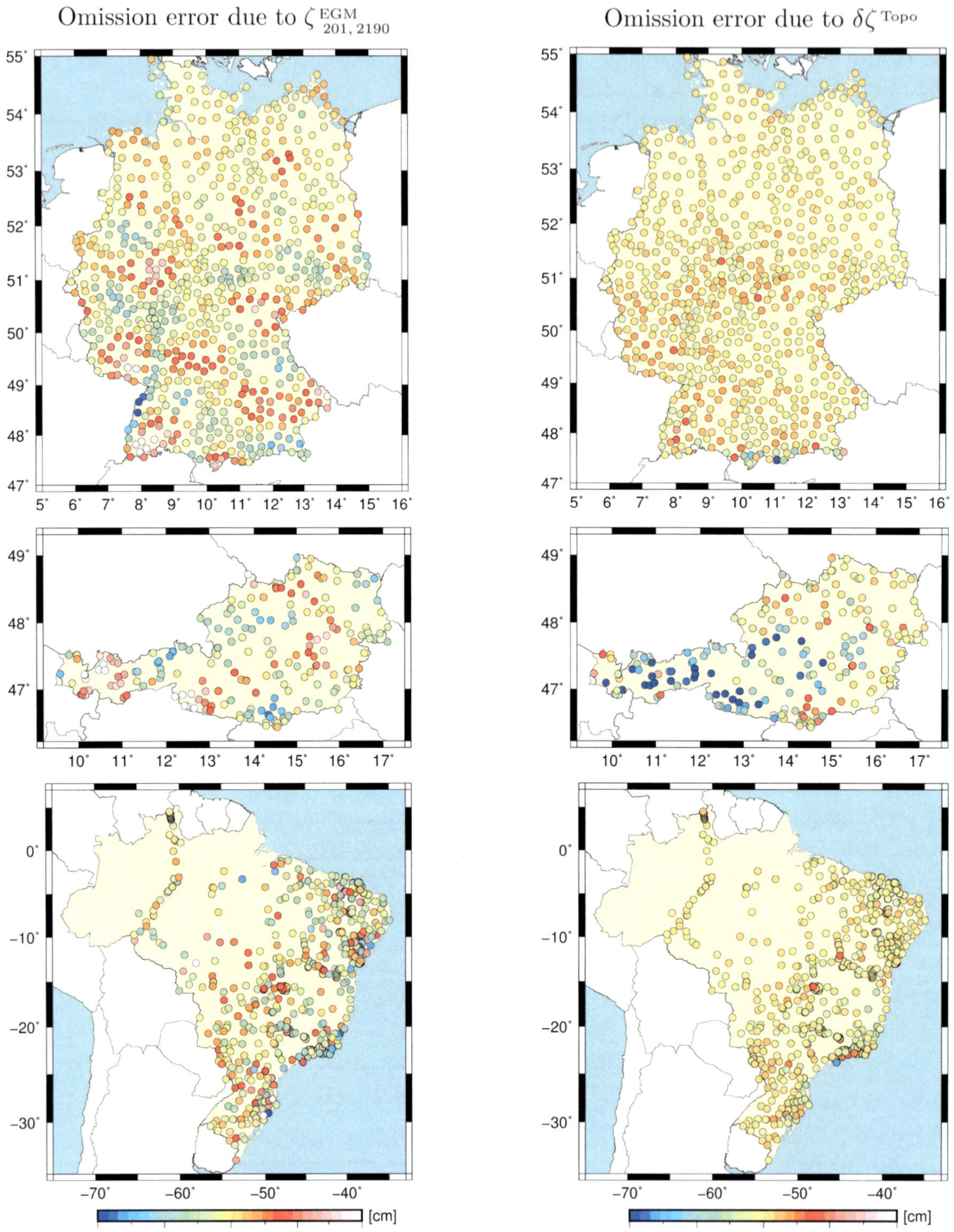

Fig. 3. Omission error obtained from $\zeta^{\mathrm{EGM}}_{201,\,2190}$ (*left column*) and $\delta\zeta^{\mathrm{Topo}}$ (*right column*) for the three study areas Germany, Austria, and Brazil (*first*, *second*, and *third row*, respectively). Note the different scaling of the color bars for the *left* and *right column*.

Table 3. Formal cumulative error σ_n^{GGM} of the used GGMs in terms of height anomalies ζ for different SH bands. All values are specified in [cm].

GGM	SH band	σ_n^{GGM}
TIM R5	2–200	2.24
DIR R5	2–200	0.79
GOCO05s	2–200	1.96
EGM2008	2–200	7.19
EGM2008	2–2190	8.24

variances, abbreviated as cumulative error in the following. In terms of height anomalies, the cumulative error up to a SH degree n can be calculated by

$$\sigma_n^{\mathrm{GGM}} = R\left[\sum_{k=0}^{n}\sum_{m=0}^{k}\left(\sigma^2\,\overline{C}_{km}^{\mathrm{GGM}} + \sigma^2\,\overline{S}_{km}^{\mathrm{GGM}}\right)\right]^{1/2}, \tag{26}$$

where $\sigma\,\overline{C}_{km}^{\mathrm{GGM}}$ and $\sigma\,\overline{S}_{km}^{\mathrm{GGM}}$ are the specified standard deviations of the SH coefficients.

In Fig. 4, the cumulative errors σ_n^{GGM} of the used GGMs are plotted, while Table 3 provides selected numerical values. The comparison of the results for the three GOCE GGMs shows significant differences. As the DIR R5 and GOCO05s models are supplemented by GRACE and SLR measurements, they generally provide smaller cumulative errors in the case of long wavelengths ($n < 150$) than the purely GOCE-based TIM R5 model. For $n > 200$ the cumulative errors of TIM R5 and GOCO05s strongly increase and their curves nearly coincide. Due to the limited bandwidth of the GOCE gradiometer, this behavior reflects the increasing signal-to-noise ratio of the gravity gradient measurements. Although the cumulative error of DIR R5 also becomes larger for an increasing degree n, this GGM provides significantly smaller values compared to TIM R5 and GOCO05s. Thus, one might assume that the DIR R5 model is somehow not as strongly affected by the increasing noise.

However, when analyzing the cumulative errors in Fig. 4 and Table 3, it has to be taken into account that (i) they are based on formal errors resulting from internal accuracy estimates of the used processing approach, (ii) they do not consider any correlations between the SH coefficients and, thus, might be too optimistic, and (iii) they provide global mean values and do not reflect regional variations.

Thus, in order to get reliable uncertainty estimates, the formal cumulative errors σ_n^{GGM} need to be calibrated by using suitable regional scaling factors m^{GGM}:

$$\left(\hat{\sigma}_n^{\mathrm{GGM}}\right)^2 = m^{\mathrm{GGM}} \cdot \left(\sigma_n^{\mathrm{GGM}}\right)^2. \tag{27}$$

In doing so, Bruinsma et al. (2014) specifies a scaling factor of $m^{\mathrm{DIR\ R5}} = 4$ for the DIR R5 model in Germany, which explains the differences to the other two GOCE models.

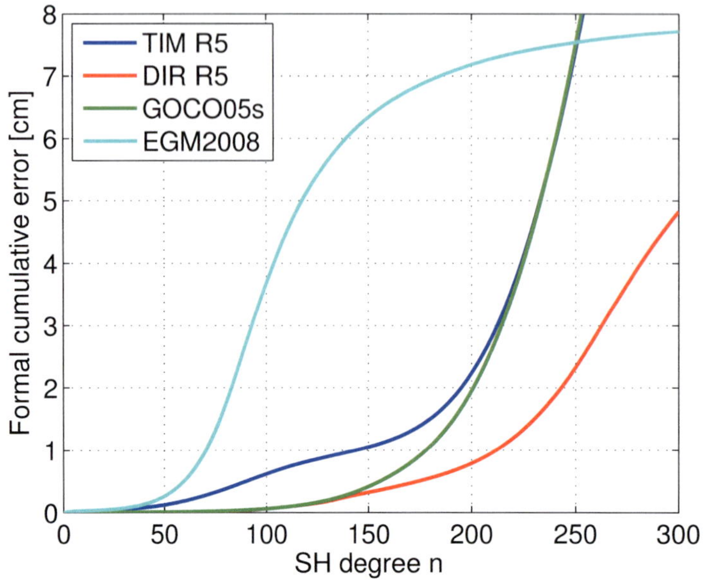

Fig. 4. Formal cumulative error $\sigma_n^{\,\mathrm{GGM}}$ of TIM R5 (*blue curve*), DIR R5 (*red curve*), GOCO05s (*green curve*), and EGM2008 (*cyan curve*) in terms of height anomalies ζ.

4.3. Estimation of uncertainties

In the following, individual scaling factors $m^{\,\mathrm{TIM\ R5}}$, $m^{\,\mathrm{DIR\ R5}}$, $m^{\,\mathrm{GOCO05s}}$, and $m^{\,\mathrm{EGM}}$ for the cumulative errors of the GGMs are estimated for each study area. For this purpose, a least squares approach suggested by Voigt and Denker (2015) is adapted to the present study. Besides the four scaling factors, this approach allows to simultaneously estimate the a priori unknown uncertainty $\sigma^{\,\mathrm{GNSS/lev}}$ of the GNSS/leveling data. For the estimation of these five unknowns, various differences are calculated in the GNSS/leveling points P_j and the resulting STD values are introduced as observations in a LSA. In contrast to the original approach of Voigt and Denker (2015), STD values are used rather than RMS (root mean square) differences, as they are not affected by the height datum offset contained in the GNSS/leveling data. Furthermore, by investigating the differences resulting from Eq. (19), as shown in Fig. 5, it seems reasonable to first remove systematic effects by means of considering the tilts a_1 and a_2, before computing the required STD values.

Altogether, three groups of observation equations are used to estimate the unknown parameters. The first group results from Eq. (19) by setting $N = 200$ and $dN = 0$:

$$\mathrm{STD}\left(\zeta^{\,\mathrm{GNSS/lev}} - \zeta^{\,\mathrm{G/E}\,(200,0)}_{\,0,\,2190} - \delta\zeta^{\,\mathrm{Topo}}\Big|_{P_j}\right)$$
$$= \left[\left(\sigma^{\,\mathrm{GNSS/lev}}\right)^2 + m^{\,\mathrm{GOCE}}\left(\sigma^{\,\mathrm{GOCE}}_{\,200}\right)^2 + m^{\,\mathrm{EGM}}\left(\left(\sigma^{\,\mathrm{EGM}}_{\,2190}\right)^2 - \left(\sigma^{\,\mathrm{EGM}}_{\,200}\right)^2\right)\right]^{1/2}, \qquad (28)$$

where $\sigma^{\,\mathrm{GOCE}}_{\,200}$, $\sigma^{\,\mathrm{EGM}}_{\,200}$, and $\sigma^{\,\mathrm{EGM}}_{\,2190}$ are the formal cumulative errors as specified in Table 3. Note that on the right-hand side of Eq. (28), the uncertainty of the topography-implied gravity signals $\delta\zeta^{\,\mathrm{Topo}}$ is omitted as it cannot be estimated independently of $\sigma^{\,\mathrm{GNSS/lev}}$.

Therefore, $\sigma^{\text{GNSS/lev}}$ in Eq. (28) contains the error budget of both, the GNSS/leveling data and the topography-implied gravity signals. However, based on the results of Voigt and Denker (2015), the uncertainty of the latter is expected to be at the sub-mm level and, hence, considered negligible.

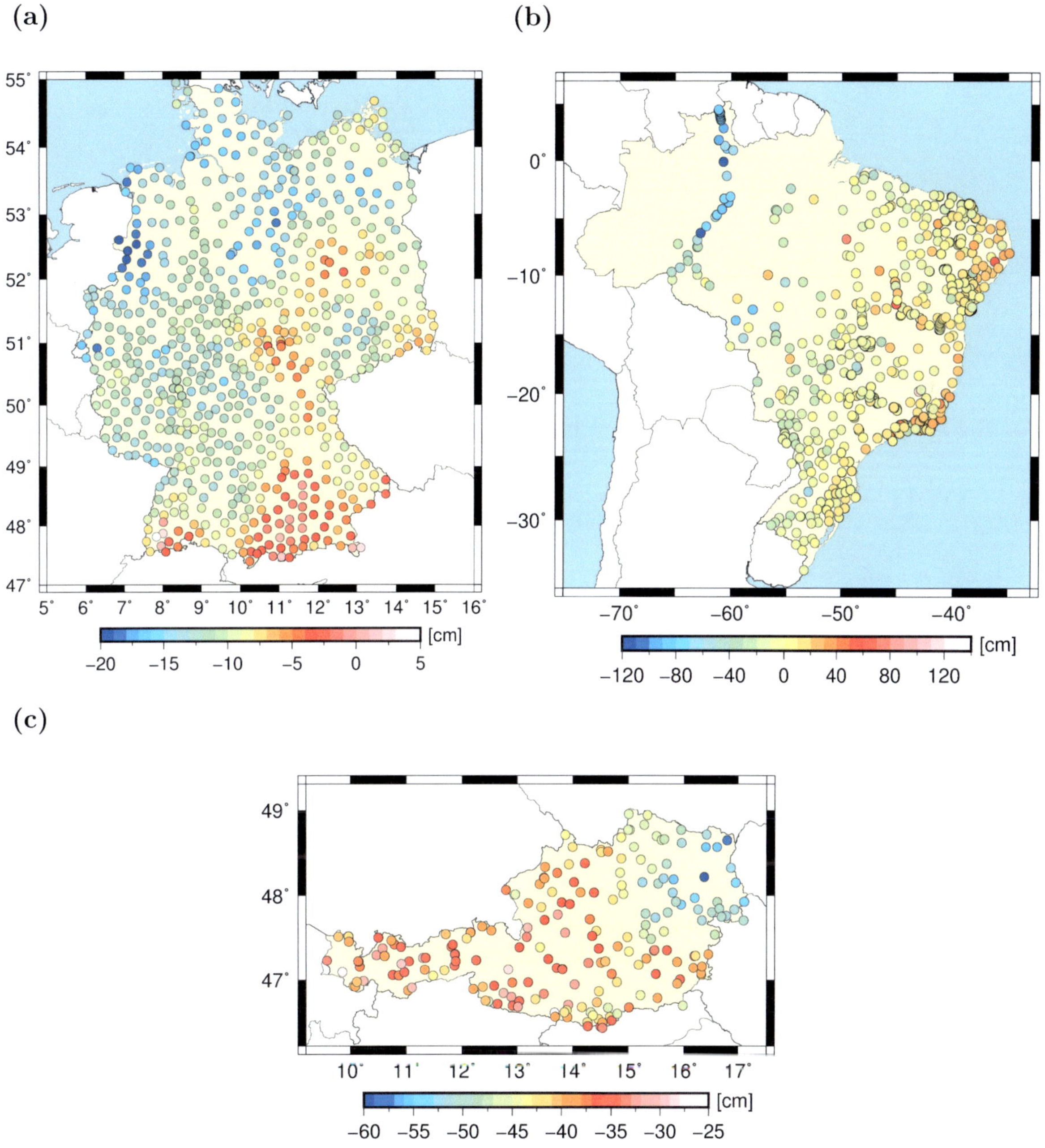

Fig. 5. Resulting values l_j of the observation equation according to Eq. (19) in **(a)** Germany, **(b)** Brazil, and **(c)** Austria, exemplarily shown for TIM R5 with $N = 200$ and $dN = 0$. Note the different scaling of the color bars.

Table 4. STD values calculated according to Eqs. (28) – (30) by using different GGMs in the three study areas Germany, Austria, and Brazil. All values are specified in [cm].

	GGM	Germany	Austria	Brazil
	TIM R5	2.79	4.67	18.03
Eq. (28)	DIR R5	2.65	4.74	18.02
	GOCO05s	2.76	4.56	18.02
Eq. (29)	EGM2008	1.84	3.97	26.26
	TIM R5	2.69	3.41	19.79
Eq. (30)	DIR R5	2.52	3.32	19.79
	GOCO05s	2.68	3.24	19.70

For the second observation equation, only the EGM2008 information is used:

$$\mathrm{STD}\left(\zeta^{\,\mathrm{GNSS/lev}} - \zeta^{\,\mathrm{EGM}}_{\,0,\,2190} - \delta\zeta^{\,\mathrm{Topo}}\Big|_{P_j}\right) = \left[\left(\sigma^{\,\mathrm{GNSS/lev}}\right)^2 + m^{\,\mathrm{EGM}}\left(\sigma^{\,\mathrm{EGM}}_{\,2190}\right)^2\right]^{1/2}. \tag{29}$$

To avoid an under-determined system, a third group of equations is needed, where the differences between the GOCE GGMs and EGM2008 in the SH band 0–200 are used:

$$\mathrm{STD}\left(\zeta^{\,\mathrm{GOCE}}_{\,0,\,200} - \zeta^{\,\mathrm{EGM}}_{\,0,\,200}\Big|_{P_j}\right) = \left[m^{\,\mathrm{GOCE}}\left(\sigma^{\,\mathrm{GOCE}}_{\,200}\right)^2 + m^{\,\mathrm{EGM}}\left(\sigma^{\,\mathrm{EGM}}_{\,200}\right)^2\right]^{1/2}. \tag{30}$$

Altogether, Eqs. (28) – (30) provide seven observation equations, whose numerical values for each study area are indicated in Table 4. As the relations between the observations and the unknown parameters are nonlinear, the observation equations have to be linearized with respect to approximate values and the LSA has to be carried out iteratively.

In Table 5, the results of the LSA in terms of the estimated parameters and their corresponding standard deviations are specified. By applying the scaling factors to the formal cumulative errors according to Eq. (27), Table 6 provides the estimated calibrated cumulative errors $\hat{\sigma}^{\,\mathrm{GGM}}_{\,n}$ for each study area.

In contrast to Table 3, all three GOCE GGMs now provide nearly the same performance within one study area, where variations are on the mm level. Moreover, also in the different study areas the estimated uncertainties of the GOCE GGMs are comparable and stay in the same order of magnitude between 2 to 3 cm. The differences can be explained as follows. Due to the orbital inclination of the GOCE satellite of 96.7°, the density of its ground tracks varies with latitude. In the case of higher latitudes, more observations are collected, which generally results in a more accurate recovery of the gravity field (cf. Klokočník et al., 2008). Therefore, a predominant latitude dependency in the accuracy of the GOCE GGMs

Table 5. Estimated parameters (regional scaling factors $m^{\,\mathrm{GGM}}$ and uncertainties of GNSS/leveling data $\sigma^{\,\mathrm{GNSS/lev}}$) and their corresponding standard deviations for the three study areas Germany, Austria, and Brazil.

	Germany	Austria	Brazil
$m^{\,\mathrm{TIM\ R5}}$	1.16 ± 0.01	1.77 ± 0.08	2.56 ± 0.27
$m^{\,\mathrm{DIR\ R5}}$	8.03 ± 0.10	14.00 ± 0.65	20.67 ± 2.21
$m^{\,\mathrm{GOCO05s}}$	1.49 ± 0.02	2.04 ± 0.10	2.91 ± 0.36
$m^{\,\mathrm{EGM}}$	0.03 ± 0.00	0.05 ± 0.01	7.31 ± 0.02
$\sigma^{\,\mathrm{GNSS/lev}}$ [cm]	1.23 ± 0.02	3.52 ± 0.04	13.89 ± 0.04

Table 6. Estimated calibrated cumulative errors $\hat{\sigma}_n^{\,\mathrm{GGM}}$ for GGM-derived height anomalies in different SH bands and uncertainties of GNSS/leveling data for the three study areas Germany, Austria, and Brazil. All values are specified in [cm].

	SH band	Germany	Austria	Brazil
TIM R5	2–200	2.41 ± 0.01	2.98 ± 0.07	3.59 ± 0.19
DIR R5	2–200	2.23 ± 0.01	2.95 ± 0.07	3.58 ± 0.19
GOCO05s	2–200	2.39 ± 0.01	2.79 ± 0.07	3.34 ± 0.21
EGM2008	2–200	1.19 ± 0.02	1.61 ± 0.10	19.45 ± 0.03
EGM2008	2–2190	1.36 ± 0.02	1.84 ± 0.11	22.29 ± 0.03
GNSS/lev	–	1.23 ± 0.02	3.52 ± 0.04	13.89 ± 0.04

is to be expected. In this context, the variations of the uncertainties in Table 6 seem to be reasonable as they are in accordance with the geographical latitudes of the study areas.

In the case of EGM2008, the necessity of using regional scaling factors for the cumulative errors is even more apparent. While EGM2008 is of high precision for Germany and Austria with uncertainties that are lower than those of the GOCE GGMs, it is the opposite in the case of Brazil, where a low accuracy is observed for EGM2008. These regional variations can be attributed to the different quality and availability of terrestrial gravity data incorporated into EGM2008. For example, in the case of Brazil, gravity data was of proprietary nature or unavailable for larger areas (Pavlis et al., 2012). Thus, the performance of EGM2008 is expected to be considerably worse over Brazil.

As can further be seen in Table 6, the estimated uncertainties of the GNSS/leveling data strongly vary between the three study areas. While an uncertainty of 1.2 cm is estimated for the German data set, the corresponding value for Austria is about three times larger (3.5 cm). In the case of Brazil, a significantly larger uncertainty of 13.9 cm is obtained.

This may reflect remaining distortions in the leveling network that are not completely compensated by the applied functional model in Eq. (19). The order of magnitude is in accordance with the roughly estimated uncertainty value of 10.4 cm as specified in Ferreira and de Freitas (2011) for 97 GNSS/leveling points in South Brazil.

5. Results and discussion

In this section, the results are presented and discussed with respect to an optimized combination of GOCE GGMs and EGM2008 (Sect. 5.1), and the benefit and contribution of topography-implied gravity signals (Sect. 5.2).

5.1. Optimized combination of GOCE GGMs and EGM2008

In the following, optimal parameters for the combination of the GOCE GGMs and EGM2008 are derived, according to the used Hanning transition window as specified in Eq. (7). For this purpose, the SH degree of combination N and the transition bandwidth dN are systematically varied. Each resulting combination is then used to derive the height anomaly $\zeta^{\,\mathrm{G/E}\,(N,\,dN)}_{\,0,\,2190}$ that is processed in the LSA, as described in Sect. 2.4. The accuracy of the combination is evaluated in terms of the STD calculated from the residuals v_j of the LSA.

Taking the TIM R5 model as a representative example, Fig. 6 (*left column*) displays the resulting STD in terms of 2D plots for each study area, where the SH degree of combination N is shown on the *horizontal axis* and the transition bandwidth dN on the *vertical axis*. Note that the triangle shape results from the fact that not all combinations of N and dN are possible, as otherwise the transition window would range outside the minimum or maximum degree of the GOCE GGM, cf. Eq. (7). In addition to the STD values, Fig. 6 (*right column*) shows the corresponding estimated height datum offsets δH.

In the case of the German data set (Fig. 6, *first row*), most of the combinations provide STD values of about 2 cm, with only smaller variations on the mm level. However, for $N > 220$, the STDs strongly increase toward values larger than 4 cm. This is a consequence of the GOCE-derived SH coefficients of higher degrees that cannot be properly estimated due to an increasing signal-to-noise ratio in the gravity gradient measurements (cf. Brockmann et al., 2014). The advantage of using a transition window is particularly visible for combination degrees $125 < N < 220$. For this SH band, the pure concatenation of the SH coefficients, i.e., $dN = 0$, provides highly variable STD values with sporadic amplitudes above 3.5 cm. In contrast, a significantly smoother behavior can be detected for combinations with $dN > 5$, where the exact value of N is not crucial for achieving a low STD of about 2 cm.

Numerically, the minimum standard deviation of 1.78 cm is reached for $N = 64$ and $dN = 0$, which means that the influence of the GOCE GGM would be very limited. This supports the well-known fact that, in the case of Germany (or Europe), EGM2008 outperforms GOCE-based models due to highly precise terrestrial gravity data incorporated in EGM2008. However, from the practical view of a height datum estimation, it is not reasonable to choose such a low degree of combination as will be explained in the following.

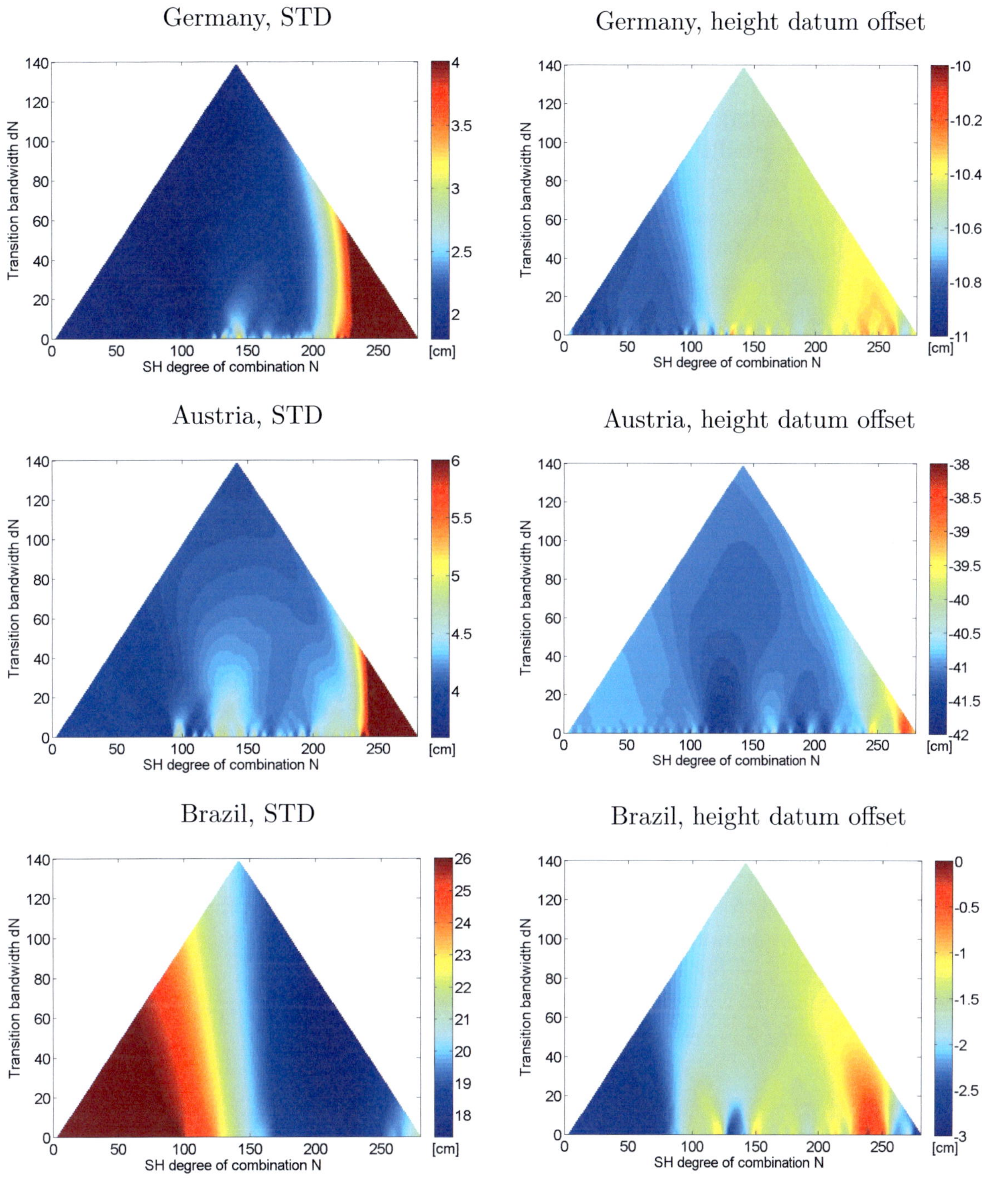

Fig. 6. Results for different combinations of TIM R5 and EGM2008 with respect to the used SH degree of combination N and the transition bandwidth dN. The achieved STD values of the residuals v_j are shown in the *left column*, while the estimated height datum offsets δH are presented in the *right column*. The different results for the three study areas Germany, Austria, and Brazil are displayed in the *first*, *second*, and *third row*, respectively. Note the different scaling of the color bars.

In the case of combined high-resolution GGMs like EGM2008, it has to be taken into consideration that they are based on terrestrial gravity anomalies and might therefore also be (indirectly) affected by height datum offsets (cf. Heck, 1990).

For the case of EGM2008, Gatti et al. (2013) specified a SH band of 100–200, where the SH coefficients are partially affected by discrepancies in the LVD. However, recent studies have demonstrated that the indirect influence of terrestrial gravity data can be reduced to below 1 cm, when a satellite-based GGM is used for representing the long wavelengths, e.g., Gerlach and Rummel (2013), Amjadiparvar et al. (2016). This statement can generally be supported for Germany, when having a look at the estimated height datum offsets δH, plotted for the different combinations of N and dN. Here, the variations in the offset stay in a range of 1 cm. Nevertheless, as there are only minimal differences in the provided STD values, a combination degree of $N = 180$ and a transition bandwidth of $dN = 20$ are used for the German study area.

The results for the Austrian data set, as illustrated in the *second row* of Fig. 6, generally provide comparable findings, but on the basis of higher STD values and more variations in the estimated height datum offset δH. Hence, keeping in mind the aforementioned indirect bias effect, a local minimum of the STD values around degree 200 is used by selecting $N = 195$ and $dN = 0$.

In contrast to the mostly comparable results in Germany and Austria, Fig. 6 (*third row*) demonstrates a totally different behavior in the case of the Brazilian data set. As the performance of EGM2008 is considerably worse in Brazil, a higher proportion of GOCE data generally leads to smaller STD values. With an increasing degree of combination N, the STD decreases from about 30 cm down to 17 cm, which impressively shows the gained improvement in the gravity information over Brazil due to the use of GOCE satellite data. Only for the highest degrees, i.e., $N > 260$, the STD values slightly increase again. The minimum of the STD values can be detected around $N = 220$ and $dN = 50$, which will be used as parameters for the Brazilian data set. The estimated height datum offsets range between −3 to 0 cm and show a strong variability with respect to different combinations. Particularly, for combination degrees around $N = 130$ and $N = 240$, conspicuous structures with rapidly changing values can be detected.

Note that comparable results are obtained for the use of DIR R5 and GOCO05s. The corresponding figures are provided in Figs. 8 and 9 in the Appendix.

5.2. Contribution of topography-implied gravity signals

To analyze the contribution and benefit of topographic-implied gravity signals, the results of the LSA with and without the consideration of $\delta\zeta^{\mathrm{Topo}}$ in Eq. (19) are compared in the following. For readability, these two cases will be called *topo* and *non-topo scenario*, respectively.

For the three study areas, Table 7 presents numerical results of both scenarios with respect to the different GOCE GGMs. Besides the estimated height datum offset δH and tilts a_1 and a_2, also the STD and range of the least squares adjusted residuals v_j are specified, in order to assess the performance of the respective solution.

Table 7. Comparison of the topo and non-topo scenarios in terms of the estimated height datum offset δH, tilts a_1 (north-south direction) and a_2 (east-west direction), and the STD and range of the residuals v_j. Results are shown for the three study areas Germany, Austria, and Brazil and different GOCE GGMs.

	Scenario	GGM	δH [cm]	a_1 [cm/deg]	a_2 [cm/deg]	STD [cm]	Range [cm]
Germany $N = 180$ $dN = 20$	non-topo	TIM R5	-10.4 ± 0.1	-1.28 ± 0.05	1.18 ± 0.07	2.5	22.1
		DIR R5	-10.5 ± 0.1	-1.31 ± 0.05	1.19 ± 0.07	2.3	20.8
		GOCO05s	-10.4 ± 0.1	-1.27 ± 0.05	1.13 ± 0.07	2.4	22.1
	topo	TIM R5	-10.5 ± 0.1	-1.27 ± 0.04	1.25 ± 0.06	2.2	19.0
		DIR R5	-10.6 ± 0.1	-1.30 ± 0.04	1.26 ± 0.06	2.0	17.7
		GOCO05s	-10.5 ± 0.1	-1.27 ± 0.04	1.20 ± 0.06	2.1	19.0
Austria $N = 195$ $dN = 0$	non-topo	TIM R5	-43.4 ± 0.4	-3.23 ± 0.71	-0.93 ± 0.35	6.0	43.2
		DIR R5	-43.3 ± 0.4	-3.36 ± 0.72	-0.78 ± 0.36	6.1	42.5
		GOCO05s	-43.2 ± 0.4	-3.27 ± 0.71	-0.85 ± 0.36	6.0	43.9
	topo	TIM R5	-40.4 ± 0.3	-3.99 ± 0.48	-2.53 ± 0.24	4.1	26.5
		DIR R5	-40.4 ± 0.3	-4.11 ± 0.49	-2.38 ± 0.24	4.2	23.8
		GOCO05s	-40.3 ± 0.3	-4.03 ± 0.47	-2.44 ± 0.24	4.0	27.3
Brazil $N = 220$ $dN = 50$	non-topo	TIM R5	-1.2 ± 0.7	-1.20 ± 0.09	2.70 ± 0.11	17.4	206.4
		DIR R5	-1.4 ± 0.7	-1.19 ± 0.09	2.69 ± 0.11	17.4	210.2
		GOCO05s	-1.2 ± 0.7	-1.19 ± 0.09	2.69 ± 0.11	17.4	206.1
	topo	TIM R5	-1.1 ± 0.7	-1.20 ± 0.08	2.70 ± 0.11	17.4	204.7
		DIR R5	-1.3 ± 0.7	-1.19 ± 0.08	2.69 ± 0.11	17.4	208.5
		GOCO05s	-1.1 ± 0.7	-1.19 ± 0.08	2.69 ± 0.11	17.3	204.4

In the case of Germany, all three GOCE GGMs provide comparable results, where differences in the estimated height datum offset and the achieved STD are on the mm level. On the one hand, the additional use of topographic signals in the German study area only slightly affects the value of the height datum offset. This is within the expectations due to the small mean value provided by $\delta\zeta^{\text{Topo}}$ (cf. Table 2). On the other hand, it can also be seen that the STD and range decrease in all cases, showing the positive impact of considering the topographic signals. With a STD of 2.0 cm, the DIR R5 model provides the most accurate result with an estimated height datum offset of -10.6 ± 0.1 cm.

The most significant improvements due to topography-implied gravity signals are obtained for the Austrian study area. For all analyzed GGMs, the STD decreases from about 6 to

4 cm, and the range is reduced from about 40 to 25 cm. Additionally, the impact of the topography causes a considerable change in the estimated height datum offset of about 3 cm, together with a slight improvement in the parameter's accuracy. As above, the differences between the three GOCE GGMs are marginal. Since the GOCO05s model exhibits the best performance with respect to the STD (4.0 cm), the height datum offset is estimated to be $\delta H = -40.3 \pm 0.3$ cm.

The Brazilian data set shows the limitations of the applied method. Although the topography-implied gravity signals $\delta\zeta^{\text{Topo}}$ provide values with magnitudes on the level of several centimeters (cf. Table 2), there seems to be no noticeable influence on the results of the LSA. In contrast to the other two study areas, the STD and range attain significantly larger values of about 17 cm and 200 cm, respectively. Thus, the effect of the topographic gravity signals is apparently dominated (or over-modulated) by the generally large error level of the Brazilian GNSS/leveling data. According to the results of the GOCO05s model, the height datum offset is estimated to be $\delta H = -1.1 \pm 0.7$ cm.

Note that the derived height datum offsets are not defined with respect to the unknown global vertical datum. Thus, for a comparison with offsets estimated in other studies, it is reasonable to make use of relative height datum offsets between two countries. For example, the relative offset between Germany and Austria is estimated to be 29.7 ± 0.3 cm in this study. Although different GNSS/leveling benchmarks have been used, this value is in good agreement with the offset of 28.6 cm derived from Rülke et al. (2012).

For a more detailed analysis of the obtained results, the residuals v_j of the respective GNSS/leveling benchmarks P_j are plotted in Fig. 7 as a function of their physical height. While in the *left column* the non-topo scenario is shown for the three study areas, the *right column* displays the corresponding topo scenarios. Each residual is indicated by a blue cross. Additionally, residuals that would be classified as outliers by iterative data snooping are marked by a red circle. In this context, a Student's t-test with a conservative confidence level of 99.99 % has been used for the detection of outliers (cf. Heck, 1981). Corresponding to the optimum values of Table 7, the results in Fig. 7 are exemplarily shown for the DIR R5 model in the case of Germany, and the GOCO05s model in the case of the two other study areas. As a supplement, Fig. 10 in the Appendix presents the corresponding spatial distributions of the residuals v_j.

For the German data set (Fig. 7, *first row*), GNSS/leveling points in lower elevation generally provide smaller residuals and are more condensed around zero. In contrast, for points with heights above 500 m, the residuals are more loosely scattered and have larger magnitudes. In the case of the non-topo scenario, four points with heights above 1200 m exhibit residuals of more than 10 cm in absolute value, which would be classified as outliers. However, the residuals of these benchmarks are particularly reduced in the topo scenario. Moreover, also the magnitude of a few other residuals can be significantly decreased. As a consequence, only two of the four points still remain as outliers in the topo scenario.

In Austria (*second row*), the non-topo scenario shows a high variation of the residuals, ranging between −27 and 17 cm. As for Germany, the benchmarks in the highest elevations provide residuals with the largest magnitudes, where one outlier is detected at a height of about 2150 m. In contrast, the residuals in the topo scenario are significantly reduced

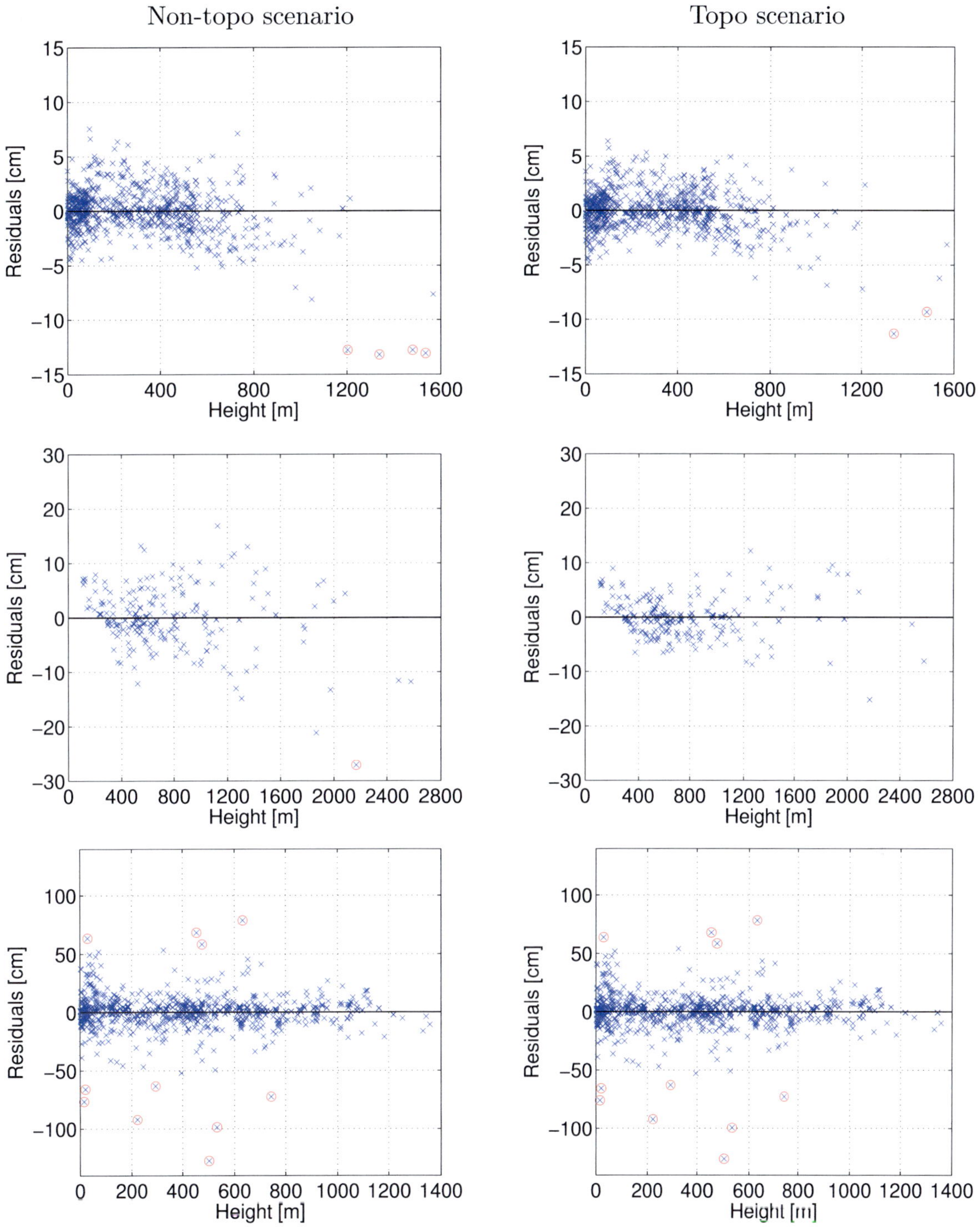

Fig. 7. Least squares adjusted residuals v_j at the respective GNSS/leveling benchmarks P_j as a function of their physical height, shown for the topo and non-topo scenario (*right* and *left column*, respectively), and the three study areas Germany, Austria, and Brazil (*first*, *second*, and *third row*, respectively). Each residual is represented by a blue cross and detected outliers are marked by a red circle.

and are more homogeneously distributed within a narrowed band of about ±12 cm. In this case, no benchmark is classified as an outlier.

As already evident from Table 7, these remarkable results cannot be reproduced for the case of Brazil (*third row*). Contrary to the other study areas, residuals have a strong variation and there seems to be no correlation between benchmark heights and residual magnitudes. Here, a large number of residuals that are not geographically clustered even exceed values of ±50 cm, with most of them being marked as outliers. Comparing the residuals of the topo and the non-topo scenarios, only slight changes are visible and the same outliers are detected.

Finally, to highlight the benefit of using topography-implied gravity signals, Table 8 specifies improvement rates in terms of percentage changes in the STD and range between the topo and the non-topo scenarios. While, for the German study area, the STD can be reduced by about 12.5 %, impressive improvement rates up to 33.3 % are reached in the case of Austria. The corresponding values for the range indicate an improvement of about 14 % for Germany and 40 % for Austria. In contrast, the obtained improvement rates for the Brazilian data set are insignificant with values below 1 %.

Table 8. Improvement rate due to topography-implied gravity signals $\delta\zeta^{\mathrm{Topo}}$ in terms of percentage changes in the STD and range between the topo and the non-topo scenarios.

	GGM	Improvement rate [%]	
		STD	Range
Germany	TIM R5	12.0	14.0
	DIR R5	13.0	14.9
	GOCO05s	12.5	14.0
Austria	TIM R5	31.7	38.7
	DIR R5	31.1	44.0
	GOCO05s	33.3	37.8
Brazil	TIM R5	0.0	0.8
	DIR R5	0.0	0.8
	GOCO05s	0.6	0.8

6. Conclusions and outlook

In this paper, a satellite-based method for global height system unification has been presented, using GGMs derived from ESA's gravity field mission GOCE. These GOCE GGMs provide a global homogeneous reference surface that is not affected by discrepancies of the local vertical datum. By comparing the GGM information with measured GNSS/leveling data, the datum offset of the local height system is estimated within a least squares adjustment, using a 3-parameter (plane) model. Due to the limited spectral resolution of GOCE GGMs, this approach suffers from an omission error.

In terms of a spectral extension, the GOCE information is therefore combined with the high-resolution EGM2008 and additionally augmented by high-frequency topography-implied gravity signals. In this context, a novel (residual) gravity forward modeling approach is suggested that is based on a high-pass filtering in the gravity domain. To this end, the forward-modeled RWI-based topographic effects of the $1' \times 1'$ Earth2014 model have been reduced by the information of a consistent SH harmonic expansion of the topographic potential. In contrast to residual terrain modeling (RTM), the benefit of this method is that it does not rely on the generally assumed – but not generally valid – spectral consistency of topographic heights and implied gravity. While in principle the proposed procedure is computationally more expensive, the availability of suitable topographic gravity field models, such as RWI_TOPO_2015, made this approach feasible.

By using study areas in Germany, Austria, and Brazil, the investigations in this paper have focused on (i) the spectral combination of GOCE GGMs and EGM2008 using a Hanning transition window and (ii) a detailed analysis of the benefit and contribution of topography-implied gravity signals on the estimation of height datum offsets. Moreover, for each study area, the uncertainties of all involved terms have been estimated in a separate least squares approach.

In this context, regional scaling factors have been derived to calibrate the cumulative error of GOCE GGMs and EGM2008, leading to more realistic error predictions. While the performance of the used GOCE GGMs (TIM R5, DIR R5, and GOCO05s) is comparable in all three study areas, larger regional differences in the accuracy of EGM2008 as well as the GNSS/leveling data become visible.

For the combination of GOCE GGMs and EGM2008, optimized parameters in terms of combination degree N and transition bandwidth dN have been derived for each study area. To assess the performance of topography-implied gravity signals, an estimation of height datum offsets has been performed with and without the consideration of these high-frequency signals. In the case of the Austrian study area, topography-implied gravity signals lead to impressive improvements of 30 – 40 % in STD and range of the residuals. Moreover, the estimated height datum offset changes considerably by about 3 cm.

While significant improvement rates of 12 – 15 % have also been achieved for Germany, this is not the case for Brazil, where a large error level in the GNSS/leveling data dominates the effect of topography-implied gravity signals. To summarize, the investigations in this paper have proved the importance of topography-implied gravity signals for an accurate estimation of height datum offsets. Therefore, it is strongly recommended to take these high-frequency components into account, in particular in regions with highly variable topography, where the remaining omission error of EGM2008 can still be on a significant cm–dm level.

Although the obtained results are already quite promising, it can be expected that by using a DTM of higher spatial resolution than the $1' \times 1'$ Earth2014 model, the benefit of topography-implied gravity signals can be even increased. However, this also requires a consistent topographic gravity field model that is linked to this DTM.

In the next step, a detailed comparison of the novel (residual) gravity forward modeling approach with the classical RTM method will be conducted.

Acknowledgements. The authors acknowledge the financial support provided by the German Research Foundation (DFG) under grant number HE1433/20-2. Furthermore, we would like to thank the German Federal Agency for Cartography and Geodesy (BKG), the Austrian Federal Office for Metrology and Surveying (BEV), and the Brazilian Institute of Geography and Statistics (IBGE) for kindly providing the GNSS/leveling data sets. Finally, two anonymous reviewers as well as the Editor-in-Chief are acknowledged for their valuable comments, which helped to improve the manuscript.

Appendix

In addition to the results presented for the TIM R5 model in Fig. 6 (Sect. 5.1), this appendix provides corresponding plots for the DIR R5 and GOCO05s model as displayed in Figs. 8 and 9, respectively. Furthermore, as a supplement to Fig. 7 (Sect. 5.2), the spatial distributions of the least squares residuals are displayed in Fig. 10.

References

Amjadiparvar, B., Rangelova, E., and Sideris, M. G. (2016): The GBVP approach for vertical datum unification: recent results in North America. *Journal of Geodesy* 90(1):45–63. DOI: 10.1007/s00190-015-0855-8.

Amjadiparvar, B., Rangelova, E., Sideris, M. G., and Véronneau, M. (2013): North American height datums and their offsets: the effect of GOCE omission errors and systematic levelling effects. *Journal of Applied Geodesy* 7(1):39–50. DOI: 10.1515/jag-2012-0034.

Barzaghi, R., Carrion, D., Reguzzoni, M., and Venuti, G. (2016): A feasibility study on the unification of the Italian height systems using GNSS-Leveling data and global satellite gravity models. In: Rizos, C., and Willis, P. (eds.) IAG 150 years. Proceedings of the IAG Scientific Assembly, Postdam, Germany, Sept. 1–6, 2013. *International Association of Geodesy Symposia*, vol. 143. Springer Berlin Heidelberg, pp. 281–288. DOI: 10.1007/1345_2015_35.

Blackman, R. B. and Tukey, J. W. (1958): The measurement of power spectra from the point of view of communications engineering – Part I. *The Bell System Technical Journal* 37(1):185–282. DOI: 10.1002/j.1538-7305.1958.tb03874.x.

Blewitt, G., Altamimi, Z., Davis, J., Gross, R., Kuo, C.-Y., Lemoine, F. G., Moore, A. W., Neilan, R. E., Plag, H.-P., Rothacher, M., Shum, C. K., Sideris, M. G., Schöne, T., Tregoning, P., and Zerbini, S. (2010): Geodetic observations and global reference frame contributions to understanding sea-level rise and variability. In: Church, J. A., Woodworth, P. L., Aarup, T., and Wilson, W. S. (eds.) Understanding sea-level rise and variability, Wiley-Blackwell, pp. 256–284. DOI: 10.1002/9781444323276.ch9.

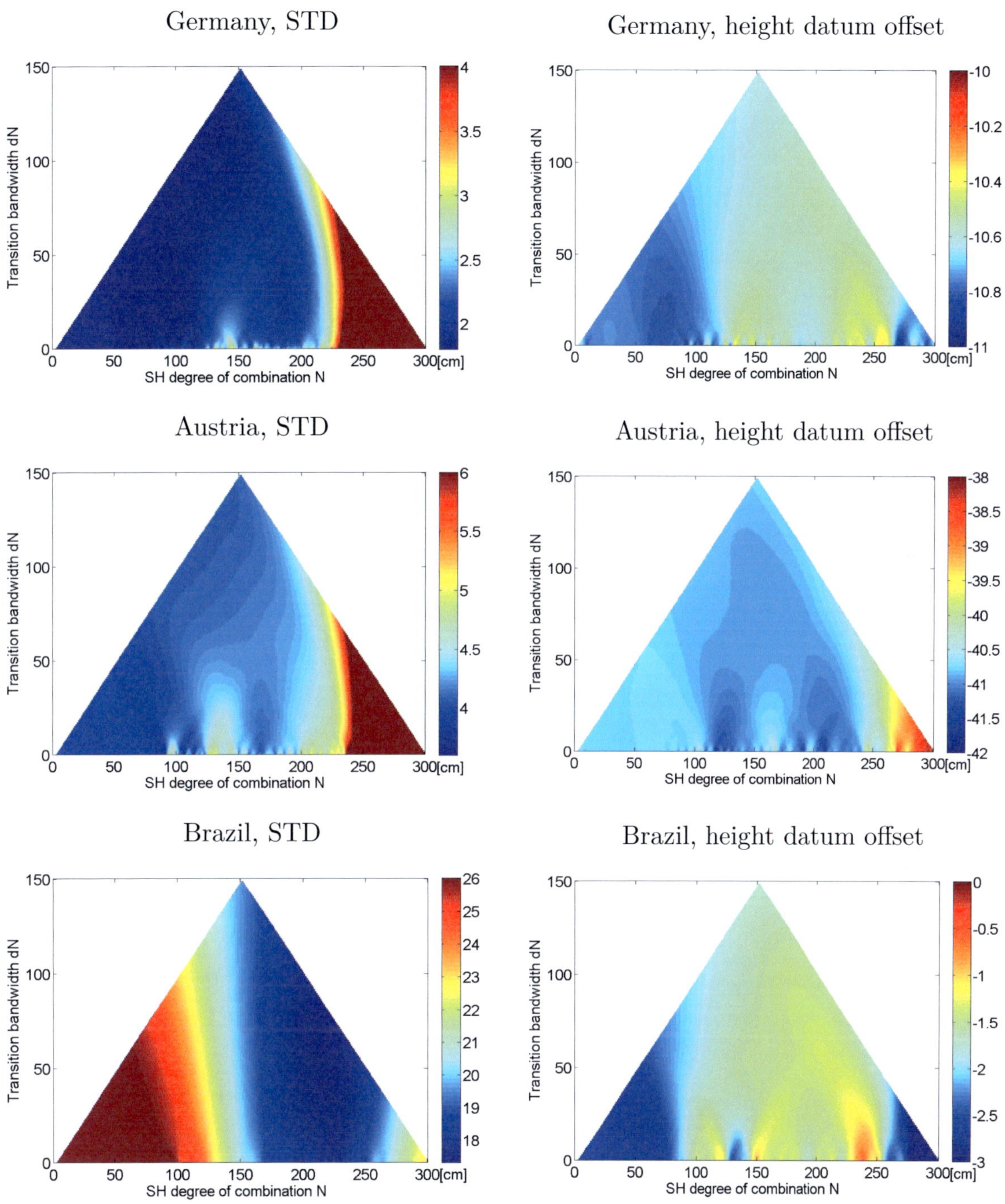

Fig. 8. Results for different combinations of DIR R5 and EGM2008 with respect to the used SH degree of combination N and the transition bandwidth dN. The achieved STD values of the residuals v_j are shown in the *left column*, while the estimated height datum offsets δH are presented in the *right column*. The different results for the three study areas Germany, Austria, and Brazil are displayed in the *first*, *second*, and *third row*, respectively. Note the different scaling of the color bars.

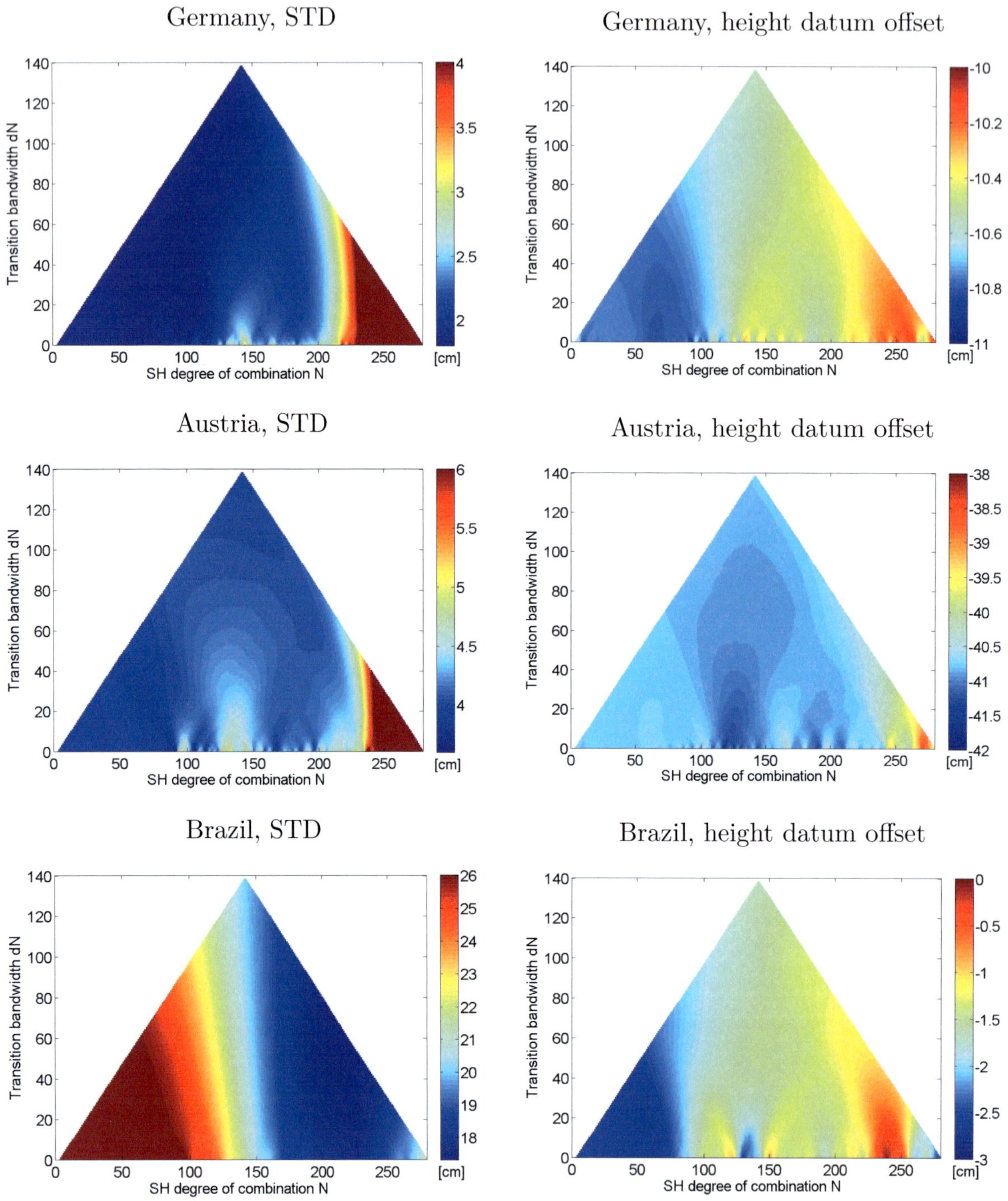

Fig. 9. Results for different combinations of GOCO05s and EGM2008 with respect to the used SH degree of combination N and the transition bandwidth dN. The achieved STD values of the residuals v_j are shown in the *left column*, while the estimated height datum offsets δH are presented in the *right column*. The different results for the three study areas Germany, Austria, and Brazil are displayed in the *first*, *second*, and *third row*, respectively. Note the different scaling of the color bars.

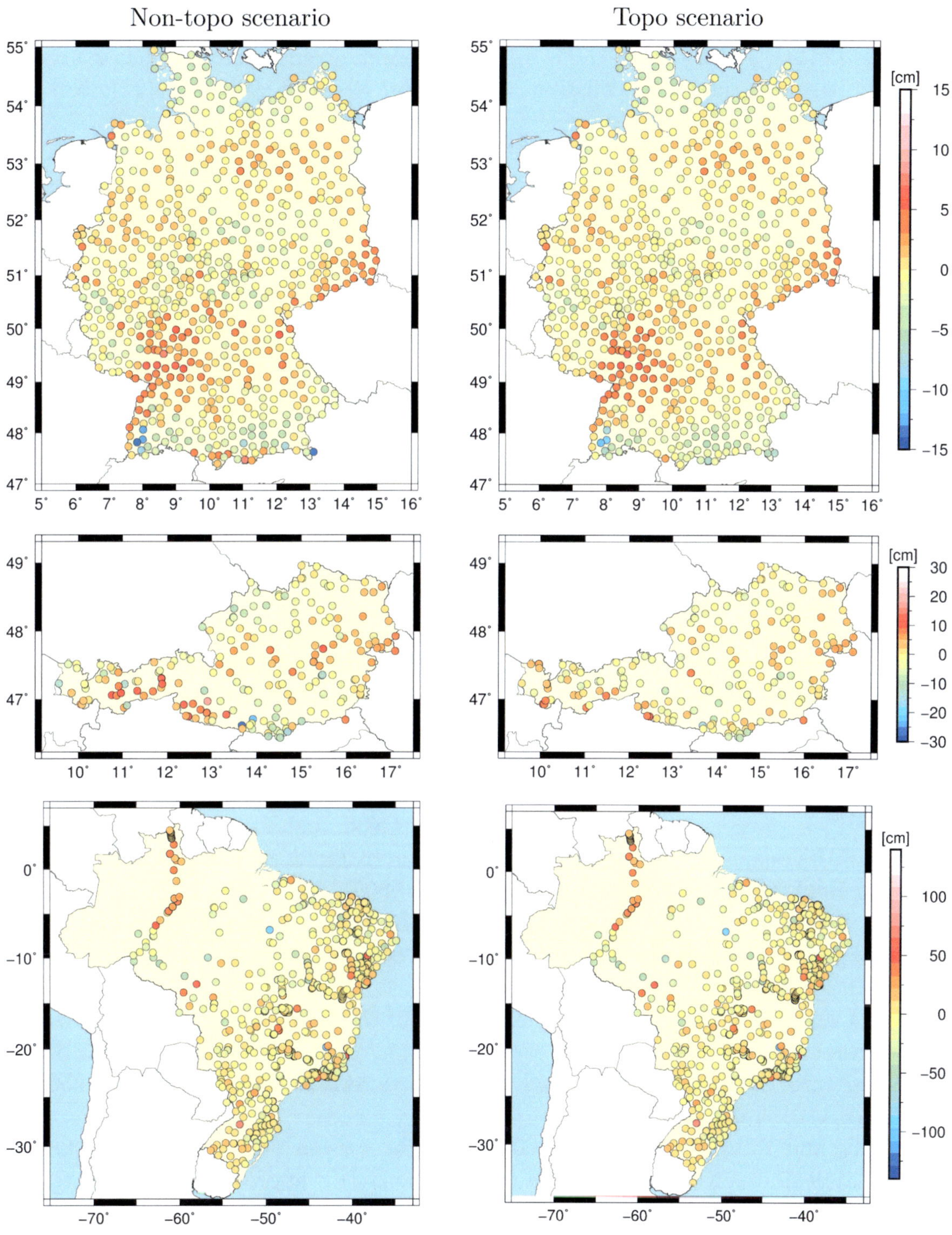

Fig. 10. Least squares adjusted residuals v_j at the respective GNSS/leveling benchmarks P_j, shown for the topo and non-topo scenario (*right* and *left column*, respectively), and the three study areas Germany, Austria, and Brazil (*first*, *second*, and *third row*, respectively). Note the different scaling of the color bars.

Brockmann, J. M., Zehentner, N., Höck, E., Pail, R., Loth, I., Mayer-Gürr, T., and Schuh, W.-D. (2014): EGM_TIM_RL05: an independent geoid with centimeter accuracy purely based on the GOCE mission. *Geophysical Research Letters* 41(22):8089–8099. DOI: 10.1002/2014GL061904.

Bruinsma, S. L., Förste, C., Abrikosov, O., Lemoine, J.-M., Marty, J.-C., Mulet, S., Rio, M.-H., and Bonvalot, S. (2014): ESA's satellite-only gravity field model via the direct approach based on all GOCE data. *Geophysical Research Letters* 41(21):7508–7514. DOI: 10.1002/2014GL062045.

Colombo, O. L. (1980): A world vertical network. Report 296. Department of Geodetic Science and Surveying, The Ohio State University, Columbus, USA.

Ekman, M. (1989): Impacts of geodynamic phenomena on systems for height and gravity. *Bulletin Géodésique* 63(3):281–296. DOI: 10.1007/BF02520477.

ESA (1999): The four candidate Earth explorer core missions – Gravity field and steady-state ocean circulation. In: Battrick, B. (ed.) Reports for mission selection, *ESA Special Publication*, vol. 1233(1). ESA Publications Division, ESTEC, Noordwijk, The Netherlands.

Ferreira, V. G. and de Freitas, S. R. C. (2011): Geopotential numbers from GPS satellite surveying and disturbing potential model: a case study of Parana, Brazil. *Journal of Applied Geodesy* 5(3–4):155–162. DOI: 10.1515/JAG.2011.016.

Ferreira, V. G., de Freitas, S. R. C., and Heck, B. (2016): Analysis of the discrepancy between the Brazilian vertical reference frame and GOCE-based geopotential models. In: Rizos, C., and Willis, P. (eds.) IAG 150 years. Proceedings of the IAG Scientific Assembly, Postdam, Germany, Sept. 1–6, 2013. *International Association of Geodesy Symposia*, vol. 143. Springer Berlin Heidelberg, pp. 227–232. DOI: 10.1007/1345_2015_20.

Forsberg, R. and Tscherning, C. C. (1997): Topographic effects in gravity field modelling for BVP. In: Sansò, F., and Rummel, R. (eds.) Geodetic boundary value problems in view of the one centimeter geoid, *Lecture Notes in Earth Sciences*, vol. 65. Springer Berlin Heidelberg, pp. 239–272. DOI: 10.1007/BFb0011707.

Gatti, A., Reguzzoni, M., and Venuti, G. (2013): The height datum problem and the role of satellite gravity models. *Journal of Geodesy* 87(1):15–22. DOI: 10.1007/s00190-012-0574-3.

Gerlach, C. and Fecher, T. (2012): Approximations of the GOCE error variance-covariance matrix for least-squares estimation of height datum offsets. *Journal of Geodetic Science*, Special issue on regional and global geoid-based vertical datums 2(4):247–256. DOI: 10.2478/v10156-011-0049-0.

Gerlach, C. and Rummel, R. (2013): Global height system unification with GOCE: a simulation study on the indirect bias term in the GBVP approach. *Journal of Geodesy* 87(1):57–67. DOI: 10.1007/s00190-012-0579-y.

Gomez, M. E., Pereira, R. A. D., Ferreira, V. G., Cogliano, D. D., Luz, R. T., de Freitas, S. R. C., Farias, C., Perdomo, R., Tocho, C., Lauria, E., and Cimbaro, S. (2016): Analysis of the discrepancies between the vertical reference frames of Argentina and Brazil. In: Rizos, C., and Willis, P. (eds.) IAG 150 years. Proceedings of the IAG Scientific Assembly, Postdam, Germany, Sept. 1–6, 2013. *International Association of*

Geodesy Symposia, vol. 143. Springer Berlin Heidelberg, pp. 289–295. DOI: 10.1007/1345_2015_75.

Grombein, T., Luo, X., Seitz, K., and Heck, B. (2014): A wavelet-based assessment of topographic-isostatic reductions for GOCE gravity gradients. *Surveys in Geophysics* 35(4):959–982. DOI: 10.1007/s10712-014-9283-1.

Grombein, T., Seitz, K., and Heck, B. (2013): Optimized formulas for the gravitational field of a tesseroid. *Journal of Geodesy* 87(7):645–660. DOI: 10.1007/s00190-013-0636-1.

Grombein, T., Seitz, K., and Heck, B. (2016a): The Rock-Water-Ice topographic gravity field model RWI_TOPO_2015 and its comparison to a conventional rock-equivalent version. *Surveys in Geophysics* 37(5):937–976. DOI: 10.1007/s10712-016-9376-0.

Grombein, T., Seitz, K., and Heck, B. (2016b): Height system unification based on the fixed GBVP approach. In: Rizos, C., and Willis, P. (eds.) IAG 150 years. Proceedings of the IAG Scientific Assembly, Postdam, Germany, Sept. 1–6, 2013. *International Association of Geodesy Symposia*, vol. 143. Springer Berlin Heidelberg, pp. 305–311. DOI: 10.1007/1345_2015_104.

Gruber, T., Gerlach, C., and Haagmans, R. (2012): Intercontinental height datum connection with GOCE and GPS-levelling data. *Journal of Geodetic Science*, Special issue on regional and global geoid-based vertical datums 2(4):270–280. DOI: 10.2478/v10156-012-0001-y.

Gruber, T., Visser, P. N. A. M., Ackermann, C., and Hosse, M. (2011): Validation of GOCE gravity field models by means of orbit residuals and geoid comparisons. *Journal of Geodesy* 85(11):845–860. DOI: 10.1007/s00190-011-0486-7.

Haagmans, R. H. N. and van Gelderen, M. (1991): Error variances-covariances of GEM-T1: their characteristics and implications in geoid computation. *Journal of Geophysical Research* 96(B12):20011–20022. DOI: 10.1029/91JB01971.

Heck, B. (1981): Der Einfluß einzelner Beobachtungen auf das Ergebnis einer Ausgleichung und die Suche nach Ausreißern in den Beobachtungen. *Allgemeine Vermessungs-Nachrichten* 88(1981):17–34.

Heck, B. (1990): An evaluation of some systematic error sources affecting terrestrial gravity anomalies. *Bulletin Géodésique* 64(1):88–108. DOI: 10.1007/BF02530617.

Heck, B. (2004): Problems in the definition of vertical reference frames. In: Sansò, F. (ed.) V Hotine-Marussi Symposium on Mathematical Geodesy, Matera, Italy, June 17–21, 2003. *International Association of Geodesy Symposia*, vol. 127. Springer Berlin Heidelberg, pp. 164–173. DOI: 10.1007/978-3-662-10735-5_22.

Heck, B. and Rummel, R. (1990): Strategies for solving the vertical datum problem using terrestrial and satellite geodetic data. In: Sünkel, H., and Baker, T. (eds.) Sea surface topography and the geoid. Proceedings of the IAG General Assembly, Edinburgh, Scotland, Aug. 10–11, 1989. *International Association of Geodesy Symposia*, vol. 104. Springer Berlin Heidelberg, pp. 116–128. DOI: 10.1007/978-1-4684-7098-7_14.

Heck, B. and Seitz, K. (2007): A comparison of the tesseroid, prism and point-mass approaches for mass reductions in gravity field modelling. *Journal of Geodesy* 81(2):121–136. DOI: 10.1007/s00190-006-0094-0.

Heiskanen, W. A. and Moritz, H. (1967): Physical geodesy. W. H. Freeman & Co., San Francisco, USA.
Hirt, C. (2013): RTM gravity forward-modeling using topography/bathymetry data to improve high-degree global geopotential models in the coastal zone. *Marine Geodesy* 36(2):183–202. DOI: 10.1080/01490419.2013.779334.
Hirt, C., Featherstone, W. E., and Marti, U. (2010): Combining EGM2008 and SRTM/DTM-2006.0 residual terrain model data to improve quasigeoid computations in mountainous areas devoid of gravity data. *Journal of Geodesy* 84(9):557–567. DOI: 10.1007/s00190-010-0395-1.
Hirt, C. and Kuhn, M. (2014): Band-limited topographic mass distribution generates full-spectrum gravity field: gravity forward modeling in the spectral and spatial domains revisited. *Journal of Geophysical Research* 119(4):3646–3661. DOI: 10.1002/2013JB010900.
Hirt, C. and Rexer, M. (2015): Earth2014: 1 arc-min shape, topography, bedrock and ice-sheet models – available as gridded data and degree-10,800 spherical harmonics. *International Journal of Applied Earth Observation and Geoinformation* 39:103–112. DOI: 10.1016/j.jag.2015.03.001. URL: http://ddfe.curtin.edu.au/models/Earth2014.
Höggerl, N. (1986): Die Ausgleichung des österreichischen Präzisionsnivellementnetzes. *Österreichische Zeitschrift für Vermessungswesen und Photogrammetrie* 74(4):216–249.
Höggerl, N. and Ruess, D. (2004): The new orthometric height system in Austria. In: Proceedings of the EUREF 2003 Symposium, Toledo, Spain, June 4–6, 2003. *EUREF Publication*, no. 13. Mitteilungen des Bundesamtes für Kartographie und Geodäsie, vol. 33. Frankfurt, pp. 202–206.
Holmes, S. A. and Pavlis, N. K. (2006): Spherical harmonic synthesis software harmonic_synth_v02.f. URL: http://earth-info.nga.mil/GandG/wgs84/gravitymod/new_egm/new_egm.html.
Huang, J. and Véronneau, M. (2013): Canadian gravimetric geoid model 2010. *Journal of Geodesy* 87(8):771–790. DOI: 10.1007/s00190-013-0645-0.
IAG Resolutions (2015): IAG resolution (No. 1) for the definition and realization of an International Height Reference System (IHRS). International Association of Geodesy. URL: http://iag.dgfi.tum.de/fileadmin/IAG-docs/IAG_Resolutions_2015.pdf.
IERS Conventions (2010). In: Petit, G., and Luzum, B. (eds.) IERS Technical Note, no. 36. Verlag des Bundesamts für Kartographie und Geodäsie, Frankfurt am Main, Germany.
Ihde, J. (1995): Geoid determination by GPS and levelling. In: Sünkel, H., and Marson, I. (eds.) Gravity and geoid, Joint symposium of the International Gravity Commission and the International Geoid Commission. Graz, Austria, Sept. 11–17, 1994. *International Association of Geodesy Symposia*, vol. 113. Springer Berlin Heidelberg, pp. 519–528. DOI: 10.1007/978-3-642-79721-7_55.
Ihde, J. and Sánchez, L. (2005): A unified global height reference system as a basis for IGGOS. *Journal of Geodynamics* 40(4–5):400–413. DOI: 10.1016/j.jog.2005.06.015.
Klokočník, J., Wagner, C. A., Kostelecký, J., Bezděk, A., Novák, P., and McAdoo, D. (2008): Variations in the accuracy of gravity recovery due to ground track variability: GRACE,

CHAMP, and GOCE. *Journal of Geodesy* 82(12):917–927. DOI: 10.1007/s00190-008-0222-0.

Kotsakis, C., Katsambalos, K., and Ampatzidis, D. (2012): Estimation of the zero-height geopotential level W_0^{LVD} in a local vertical datum from inversion of co-located GPS, leveling and geoid heights: a case study in the Hellenic islands. *Journal of Geodesy* 86(6):423–439. DOI: 10.1007/s00190-011-0530-7.

Kutterer, H. and Neilan, R. (2016): Global Geodetic Observing System (GGOS). *The Geodesists Handbook 2016, Journal of Geodesy* 90(10):1079–1094. DOI: 10.1007/s00190-016-0948-z.

Lemoine, F. G., Kenyon, S. C., Factor, J. K., Trimmer, R. G., Pavlis, N. K., Chinn, D. S., Cox, C. M., Klosko, S. M., Luthcke, S. B., Torrence, M. H., Wang, Y. M., Williamson, R. G., Pavlis, E. C., Rapp, R. H., and Olson, T. R. (1998): The development of the joint NASA GSFC and the National Imagery and Mapping Agency (NIMA) geopotential model EGM96. Technical report 1998-206861. NASA Goddard Space Flight Center, Greenbelt, Maryland, USA. URL: http://ntrs.nasa.gov/search.jsp?R=19980218814.

Luz, R. T., Bosch, W., de Freitas, S. R. C., Heck, B., and Dalazoana, R. (2009a): Evaluating the Brazilian vertical datum through improved coastal satellite altimetry data. In: Sideris, M. G. (ed.) Observing our changing Earth. Proceedings of the IAG General Assembly, Perugia, Italy, July 2–13, 2007. *International Association of Geodesy Symposia*, vol. 133. Springer Berlin Heidelberg, pp. 735–741. DOI: 10.1007/978-3-540-85426-5_84.

Luz, R. T., de Freitas, S. R. C., Heck, B., and Bosch, W. (2009b): Challenges and first results towards the realization of a consistent height system in Brazil. In: Drewes, H. (ed.) Geodetic reference frames, IAG Symposium. Munich, Germany, Oct. 9–14, 2006. *International Association of Geodesy Symposia*, vol. 134. Springer Berlin Heidelberg, pp. 291–296. DOI: 10.1007/978-3-642-00860-3_45.

Luz, R. T., Guimarães, V. M., Rodrigues, A. C., and Correia, J. D. (2002): Brazilian first order levelling network. In: Drewes, H., Dodson, A. H., Fortes, L. P. S., Sánchez, L., and Sandoval, P. (eds.) Vertical reference systems, IAG Symposium. Cartagena, Colombia, Feb. 20–23, 2001. *International Association of Geodesy Symposia*, vol. 124. Springer Berlin Heidelberg, pp. 20–22. DOI: 10.1007/978-3-662-04683-8_5.

Mäkinen, J. and Ihde, J. (2009): The permanent tide in height systems. In: Sideris, M. G. (ed.) Observing our changing Earth. Proceedings of the IAG General Assembly, Perugia, Italy, July 2–13, 2007. *International Association of Geodesy Symposia*, vol. 133. Springer Berlin Heidelberg, pp. 81–87. DOI: 10.1007/978-3-540-85426-5_10.

Mayer-Gürr, T., Pail, R., Gruber, T., Fecher, T., Rexer, M., Schuh, W.-D., Kusche, J., Brockmann, J.-M., Rieser, D., Zehentner, N., Kvas, A., Klinger, B., O., B., Höck, E., Krauss, S., and Jäggi, A. (2015): The combined satellite gravity field model GOCO05s. In: General Assembly of the European Geosciences Union 2015. Vienna, Austria, Apr. 12–17, 2015. *Geophysical Research Abstracts*, vol. 17. EGU2015-12364.

Mayer-Gürr, T., Zehentner, N., Klinger, B., and Kvas, A. (2014): ITSG-Grace2014: a new GRACE gravity field release computed in Graz. In: GRACE Science Team Meeting (GSTM). Potsdam, Germany, Sept. 29–Oct. 1, 2014. URL: https://pure.tugraz.at/portal/files/3412370/2014-09-30_mayer-guerr_etal_ITG-Grace2014_GSTM_Potsdam.pdf.

Montecino, H. D. and de Freitas, S. R. C. (2014): Strategies for connecting Imbituba and Santana Brazilian datums based on satellite gravimetry and residual terrain model. In: Rizos, C., and Willis, P. (eds.) Earth on the edge: science for a sustainable planet. Proceedings of the IAG General Assembly, Melbourne, Australia, June 28–July 2, 2011. *International Association of Geodesy Symposia*, vol. 139. Springer Berlin Heidelberg, pp. 543–549. DOI: 10.1007/978-3-642-37222-3_72.

Moritz, H. (1980): Geodetic Reference System 1980. *Bulletin Géodésique* 54(3):395–405. DOI: 10.1007/BF02521480.

Pail, R., Bruinsma, S. L., Migliaccio, F., Förste, C., Goiginger, H., Schuh, W.-D., Höck, E., Reguzzoni, M., Brockmann, J. M., Abrikosov, O., Veicherts, M., Fecher, T., Mayrhofer, R., Krasbutter, I., Sansò, F., and Tscherning, C. C. (2011): First GOCE gravity field models derived by three different approaches. *Journal of Geodesy* 85(11):819–843. DOI: 10.1007/s00190-011-0467-x.

Pail, R., Goiginger, H., Schuh, W.-D., Höck, E., Brockmann, J. M., Fecher, T., Gruber, T., Mayer-Gürr, T., Kusche, J., Jäggi, A., and Rieser, D. (2010): Combined satellite gravity field model GOCO01S derived from GOCE and GRACE. *Geophysical Research Letters* 37(20):L20314. DOI: 10.1029/2010GL044906.

Pavlis, N. K., Holmes, S. A., Kenyon, S. C., and Factor, J. K. (2012): The development and evaluation of the Earth Gravitational Model 2008. *Journal of Geophysical Research* 117:B04406. DOI: 10.1029/2011JB008916.

Rapp, R. H. (1983): The need and prospects for a world vertical datum. In: Proceedings of the 18th IAG General Assembly, XVII IUGG General Assembly, Hamburg, Germany, Aug. 15–27, 1983. *International Association of Geodesy Symposia*, vol. 2, pp. 432–445.

Rapp, R. H., Nerem, R. S., Shum, C., Klosko, S. M., and Williamson, R. G. (1991): Consideration of permanent tidal deformation in the orbit determination and data analysis for the Topex/Poseidon mission. Technical report NASA-TM-100775. NASA Goddard Space Flight Center, Greenbelt, Maryland, USA. URL: http://ntrs.nasa.gov/search.jsp?R=19910021305.

Ruess, D. and Mitterschiffthaler, P. (2015): Rezente Höhenänderungen in Österreich abgeleitet aus geodätischen Wiederholungsmessungen. In: Hanke, K., and Weinold, T. (eds.), 18. Internationale Geodätische Woche. Obergurgl, Austria, Feb. 8–14, 2015. Wichmann-Verlag, Heidelberg, pp. 111–123.

Rülke, A., Liebsch, G., Sacher, M., Schäfer, U., Schirmer, U., and Ihde, J. (2012): Unification of European height system realizations. *Journal of Geodetic Science*, Special issue on regional and global geoid-based vertical datums 2(4):343–354. DOI: 10.2478/v10156-011-0048-1.

Rülke, A., Liebsch, G., Sacher, M., Schäfer, U., Ihde, J., and Woodworth, P. L. (2016): Practical aspects of the unification of height system realizations in Europe. In: Rizos, C., and Willis, P. (eds.) IAG 150 years. Proceedings of the IAG Scientific Assembly, Postdam, Germany, Sept. 1–6, 2013. *International Association of Geodesy Symposia*, vol. 143. Springer Berlin Heidelberg, pp. 367–373. DOI: 10.1007/1345_2015_168.

Rummel, R. (2002): Global unification of height systems and GOCE. In: Sideris, M. G. (ed.) Gravity, geoid and geodynamics 2000, GGG2000 IAG International Symposium. Banff,

Alberta, Canada, July 31–Aug. 4, 2000. *International Association of Geodesy Symposia*, vol. 123. Springer Berlin Heidelberg, pp. 13–20. DOI: 10.1007/978-3-662-04827-6_3.

Rummel, R. and Teunissen, P. (1988): Height datum definition, height datum connection and the role of the geodetic boundary value problem. *Bulletin Géodésique* 62(4):477–498. DOI: 10.1007/BF02520239.

Rummel, R., Yi, W., and Stummer, C. (2011): GOCE gravitational gradiometry. *Journal of Geodesy* 85(11):777–790. DOI: 10.1007/s00190-011-0500-0.

Sacher, M., Ihde, J., Liebsch, G., and Mäkinen, J. (2009): EVRF2007 as realization of the European Vertical Reference System. *Bollettino di Geodesia e Scienze Affini* 68(1):35–50.

Sánchez, L. (2009): Strategy to establish a global vertical reference system. In: Drewes, H. (ed.) Geodetic reference frames, IAG Symposium. Munich, Germany, Oct. 9–14, 2006. *International Association of Geodesy Symposia*, vol. 134. Springer Berlin Heidelberg, pp. 273–278. DOI: 10.1007/978-3-642-00860-3_42.

Sánchez, L. (2015): Ein einheitliches vertikales Referenzsystem für Südamerika im Rahmen eines globalen Höhensystems. *Deutsche Geodätische Kommission, Reihe C*, no. 748. Verlag der Bayerischen Akademie der Wissenschaften in Kommission beim Verlag C. H. Beck, Munich, Germany. URL: https://www.dgk.badw.de.devweb.mwn.de/fileadmin/docs/c-748.pdf.

Sansò, F. and Venuti, G. (2002): The height datum/geodetic datum problem. *Geophysical Journal International* 149(3):768–775. DOI: 10.1046/j.1365-246X.2002.01680.x.

Sjöberg, L. (2011): On the definition and realization of a global vertical datum. *Journal of Geodetic Science* 1(2):154–157. DOI: 10.2478/v10156-010-0018-z.

Šprlák, M., Gerlach, C., and Pettersen, B. R. (2015): Validation of GOCE global gravitational field models in Norway. In: Huang, J., Reguzzoni, M., and Gruber, T. (eds.) Assessment of GOCE geopotential models, *Newton's Bulletin*, no. 5. International Association of Geodesy and International Gravity Field Service, pp. 13–24. URL: http://www.isgeoid.polimi.it/Newton/Newton_5/03_Sprlak_13_24.html.

Tapley, B. D., Bettadpur, S., Ries, J. C., Thompson, P. F., and Watkins, M. M. (2004): GRACE measurements of mass variability in the Earth system. *Science* 305(5683):503–505. DOI: 10.1126/science.1099192.

Voigt, C. and Denker, H. (2015): Validation of GOCE gravity field models in Germany. In: Huang, J., Reguzzoni, M., and Gruber, T. (eds.) Assessment of GOCE geopotential models, *Newton's Bulletin*, no. 5. International Association of Geodesy and International Gravity Field Service, pp. 37–48. URL: http://www.isgeoid.polimi.it/Newton/Newton_5/05_Voigt_37_48.html.

Weber, D. (1994): Das neue gesamtdeutsche Haupthöhennetz DHHN 92. *Allgemeine Vermessungs-Nachrichten* 101(5):179–193.

Wolf, H. (1974): Über die Einführung von Normalhöhen. *Zeitschrift für Vermessungswesen* 99:1–5.

Woodworth, P. L., Hughes, C. W., J., B. R., and Gruber, T. (2012): Towards worldwide height system unification using ocean information. *Journal of Geodetic Science*, Special issue on regional and global geoid-based vertical datums 2(4):302–318. DOI: 10.2478/v10156-012-0004-8.

Xu, P. (1992): A quality investigation of global vertical datum connection. *Geophysical Journal International* 110(2):361–370. DOI: 10.1111/j.1365-246X.1992.tb00880.x.

Chapter VI.

Height system unification based on the fixed GBVP approach

Thomas Grombein, Kurt Seitz, Bernhard Heck
IAG 150 Years. International Association of Geodesy Symposia, vol. 143, pp. 305–311, 2016. DOI: 10.1007/1345_2015_104.

Submitted: 14 November 2013 / Accepted: 26 April 2014 / Published online: 07 July 2015

Abstract. In general, any national or regional height reference system is related to an individual vertical datum, defined by one or several tide gauges. The discrepancies of these local vertical datums cause height datum offsets in a range of about ± 1–2 m at a global scale. For the purpose of height system unification, global geopotential models derived from homogeneous satellite data provide an important contribution. However, to achieve a unification of high precision, the use of local terrestrial gravity data in the framework of a Geodetic Boundary Value Problem (GBVP) is required. By solving the GBVP at GNSS/leveling benchmarks, the unknown height datum offsets can be estimated in a least squares adjustment. In contrast to previous studies, related to the scalar free GBVP based on gravity anomalies, this paper discusses the alternative use and benefit of the fixed GBVP. This modern formulation of the GBVP is related to gravity disturbances, using the surface of the Earth as boundary surface. In contrast to gravity anomalies, gravity disturbances are not affected by the discrepancies of the local height datum. Therefore, in comparison to a scalar free GBVP approach, the proposed method is not affected by indirect bias terms, which will simplify a height system unification. In this paper, the theory of the fixed GBVP approach is developed and formulas in spherical approximation are derived. Moreover, the method is validated using a closed loop simulation based on the global geopotential model EGM2008, showing mm-accuracy of the estimated height datum offsets.

Keywords Height system unification · Geodetic Boundary Value Problem (GBVP) · Hotine's integral formula

1. Introduction

In geodesy, there are two different types of height systems: geometrical and physical. In the former, geometrically defined ellipsoidal heights are used, related to the orthogonal distance to a reference ellipsoid. In the latter, physical heights are utilized that refer to a physically defined reference surface linked to the Earth's gravity potential W.

The ellipsoidal height $h(P)$ of a point P on the Earth's surface can directly be measured using methods of GNSS positioning (Global Navigation Satellite Systems). By combining GNSS observations with other space techniques, global three-dimensional terrestrial reference frames have been established that provide sub-cm consistency in the vertical component, e.g., ITRF2008 (Altamimi et al., 2011).

For physical (or national) height systems the situation is quite different. Physical heights are determined by a combination of spirit leveling and gravimetry with respect to a fixed datum point P_0. These observations are then used to derive geopotential numbers

$$C(P) := W_0 - W(P), \tag{1}$$

representing the difference of the gravity potential value between a leveling point P and the datum point P_0, i.e., $W_0 := W(P_0)$. In that way, the equipotential surface passing through P_0 is chosen as the reference level of the height system. Dividing Eq. (1) by the mean normal gravity value $\overline{\gamma}$ along the normal plumb line, the geopotential number $C(P)$ is transformed to the (metric) normal height $H(P)$ (Heiskanen and Moritz, 1967, p. 170 f.).

For the practical realization of a physical height system, the height reference level is conventionally linked to the mean sea level (MSL), observed at one or several tide gauges, i.e., the datum point P_0 is selected such that the zero level is fixed to the local MSL. As the leveling networks of different national surveys mostly refer to individual tide gauges, hundreds of different national height systems exist worldwide that are realized by their own local vertical datum. Due to the sea surface topography, different tide gauges do not refer to the same equipotential surface. Therefore, the reference levels of different physical height systems are inconsistent by about ± 1–$2\,\mathrm{m}$ at a global scale (Heck, 1990; Gerlach and Rummel, 2013).

On the other hand, many global and regional applications such as monitoring of sea level change, ice sheet melting, or post-glacial rebound require a high-precision and consistent global physical height system. Moreover, this is also relevant for establishing the Global Geodetic Observing System (Ihde and Sánchez, 2005). In order to overcome the problem of height datum inconsistencies, different strategies and approaches for height system unification have been discussed and proposed in various publications (e.g., Colombo, 1980; Rapp, 1983; Heck and Rummel, 1990; Sansò and Venuti, 2002; Sánchez, 2009).

Considering a local height datum zone σ^i that is linked to the gravity potential value W_0^i, the geopotential number in Eq. (1) analogously reads

$$C^i(P) = W_0^i - W(P). \tag{2}$$

Combining Eqs. (1) and (2), the relation between the local datum zone σ^i and a global datum specified by the gravity potential value W_0 is described by the height datum offset

$$\delta H^i := \frac{C(P) - C^i(P)}{\bar{\gamma}} = \frac{W_0 - W_0^i}{\bar{\gamma}}. \tag{3}$$

For the determination of δH^i, observation points that combine physical and geometrical height information are of particular interest, i.e., GNSS/leveling benchmarks. For these points, global geopotential models (GGM) can be used to determine approximated values $C(P) = W_0 - W_{\mathrm{GGM}}(P)$, which can be inserted in Eq. (3). In this context, GGM derived from recent gravity field satellite missions like GRACE and GOCE provide an important contribution, as they provide a homogeneous reference surface that is not affected by a height datum offset (Rummel, 2002; Gatti et al., 2013). Due to the limited resolution of the used GGM, such an approach suffers from an omission error. Although this error can be reduced, representing shorter wavelengths by the high-resolution EGM2008 (Pavlis et al., 2012) or regional geoid models, the expected accuracy for δH^i is limited to cm–dm level (Gruber et al., 2012; Rülke et al., 2012).

To achieve a unification at sub-cm level, the use of terrestrial gravity data in a Geodetic Boundary Value Problem is indispensable (GBVP, Heiskanen and Moritz, 1967, p. 36 f.). For this purpose, the solution of the GBVP is used to estimate height datum offsets in a least squares approach (e.g., Heck and Rummel, 1990). In contrast to previous publications, mostly related to the scalar free GBVP approach (Rummel and Teunissen, 1988; Xu, 1992; Gerlach and Rummel, 2013), this paper discusses perspectives and benefits of the alternative use of a fixed GBVP approach for height system unification. In order to reduce systematic errors, a combination with a GGM and topographic information in a remove-compute-restore approach is advisable, as frequently used in gravimetric (quasi-)geoid determination (Forsberg and Tscherning, 1997). However, such a combination is beyond the scope of this article. Therefore, the presented formulas will be restricted to the use of terrestrial gravity data.

The paper is organized as follows: in Sect. 2 the proposed fixed GBVP approach is presented and formulas in spherical approximation are derived. In order to validate the method and analyze its accuracy, a closed loop simulation based on EGM2008 is presented in Sect. 3. Finally, in Sect. 4, a summary and an outlook to ongoing research are provided.

2. Fixed GBVP approach

Let the Earth's surface S be partitioned into n disjoint local height datum zones σ^i, $i = 1, \dots, n$, i.e., $S = \bigcup_{i=1}^{n} \sigma^i$ with $\sigma^i \cap \sigma^k = \emptyset$ for $i \neq k$. Each datum zone is assumed to be linked to an individual equipotential surface defined by the gravity potential value W_0^i. Furthermore, let each datum zone σ^i contain m_i GNSS/leveling benchmarks P_j^i, $j = 1, \dots, m_i$, where the (unbiased) ellipsoidal height h and the (biased) normal height H^i are known. For these benchmarks, the (biased) height anomaly $\zeta^i = h - H^i$ can be calculated, which is linked to the disturbing potential T by the generalized Bruns' formula (Heiskanen and Moritz, 1967, p. 100):

$$\zeta^i(P_j^i) = \frac{T(P_j^i) - (W_0^i - U_0)}{\gamma} = \frac{T(P_j^i) - \Delta W_0}{\gamma} + \delta H^i, \tag{4}$$

where U_0 denotes the constant normal gravity potential value of the used reference ellipsoid, γ is the normal gravity value at the Earth's surface, and

$$\Delta W_0 := W_0 - U_0. \tag{5}$$

To determine the disturbing potential T, the fixed GBVP will be used that is based on gravity disturbances

$$\delta g := g(P) - \gamma(P) \approx -\frac{\partial T}{\partial r}\bigg|_S \tag{6}$$

resulting from the difference between the measured gravity $g(P)$ and the normal gravity $\gamma(P)$, both defined at the Earth's surface point $P \in S$. Here, $\partial/\partial r$ denotes the partial derivative with respect to the geocentric radius r. Considering the normal gravity formula (Heiskanen and Moritz, 1967, p. 79), the ellipsoidal height $h(P)$ of the gravity measurement benchmark is required to obtain $\gamma(P)$. Thus, in the case of the fixed GBVP, the geometry of the Earth's surface S is assumed to be known, e.g, by GNSS positioning.

Utilizing the analytical solution of the fixed GBVP, the disturbing potential T can be obtained in constant radius approximation by Hotine's spherical integral formula (Hotine, 1969, p. 311ff.; Heck, 2011):

$$T(\varphi, \lambda) = \frac{R}{4\pi} \iint_\sigma \delta g(\varphi', \lambda') \cdot \mathrm{H}(\psi)\, \mathrm{d}\sigma, \tag{7}$$

where

$$\mathrm{H}(\psi) = \frac{1}{\sin(\psi/2)} - \ln\left(1 + \frac{1}{\sin(\psi/2)}\right) \tag{8}$$

and ψ is the spherical distance between the position vectors of the computation point $P(r = R, \varphi, \lambda)$ and the running integration point $P'(\varphi', \lambda')$, both located on the sphere with radius R. The surface of the unit sphere is denoted by σ with the corresponding surface element $\mathrm{d}\sigma = \cos\varphi'\, \mathrm{d}\varphi'\, \mathrm{d}\lambda'$.

Applying Eq. (7) to Eq. (4) leads to

$$\zeta^i(P_j^i) = \frac{R}{4\pi\gamma} \iint_\sigma \delta g \cdot \mathrm{H}(\psi)\, \mathrm{d}\sigma - \frac{\Delta W_0}{\gamma} + \delta H^i, \tag{9}$$

which is the basic equation of the fixed GBVP approach that can already be used for the estimation of the unknown height datum offsets δH^i at GNSS/leveling benchmarks P_j^i.

However, the lacking availability of globally distributed gravity disturbances δg complicates the practical evaluation of Eq. (9). Since for most (historical) gravity measurement benchmarks of the pre-GNSS era the ellipsoidal height h has not been determined, gravity disturbances δg according to Eq. (6) could not be compiled. Instead, gravity measurements g have frequently been used to derive gravity anomalies Δg that serve as boundary values for the traditional scalar free GBVP. Taking into account the present situation, Eq. (9) will be extended by considering the transformation of gravity anomalies Δg to gravity disturbances δg.

2.1. Extension to gravity anomalies

Following the theory of Molodensky (Heiskanen and Moritz, 1967, p. 291ff.), gravity anomalies

$$\Delta g := g(P) - \gamma(Q) \approx \left(-\frac{\partial T}{\partial r} - \frac{2\gamma}{r}\zeta\right)\Big|_{\Sigma} \tag{10}$$

differ from gravity disturbances δg in the normal gravity $\gamma(Q)$, evaluated at the telluroid $\Sigma \ni Q$ instead of the Earth's surface S. Considering that $h(Q) = H^i(P)$ (Heiskanen and Moritz, 1967, p. 293), the normal gravity value $\gamma(Q)$ depends on the (biased) normal height. Thus, in contrast to gravity disturbances, gravity anomalies are affected by the height datum offset δH^i of the local datum zone σ^i (Heck, 1990). This becomes clear when inserting Eq. (4) into Eq. (10):

$$\Delta g^i = \left(-\frac{\partial T}{\partial r} - \frac{2}{r}T + \frac{2}{r}\Delta W_0 - \frac{2\gamma}{r}\delta H^i\right)\Big|_{\Sigma}. \tag{11}$$

Combining the boundary conditions of Eqs. (6) and (11), the (unbiased) gravity disturbance δg can be expressed as a function of the (biased) gravity anomaly Δg^i and the height datum offset δH^i using the linear approximation

$$\delta g = \Delta g^i + \left(\frac{2}{r}T - \frac{2}{r}\Delta W_0 + \frac{2\gamma}{r}\delta H^i\right)\Big|_{S} + \delta_{\mathrm{BS}}, \tag{12}$$

where δ_{BS} denotes the error induced by the different boundary surfaces (S and Σ), which is neglected in the following.

Splitting Eq. (12) into three components

$$\delta g_0 := \Delta g^i + \frac{2}{r}T, \quad \delta g_1 := -\frac{2}{r}\Delta W_0, \quad \delta g_2 := \frac{2\gamma}{r}\delta H^i, \tag{13}$$

and inserting them separately into Eq. (9) results in

$$\zeta^i(P^i_j) = \zeta_0 + \zeta_1 + \zeta_2 - \frac{\Delta W_0}{\gamma} + \delta H^i, \tag{14}$$

where

$$\zeta_m := \frac{R}{4\pi\gamma}\iint\limits_{\sigma} \delta g_m \cdot \mathrm{H}(\psi)\,\mathrm{d}\sigma, \quad m = 0, 1, 2. \tag{15}$$

Applying constant radius approximation, i.e., $r = R$, the evaluation of Eq. (15) leads to

$$\zeta_0 = \frac{R}{4\pi\gamma} \iint\limits_{\sigma} \left(\Delta g^i + \frac{2}{R} T \right) \cdot \mathrm{H}(\psi)\, \mathrm{d}\sigma, \tag{16}$$

$$\zeta_1 = -\frac{\Delta W_0}{2\pi\gamma} \iint\limits_{\sigma} \mathrm{H}(\psi)\, \mathrm{d}\sigma = -\frac{\Delta W_0}{2\pi\gamma} \cdot 4\pi = -\frac{2\Delta W_0}{\gamma}, \tag{17}$$

$$\zeta_2 = \sum_{i=1}^{n} \frac{\delta H^i}{2\pi} \iint\limits_{\sigma^i} \mathrm{H}(\psi)\, \mathrm{d}\sigma, \tag{18}$$

where in the case of ζ_2, the (global) integral domain σ is decomposed into the disjoint height datum zones σ^i.

Finally, inserting Eqs. (16) – (18) into Eq. (14) results in

$$\zeta^i(P^i_j) = \zeta_0(\Delta g^i, T) + \delta H^0 + \delta H^i + \sum_{k=1}^{n} \delta H^k \cdot \mathrm{G}^{i,k}_j, \tag{19}$$

where

$$\begin{aligned} \delta H^0 &:= -\frac{3\Delta W_0}{\gamma}, \\ \mathrm{G}^{i,k}_j &:= \frac{1}{2\pi} \iint\limits_{\sigma^k} \mathrm{H}(\psi)\, \mathrm{d}\sigma \bigg|_{P^i_j}. \end{aligned} \tag{20}$$

In Eq. (19), different kinds of height datum offsets occur. The height datum offset δH^i represents the direct influence of the datum zone σ^i containing P^i_j. This offset, also occurring in the basic Eq. (9), is frequently called direct bias term. Moreover, Eq. (19) also comprises the height datum offsets δH^k $(k = 1, \ldots, n)$ of all datum zones, i.e., $\delta H^1, \ldots, \delta H^n$. These offsets are a consequence of the global integration of biased gravity anomalies Δg^i and are named indirect bias terms (Gerlach and Rummel, 2013). Particularly, the evaluation of the corresponding factors $\mathrm{G}^{i,k}_j$ in Eq. (20) is complicated, as the separate integration requires the coordinates of the bounding polygon for each datum zone.

While the indirect bias terms amount to about ±1–2 m, simulation studies for the scalar free GBVP approach presented by Gerlach and Rummel (2013) demonstrate that their influence can be reduced to a level below 1 cm, when a satellite-derived GGM is employed for representing the long-wavelength parts of ζ_0. However, it is worthwhile mentioning that the basic approach in Eq. (9) is not affected by the indirect bias terms. Therefore, if gravity disturbances δg become globally available, the indirect bias terms can be avoided, demonstrating the advantage of the fixed GBVP approach in future applications.

The parameter δH^0 in Eq. (19) comprises ΔW_0, defining the reference level of the global datum. As this global offset cannot be uniquely estimated within this approach, W_0 is assumed to be equal to U_0, i.e., ΔW_0 in Eq. (5) and δH^0 in Eq. (20) are set to zero. By this procedure, an "absolute" vertical datum is defined by convention (e.g., Heck, 2004).

2.2. Least squares adjustment

Using Eq. (19) with $\delta H^0 = 0$, the observation equation for least squares adjustment (LSA) is provided by

$$L_j^i = \zeta^i - \zeta_0(\Delta g^i, T)\Big|_{P_j^i} = \delta H^i + \sum_{k=1}^{n} \delta H^k \cdot \mathrm{G}_j^{i,k}, \tag{21}$$

where Δg^i are the observed (biased) gravity anomalies and T the (unbiased) disturbing potential values, derived from an a priori model (e.g., EGM2008). The quantities on the left-hand side of Eq. (21) are the known observations and those on the right-hand side contain the unknowns to be estimated. The functional model according to Eq. (21) is specified by

$$\underbrace{\begin{pmatrix} l_1^1 + v_1^1 \\ l_2^1 + v_2^1 \\ \vdots \\ l_1^2 + v_1^2 \\ \vdots \\ l_{m_n}^n + v_{m_n}^n \end{pmatrix}}_{\boldsymbol{l}+\boldsymbol{v}} = \underbrace{\begin{pmatrix} 1+\mathrm{G}_1^{1,1} & \mathrm{G}_1^{1,2} & \cdots & \mathrm{G}_1^{1,n} \\ 1+\mathrm{G}_2^{1,1} & \mathrm{G}_2^{1,2} & \cdots & \mathrm{G}_2^{1,n} \\ \vdots & & & \\ \mathrm{G}_1^{2,1} & 1+\mathrm{G}_1^{2,2} & \cdots & \mathrm{G}_1^{2,n} \\ \vdots & & & \\ \mathrm{G}_{m_n}^{n,1} & \mathrm{G}_{m_n}^{n,2} & \cdots & 1+\mathrm{G}_{m_n}^{n,n} \end{pmatrix}}_{\mathbf{A}} \cdot \underbrace{\begin{pmatrix} \delta H^1 \\ \delta H^2 \\ \delta H^3 \\ \vdots \\ \delta H^n \end{pmatrix}}_{\boldsymbol{x}},$$

where $\boldsymbol{l}$ is the observation vector, $\boldsymbol{v}$ the inconsistency vector, and $\boldsymbol{x}$ the vector of unknowns. The design matrix $\mathbf{A}$ contains the partial derivatives of the observations with respect to the unknowns. Using a standard LSA, the unknown height datum offsets are estimated by

$$\hat{\boldsymbol{x}} = \mathbf{N}^{-1} \cdot \mathbf{A}^{\mathrm{T}}\mathbf{P} \cdot \boldsymbol{l}, \tag{22}$$

where $\mathbf{N} = \mathbf{A}^{\mathrm{T}}\mathbf{P}\mathbf{A}$ is the normal matrix and $\mathbf{P}$ is the weight matrix of the observations, which can be specified by an additional stochastic model.

3. Closed loop simulation

Using the presented fixed GBVP approach, a closed loop simulation is performed following a four-step sequence:

1. Definition of eight height datum zones σ^i with individual height datum offsets δH^i ($i = 1, \ldots, 8$).
2. Addition of δH^i to EGM2008-derived observations.
3. Estimation of $\hat{\boldsymbol{x}}$ by Eq. (22) with $\mathbf{P} = \mathbf{I}$ (identity matrix).
4. Comparison of estimated and reference values.

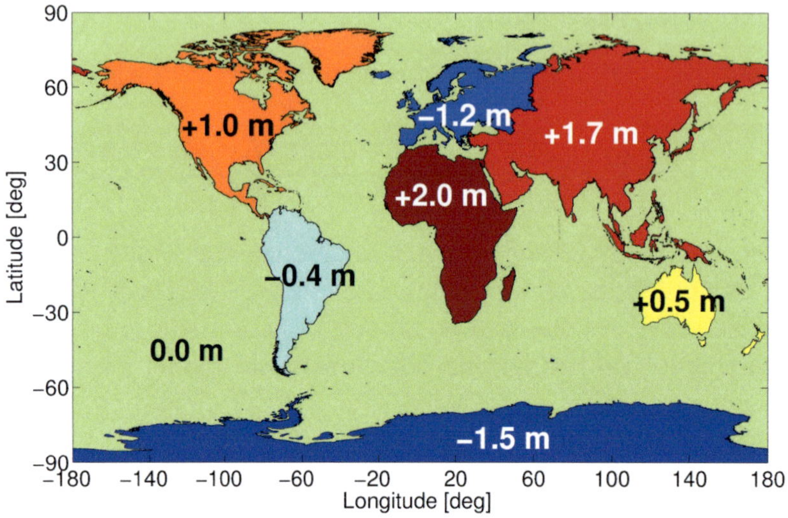

Fig. 1. Visualization of the height datum zones σ^i and their assumed height datum offsets δH^i used for the closed loop simulation.

As illustrated in Fig. 1 and specified by Table 1, the Earth's continents and oceans are utilized as height datum zones σ^i, where height datum offsets δH^i are assumed that cover the range of ± 1–$2\,\text{m}$. Using EGM2008 to degree and order 2190, global grids of consistent height anomalies ζ_{EGM} $(5^\circ \times 5^\circ)$, gravity anomalies Δg_{EGM} $(5' \times 5')$ and disturbing potential values T_{EGM} $(5' \times 5')$ are generated on a sphere with radius $R = 6371\,\text{km}$ and normal gravity $\bar{\gamma} = \gamma = 9.81\,\text{m}\,\text{s}^{-2}$. Applying the height datum offsets δH^i, simulated observations according to Eq. (21) are calculated by

$$L_j^i = \underbrace{\zeta_{\text{EGM}} + \delta H^i}_{\zeta^i} - \zeta_0(\underbrace{\Delta g_{\text{EGM}} - \frac{2\gamma}{R}\delta H^i}_{\Delta g^i}, \underbrace{T_{\text{EGM}}}_{T})\Big|_{P_j^i}, \tag{23}$$

where the integration is performed by Gauss–Legendre quadrature (e.g., Schwarz, 1989, p. 361ff.).

To analyze the impact of the global distribution of the used benchmarks P_j^i, four different scenarios (a) – (d) are considered as displayed in Fig. 2. In scenario (a), all 2592 observations L_j^i of the $5^\circ \times 5^\circ$ global grid are used in the LSA. In scenario (b), observations are restricted to continental areas (879 benchmarks), while in scenario (c) only observations in Europe, South America, and Australia are included (161 benchmarks). Scenario (d) is similar to (c), but additionally at least one benchmark is included in each datum zone (166 benchmarks). In each scenario, the height datum offsets of all datum zones are estimated.

In Table 1, the numerical results for the scenarios (a) – (d) are presented in terms of error values

$$\varepsilon^i = \delta H^i - \hat{x}^i, \tag{24}$$

where the estimated height datum offsets are denoted by $\hat{x}^i$, i.e., the components of $\hat{\boldsymbol{x}}$. Moreover, to quantify the stability of the LSA, Table 2 specifies the spectral condition

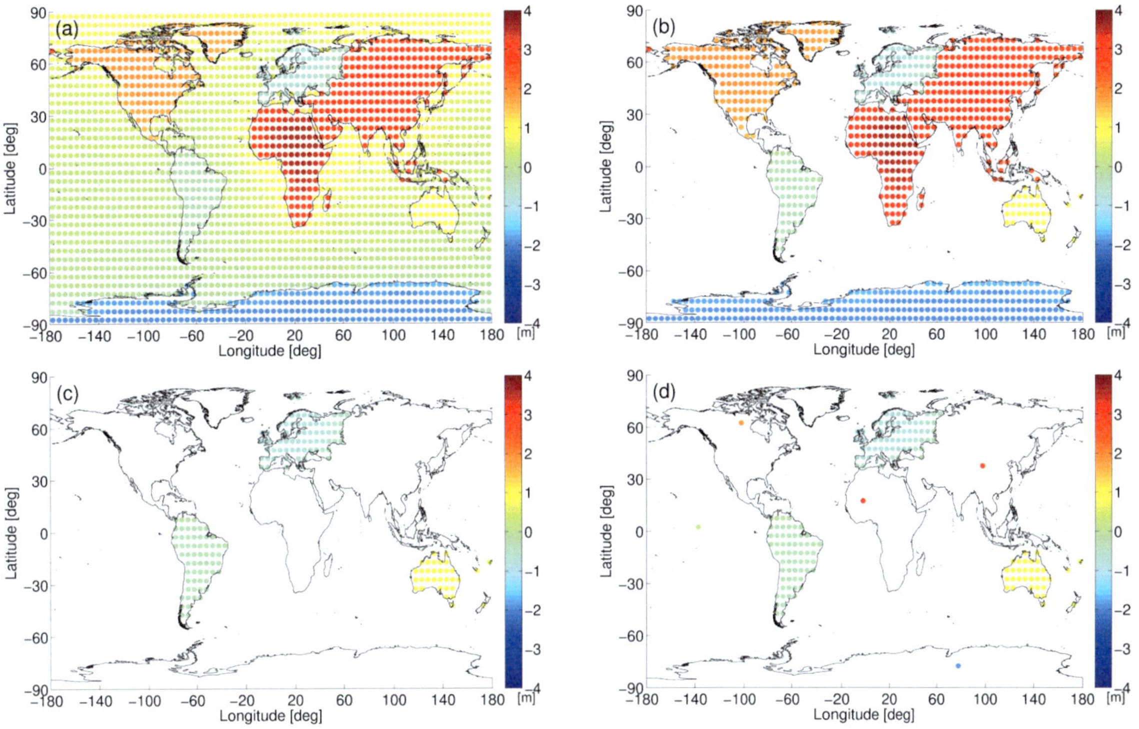

Fig. 2. Visualization of the global distribution of benchmarks P_j^i used in the scenarios (a) – (d). Each dot represents the value of an observation equation L_j^i according to Eq. (23).

number κ_2 of the normal matrix $\mathbf{N}$, i.e., the ratio of the largest to the smallest eigenvalue of the matrix (Schwarz, 1989, p. 24f.).

In the ideal scenario (a), the error values attain a sub-mm level, only in South America and Antarctica slightly larger values occur. Excluding the observations of the oceans, scenario (b) produces error values at lower mm level. Going a step further toward a realistic scenario, case (c) demonstrates that the error values are increased to cm level or even dm level in North America. In contrast to the other scenarios, the large condition number of (c) indicates the instability of the LSA. Concerning scenario (d), it is demonstrated that if at least one observation is added in each datum zone, this instability can be mitigated. Thus, scenario (d) provides an error level comparable to (b), showing that mm-accuracy can be achieved in principle. However, these accuracy values are quite optimistic and must be seen in the context of the assumed error-free observation data of the closed loop simulation. To obtain realistic values for practical applications, a formal error propagation procedure would have to be taken into account.

4. Conclusions and outlook

In contrast to geometrically defined global terrestrial reference systems, physical height systems suffer from discrepancies of about ±1–2 m due to the individual definition of their local vertical datum. In order to realize a comparison of physical heights, a height system unification is required.

In this paper, a method based on the solution of a fixed GBVP has been presented, where height datum offsets are estimated in a least squares adjustment. In contrast to previous approaches using the traditional scalar free GBVP, the formulation of the proposed method is based on (unbiased) gravity disturbances that do not cause indirect bias terms. Therefore, the fixed GBVP approach simplifies the estimation of height datum offsets, when gravity disturbances become globally available in the future. However, considering the current situation of the global gravity data base, the approach is extended by a transformation of gravity anomalies to gravity disturbances also comprising indirect bias terms. By conducting a closed loop simulation based on eight height datum zones and EGM2008-derived observations, the fixed GBVP approach has been validated, showing mm-accuracy of the estimated height datum offsets. Furthermore, the stability of the adjustment has been analyzed showing a dependency on the global distribution of the observations; at least one observation should be located in each datum zone.

Table 1. Specification of the height datum zones σ^i, their assumed height datum offsets δH^i, and the error values ε^i according to Eq. (24) for the scenarios (a) – (d).

i	Datum zone σ^i	δH^i [m]	Error values ε^i [mm]			
			(a)	(b)	(c)	(d)
1	Asia	1.7	−0.2	1.3	−19.9	−0.6
2	North America	1.0	−0.4	1.5	−177.8	−4.1
3	Europe	−1.2	0.3	1.7	22.0	2.1
4	Africa	2.0	0.3	1.7	−24.4	2.2
5	South America	−0.4	−2.7	−0.9	10.6	−0.7
6	Australia	0.5	−0.5	2.2	10.1	2.2
7	Antarctica	−1.5	2.4	4.5	−25.7	3.8
8	Ocean	0.0	0.0	−2.7	−2.1	−2.5

Table 2. Spectral condition number κ_2 of the normal matrix $\mathbf{N}$, quantifying the stability of the LSA for the scenarios (a) – (d).

Scenario	(a)	(b)	(c)	(d)
Condition number $\kappa_2(\mathbf{N})$	221	404	118 480	246

As future work, the impact of approximation errors on the presented spherical solution will be analyzed and taken into account by suitable reductions. First results concerning the fixed GBVP are presented by Müßle et al. (2014). In addition, the combination of terrestrial gravity data with a GGM and topographic information will be investigated as well as a modification of Hotine's integral kernel to restrict the global integration area (Featherstone, 2013).

Acknowledgements. The authors acknowledge the financial support provided by the German Research Foundation (DFG) under grant number HE1433/20-1. Furthermore, we would like to thank three anonymous reviewers as well as the associated editor and the Editor-in-Chief for their valuable comments, which helped to improve the manuscript.

References

Altamimi, Z., Collilieux, X., and Métivier, L. (2011): ITRF2008: an improved solution of the international terrestrial reference frame. *Journal of Geodesy* 85(8):457–473. DOI: 10.1007/s00190-011-0444-4.

Colombo, O. L. (1980): A world vertical network. Report 296. Department of Geodetic Science and Surveying, The Ohio State University, Columbus, USA.

Featherstone, W. E. (2013): Deterministic, stochastic, hybrid and band-limited modifications of Hotine's integral. *Journal of Geodesy* 87(5):487–500. DOI: 10.1007/s00190-013-0612-9.

Forsberg, R. and Tscherning, C. C. (1997): Topographic effects in gravity field modelling for BVP. In: Sansò, F., and Rummel, R. (eds.) Geodetic boundary value problems in view of the one centimeter geoid, *Lecture Notes in Earth Sciences*, vol. 65. Springer Berlin Heidelberg, pp. 239–272. DOI: 10.1007/BFb0011707.

Gatti, A., Reguzzoni, M., and Venuti, G. (2013): The height datum problem and the role of satellite gravity models. *Journal of Geodesy* 87(1):15–22. DOI: 10.1007/s00190-012-0574-3.

Gerlach, C. and Rummel, R. (2013): Global height system unification with GOCE: a simulation study on the indirect bias term in the GBVP approach. *Journal of Geodesy* 87(1):57–67. DOI: 10.1007/s00190-012-0579-y.

Gruber, T., Gerlach, C., and Haagmans, R. (2012): Intercontinental height datum connection with GOCE and GPS-levelling data. *Journal of Geodetic Science*, Special issue on regional and global geoid-based vertical datums 2(4):270–280. DOI: 10.2478/v10156-012-0001-y.

Heck, B. (1990): An evaluation of some systematic error sources affecting terrestrial gravity anomalies. *Bulletin Géodésique* 64(1):88–108. DOI: 10.1007/BF02530617.

Heck, B. (2004): Problems in the definition of vertical reference frames. In: Sansò, F. (ed.) V Hotine-Marussi Symposium on Mathematical Geodesy, Matera, Italy, June 17–21, 2003. *International Association of Geodesy Symposia*, vol. 127. Springer Berlin Heidelberg, pp. 164–173. DOI: 10.1007/978-3-662-10735-5_22.

Heck, B. (2011): A Brovar-type solution of the fixed geodetic boundary-value problem. *Studia Geophysica et Geodætica* 55(3):441–454. DOI: 10.1007/s11200-011-0025-2.

Heck, B. and Rummel, R. (1990): Strategies for solving the vertical datum problem using terrestrial and satellite geodetic data. In: Sünkel, H., and Baker, T. (eds.) Sea surface topography and the geoid. Proceedings of the IAG General Assembly, Edinburgh, Scotland, Aug. 10–11, 1989. *International Association of Geodesy Symposia*, vol. 104. Springer Berlin Heidelberg, pp. 116–128. DOI: 10.1007/978-1-4684-7098-7_14.

Heiskanen, W. A. and Moritz, H. (1967): Physical geodesy. W. H. Freeman & Co., San Francisco, USA.

Hotine, M. (1969): Mathematical geodesy. ESSA Monograph 2, US Departement of Commerce, Washington, USA.

Ihde, J. and Sánchez, L. (2005): A unified global height reference system as a basis for IGGOS. *Journal of Geodynamics* 40(4–5):400–413. DOI: 10.1016/j.jog.2005.06.015.

Müßle, M., Heck, B., Seitz, K., and Grombein, T. (2014): On the effect of planar approximation in the geodetic boundary value problem. *Studia Geophysica et Geodætica* 58(4):536–555. DOI: 10.1007/s11200-013-0249-4.

Pavlis, N. K., Holmes, S. A., Kenyon, S. C., and Factor, J. K. (2012): The development and evaluation of the Earth Gravitational Model 2008. *Journal of Geophysical Research* 117:B04406. DOI: 10.1029/2011JB008916.

Rapp, R. H. (1983): The need and prospects for a world vertical datum. In: Proceedings of the 18th IAG General Assembly, XVII IUGG General Assembly, Hamburg, Germany, Aug. 15–27, 1983. *International Association of Geodesy Symposia*, vol. 2, pp. 432–445.

Rülke, A., Liebsch, G., Sacher, M., Schäfer, U., Schirmer, U., and Ihde, J. (2012): Unification of European height system realizations. *Journal of Geodetic Science*, Special issue on regional and global geoid-based vertical datums 2(4):343–354. DOI: 10.2478/v10156-011-0048-1.

Rummel, R. (2002): Global unification of height systems and GOCE. In: Sideris, M. G. (ed.) Gravity, geoid and geodynamics 2000, GGG2000 IAG International Symposium. Banff, Alberta, Canada, July 31–Aug. 4, 2000. *International Association of Geodesy Symposia*, vol. 123. Springer Berlin Heidelberg, pp. 13–20. DOI: 10.1007/978-3-662-04827-6_3.

Rummel, R. and Teunissen, P. (1988): Height datum definition, height datum connection and the role of the geodetic boundary value problem. *Bulletin Géodésique* 62(4):477–498. DOI: 10.1007/BF02520239.

Sánchez, L. (2009): Strategy to establish a global vertical reference system. In: Drewes, H. (ed.) Geodetic reference frames, IAG Symposium. Munich, Germany, Oct. 9–14, 2006. *International Association of Geodesy Symposia*, vol. 134. Springer Berlin Heidelberg, pp. 273–278. DOI: 10.1007/978-3-642-00860-3_42.

Sansò, F. and Venuti, G. (2002): The height datum/geodetic datum problem. *Geophysical Journal International* 149(3):768–775. DOI: 10.1046/j.1365-246X.2002.01680.x.

Schwarz, H. R. (1989): Numerical analysis: a comprehensive introduction. John Wiley & Sons Ltd., Chichester, UK.

Xu, P. (1992): A quality investigation of global vertical datum connection. *Geophysical Journal International* 110(2):361–370. DOI: 10.1111/j.1365-246X.1992.tb00880.x.

Acknowledgments

First of all, I would like to express my deepest gratitude to my supervisor, Prof. Bernhard Heck, for his continuous support, his motivation and enthusiasm. His guidance and experience was of constant help during my research and the writing of this thesis and the associated publications. I am also deeply grateful to Prof. Reiner Rummel for his continued interest in my research. It is a great honor for me that he served as a second reviewer.

A very special thanks goes to Dr. Kurt Seitz for all his assistance and advice, and the immeasurable amount of time he spent to accompany my research activities. I highly appreciate our fruitful discussions and that I could talk to him whenever necessary. I would also like to extend my thanks to all my present and former colleagues at the "Außenstelle Post" for contributing to a pleasant working atmosphere, for constructive criticism, and for having a lot of fun during tea breaks and unforgettable conference trips around the world.

At the beginning of my PhD, it was an excellent experience to be part of the REAL GOCE project. Our joint project meetings have always been of great value to me. Representative for all the project members, I would like to thank the coordinator Prof. Wolf-Dieter Schuh. I am very thankful to lots of people for interesting and valuable discussions about my research on conferences, workshops, and in cooperations. In particular, I would like to thank Prof. Hussein Abd-Elmotaal, Prof. Joseph Awange, Dr. Johannes Bouman, Prof. Carla Braitenberg, Dr. Jan Martin Brockmann, Dr. Sten Claessens, Prof. Vagner Ferreira, Dr. Martin Fuchs, Dr. Thomas Gruber, Dr. Christian Hirt, Prof. Michael Kuhn, Dr. Gunter Liebsch, Prof. Torsten Mayer-Guerr, Christian Pock, Moritz Rexer, Dr. Axel Rülke, Judith Schall, Dr. Joachim Schwabe, Dr. Christian Voigt, and Dr. Franziska Wild-Pfeiffer.

The following institutions are acknowledged for providing different kinds of geodetic data sets: Austrian Federal Office for Metrology and Surveying (BEV), Brazilian Institute of Geography and Statistics (IBGE), Curtin University Perth (Australia), European Space Agency (ESA), Federal University of Paraná (Brazil), Geoscience Australia and the other Intergovernmental Committee on Surveying and Mapping (ICSM) member agencies, German Federal Agency for Cartography and Geodesy (BKG), GOCE High-Level Processing Facility (HPF), Gravity Observation Combination (GOCO) consortium, National Geospatial-Intelligence Agency (NGA, USA).

Parts of the research that contributed to this thesis were funded by the German Federal Ministry of Education and Research (BMBF) under grant number 03G0726F within the REAL GOCE project of the GEOTECHNOLOGIEN programme, and by the German

Research Foundation (DFG) under grant numbers HE1433/20-1 and HE1433/20-2 within the project “Methodical and numerical investigations into the establishment of a unified global vertical datum”. The financial support for these research projects is gratefully acknowledged.

I am especially thankful to Alexander Koch for his precise proofreading of the manuscript and his kind support to ensure a uniform appearance of this thesis. Last but not least, I owe my sincere gratitude to my family, which has always been a constant source of love, concern, support, and strength in all these years.

SCHRIFTENREIHE DES STUDIENGANGS GEODÄSIE UND GEOINFORMATIK

(ISSN 1612-9733)

2003/1 **Daniel Koenig, Kurt Seitz**
Numerische Integration von Satellitenbahnen unter Berücksichtigung der Anisotropie des Gravitationsfeldes der Erde. 2003
ISBN 3-937300-00-7

2004/1 **Marcus Ulrich Schmidt**
Objektorientierte Modellierung zur geodätischen Deformationsanalyse. 2004
ISBN 3-937300-06-6

2005/1 **Stephan Kupferer**
Anwendung der Total-Least-Squares-Technik bei geodätischen Problemstellungen. 2005
ISBN 3-937300-67-8

2007/1 **Andreas Knöpfler, Michael Mayer, André Nuckelt, Bernhard Heck, Günter Schmitt**
Untersuchungen zum Einfluss von Antennenkalibrierwerten auf die Prozessierung regionaler GPS-Netze. 2007
ISBN 978-3-86644-110-1

2007/2 **Xiaoguang Luo, Michael Mayer, Bernhard Heck**
Bestimmung von hochauflösenden Wasserdampffeldern unter Berücksichtigung von GNSS-Doppeldifferenzresiduen. 2007
ISBN 978-3-86644-115-6

2007/3 **Michael Mürle**
Aufbau eines Wertermittlungsinformationssystems. 2007
ISBN 978-3-86644-116-3

2007/4 **Heinrich Derenbach, Michael Illner, Günter Schmitt, Martin Vetter, Siegfried Vielsack**
Ausgleichsrechnung – Theorie und aktuelle Anwendungen aus der Vermessungspraxis. 2007
ISBN 978-3-86644-124-8

2007/5 **André Nuckelt**
Dreidimensionale Plattenkinematik. 2007
ISBN 978-3-86644-152-1

2007/6 **Hermann Bähr, Zuheir Altamimi, Bernhard Heck**
Variance Component Estimation for Combination of Terrestrial Reference Frames. 2007
ISBN 978-3-86644-206-1

2009/1 **Cornelia Eschelbach**
Refraktionskorrekturbestimmung durch Modellierung des Impuls- und Wärmeflusses in der Rauhigkeitsschicht. 2009
ISBN 978-3-86644-307-5

2009/2 **Bernhard Heck, Michael Mayer (Hrsg.)**
Geodätische Woche 2009
22.–24. September 2009, Messe Karlsruhe, Rheinstetten im Rahmen der INTERGEO – Kongress und Fachmesse für Geodäsie, Geoinformation und Landmanagement. Abstracts. 2009
ISBN 978-3-86644-411-9

2010/1 **Thomas Grombein, Kurt Seitz, Bernhard Heck**
Untersuchungen zur effizienten Berechnung topographischer Effekte auf den Gradiententensor am Fallbeispiel der Satellitengradiometriemission GOCE. 2010
ISBN 978-3-86644-510-9 / KIT Scientific Reports 7547

2010/2 **Thomas Fuhrmann, Andreas Knöpfler, Michael Mayer, Xiaoguang Luo, Bernhard Heck**
Zur GNSS-basierten Bestimmung des atmosphärischen Wasserdampfgehalts mittels Precise Point Positioning. 2010
ISBN 978-3-86644-539-0 / KIT Scientific Reports 7561

2010/3 **Geodätisches Institut (Hrsg.)**
Vernetzt und ausgeglichen
Festschrift zur Verabschiedung von Prof. Dr.-Ing. habil. Dr.-Ing. E.h. Günter Schmitt. 2010
ISBN 978-3-86644-576-5

2012/1 **Michael Mayer, Claudia P. Krueger, Bernhard Heck (Hrsg.)**
Highly Precise Positioning and Height Determination using GPS. Results of a PROBRAL project by Universidade Federal do Paraná (UFPR, Curitiba, Brazil) and Karlsruhe Institut für Technologie. 2012
KIT Scientific Reports 7604

2013/1 **Melanie Müßle, Bernhard Heck, Kurt Seitz, Thomas Grombein**
Untersuchungen zur planaren Approximation im Geodätschen Randwertproblem. 2013
ISBN 978-3-7315-0095-7 / KIT Scientific Reports 7652

2013/2 **Hermann Bähr**
Orbital Effects in Spaceborne Synthetic Aperture Radar Interferometry. 2013
ISBN 978-3-7315-0134-3

2014/1 **Marion Heublein**
Towards a rigorous fusion of GNSS and InSAR observations for the purpose of water vapor retrieval. 2014
ISBN 978-3-7315-0270-8

2015/1 **Andreas Knöpfler**
Korrektur stationsabhängiger Fehler bei GNSS. 2015
ISBN 978-3-7315-0352-1

2017/1 **Thomas Grombein**
Gravity forward modeling with a tesseroid-based Rock-Water-Ice approach – Theory and applications in the context of the GOCE mission and height system unification. 2017
ISBN 978-3-7315-0655-3